ALLE ZEIT WACH 1842

34. Colloquium der Gesellschaft für Biologische Chemie
14. – 16. April 1983 in Mosbach/Baden

Biological Oxidations

Edited by H. Sund and V. Ullrich

With 119 Figures

Springer-Verlag
Berlin Heidelberg New York Tokyo 1983

Professor Dr. HORST SUND
Professor Dr. VOLKER ULLRICH
Fakultät für Biologie der
Universität Konstanz
Postfach 5560
D-7750 Konstanz, FRG

ISBN-13:978-3-642-69469-1 e-ISBN-13:978-3-642-69467-7
DOI: 10.1007/978-3-642-69467-7

Softcover reprint of the hardcover 1st edition 1983

2131/3130-543210

Preface

This volume contains the lectures and discussions of the 34th Mosbach Colloquium, organized by the Gesellschaft für Biologische Chemie in commemoration of Otto Warburg's 100th anniversary. Following Warburg's main scientific interests the topics of the contributions were related to the hydrogen-transferring enzymes as well as to the oxygen-reducing and oxygen-activating enzymes. Due to the detailed knowledge on the molecular structure of NAD- and flavin-dependent dehydrogenases, a better understanding of enzyme catalysis and especially enzyme evolution has emerged. In the field of oxidases and oxygenases we begin to understand the complex biochemistry of oxygen and realize that binding, reduction, and activation of this molecule have common underlying principles. Even new metabolic pathways of oxygenases, such as lipoxygenases or cyclooxygenase, have recently been discovered and promise new innovations in cell regulation and medicine.

Thanks to the cooperation of all authors, a well-balanced mixture of basic information and progress report was provided. We especially appreciate Professor Bücher's charming, thoughtful, and honest contribution on Otto Warburg as a skilful scientist and colorful personality.

The editors are indebted to the sponsoring organizations, the Deutsche Forschungsgemeinschaft and several industrial companies, and especially acknowledge the competent help of Mrs. H. Allen in the editorial work.

October 1983

H. SUND
V. ULLRICH

Contents

Oxygenases

Integrated Systems

Contributors

You will find the addresses at the beginning of the respective contribution

Otto Heinrich Warburg
3 October 1883 - 1 August 1970

Otto Warburg: A Personal Recollection

Th. Bücher[1]

Pupils as Biographers

It is to Hans Adolf Krebs that we owe an authoritative biography of Otto Warburg. Originally written as an obituary for the Royal Society of London [1], it was also published in German [2]. The first written communication by Krebs from Warburg's laboratory dates from 1927 [3]. In the 43 years preceding Warburg's death, there existed a bond between the two men profiting in its personal as well as in its professional aspect from the stimulating tension of a teacher-pupil relationship. Krebs had most probably kept a diary with the thoroughness that characterized his style of working. In addition to this, he studied the roots of Warburg's personality and investigated the sources of the publications. After this biography, which in its affectionate sincerity honors both the biographer and his subject, what can be the further purpose and sense of my contribution?

Any appraisal of Warburg within the historical scope of natural sciences should be left to the experts. Nevertheless, it so happens that I met Otto Warburg in those very years in which Krebs, due to his emigration and the war, had lost direct contact with the Dahlem Group. The lack of Krebs' personal experience leads to a less vivid representation of the decisive period in which Warburg and his coworkers laid the foundations for many of the contributions to this Colloquium. Consequently it may appear sensible to continue and complete Krebs' attempt to "let Warburg speak in his own words", to take the risk of tracing features of his personality as remembered by his pupil.

On this matter a few words on my own behalf: neither before nor after the years in Warburg's working team did I meet a teacher in whom I was able to confide so unreservedly. The span of time between autumn 1938 until spring 1945, interrupted by 14 months of military service, was decisive for me far beyond its professional aspect. I therefore see little sense in hiding my own person in the following scenes, an endeavor which incidentally would have been a striking example of what Warburg would have called an Eiertanz.[2,3]

---

1 Physiologisch-Chemisches Institut, Goethestraße 33, D-8000 München, FRG

2 This lecture was originally given in German ("Über Otto Warburg in der Erinnerung") and, at the request of the publishers, has been translated into English. I am grateful to Mrs. H. Allen and Mrs. A. Macolister, University of Constance, and to Ms. H. Carr for attempting to put my thoughts into a comprehensible English version

3 Rural folk dance amongst a pattern of fresh eggs, where the dancer tries not to break the shells. Such scenes were frequently depicted by Dutch genre painters, e.g., The Christening Feast by Jan Steen (P111 at the Wallace Collection, London)

34. Colloquium-Mosbach 1983
Biological Oxidations

Application for a Position as a Doctoral Candidate

The intellectual Berlin of the early thirties was characterized by the most lively participation in the development of the arts and sciences. The concentration of creative authorities was unparalleled. In the socially different worlds of pre- and post-war Berlin, turning to progress in the same way as assimilating it, Otto Warburg became well-known and influential. Where formerly the principle of unity of research and teaching had taken shape, he represented a new type of "pure researcher" that fascinated the young. The public of that time took less notice of Nobel prizes than today, and turned their attention far more to the lecture evenings of the scientific societies, above all to those of the Kaiser-Wilhelm-Gesellschaft which was even honoured by the visit of the Reichspräsident Ebert in his time. In 1926, Warburg lectured to the medical corps on the cancer problem [4], in 1928 in the Harnack Haus on the action spectrum of the respiratory enzyme [5]. Fascinating discoveries were quickly discussed everywhere, and were soon incorporated in the teaching syllabus of the grammar schools, at least in those of Berlin's Western suburbs in which the Kaiser-Wilhelm Institutes were situated.

I am unable to say exactly what motivated me even during my last school years at Berlin-Steglitz' Paulsen Realgymnasium, and later during my chemical studies, to work toward an apprenticeship with Otto Warburg. It was most likely the influence of our esteemed biology teacher. At his suggestion, Albert Reid - a schoolfellow 4 years my senior - had written his Ph.D. under the supervision of Warburg. At any rate, immediately after passing my final examinations in chemistry, I made enquiries as to how I could get in contact with him. In general, Warburg not only had the reputation of being a genius, but also of being arrogant. I therefore chose the shortest, but also the least compromising way, by phoning the institute to ask for acceptance as a doctoral candidate.

Without more ado, with an appointment for 6 o'clock the same afternoon, I met Otto Warburg in the beautiful, oval-shaped library of his institute. I saw him for the first time. He was of rather slight stature, but with broad shoulders. The sleeves of his elegant woollen waistcoat were slightly frayed. Gray tweed trousers and carefully polished Scottish shoes completed the picture of a nobleman of the British School.

I repeated my request in a few sentences. Warburg looked directly into my face, his mouth closed. His look did not frighten, on the contrary, but it was penetrating, as I was soon to learn. Without going into lengthy questions concerning my person, Warburg decided that the consent of the Head of the Department should be obtained. By this, the requested opportunity was probably granted in principle. He then said, unexpectedly, establishing with incomparable authority the teacher-pupil relationship: "I presume you do not know what we are working on ... I will get you a publication."

The 1938 Review

Warburg fetched the paper *Chemische Konstitution von Fermenten* [9]. It had shortly before been published in the seventh volume of the *Ergebnisse der Enzymforschung*[4] founded by F.F. Nord and R. Weidenhagen.

Most topics under discussion at our present meeting hinge upon the discoveries reviewed in this paper, which includes the following chapters: Photometric- and Manometric Methods, Action and Prosthetic Groups, Alloxazin Proteids, Pyridin Proteids, Copper- and Iron Proteids and Hydrogen-Transferring Enzymes.

The article also represented a milestone in the development of Warburg's scientific work: At the turn of the twenties, the measurement of the *absolute* action spectrum of the Atmungsferment [6] initiated the final phase of investigations on cell respiration, even of the cell physiology of energy metabolism. He was awarded the Nobel prize for medicine and physiology in 1931. Already in 1929, however, Warburg had been invited to the John Hopkins Medical School in Baltimore to give the Herter Lecture on *Enzyme Problems and Biological Oxidations* [7]. Incidentally, by demonstrating an experiment at that time, E.S.G. Barron succeeded in overcoming Warburg's aversion to the "Thunberg technique". Barron and Harrop [8] had discovered that erythrocytes - in the presence of methylene blue - brought about the partial combustion of glucose into carbon dioxide and pyruvate

$$C_6H_{12} + O_2 = 2C_3H_4O_3 + 2H_2O$$

with astonishing activity.

On his return to Dahlem Warburg observed in experiments which had immediately been started that the replacement of glucose by the Robison ester (glucose-6-phosphate) led to a cell-free system, that the system consisted of heat-resistant and heat-sensitive components and that parts of these components were more easily isolated from yeast extracts.

Although the phenomenon discovered by Barron was an artefact, it initiated the use of preparative methods. The isolation of the ferments and coferments acting in the respiratory chain and the clarification of the chemical constitution and physiological function of their Wirkungsgruppen (active groups) disclosed fascinating mechanisms. Predictions on the "dehydrogenation theory" conceived by Thunberg and Wieland were confirmed to an unexpected degree. In the introduction of the report of this work, Warburg expressed this as follows [9]: "The most important chemical discoveries in this field were the isolation of the luminoflavin in 1932 (Fig. 1) and the isolation of nicotinic acid amide in 1934 (Fig. 2). With this the nitrogen-containing rings of alloxazine and pyridine were isolated, on whose reversible hydrogenation and dehydrogenation the action of alloxazine proteids and pyridine proteids was based."

The interdependence of methodical innovation and progress of knowledge, characteristic of Warburg's research, brought about in this connection the invention of the photometric enzyme test (Figs. 3 and 4).

---

4 Continued as *Advances in Enzymology and Related Subjects*, Interscience Publishers, Inc., New York & London (Vol. I, 1941) after the emigration of F.F. Nord

Fig. 2

Fig. 1

Fig. 1. Luminoflavin (Fig. 4 in [9]). "Although the elementary composition of luminoflavin, its molecular weight (256 g/mol), and the formation of urea at alkaline hydrolysis suggested that it was trimethylated alloxazine, the constitution remained uncertain until Stern and Holiday found in 1934 that the spectrum of luminoflavin originates when alloxazin is methylated at the nitrogens. R. Kuhn, showed then by synthesis that luminoflavin is trimethyl alloxazine"

Fig. 2. Picrolonate of nicotinamide from *Coferment*. (Fig. 7 in [9]). "The first step towards the elucidation of the coenzyme's effects, was the separation of its catalytical action into two stoichiometric reactions, from which the second reverses the first". (See also Figs 3 and 4)

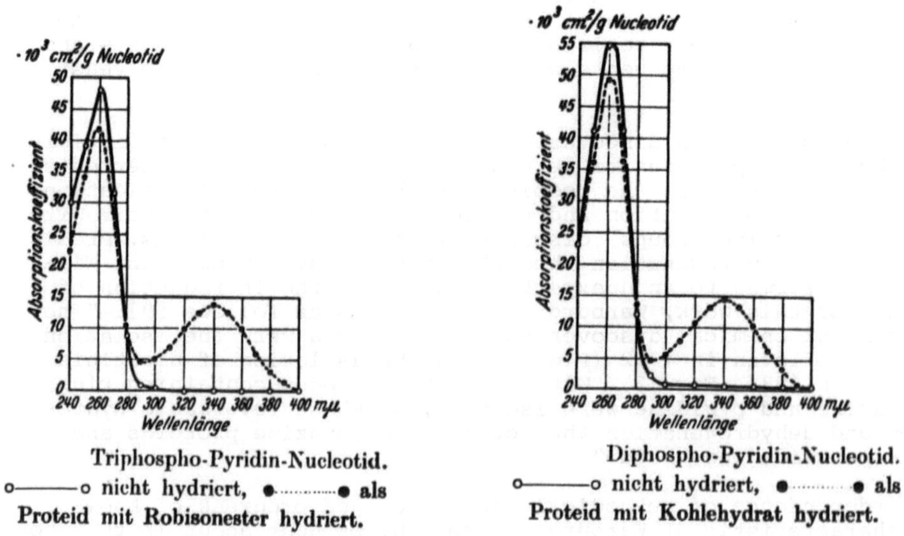

Fig. 3. Absorption spectra of NADP and NAD before and after enzymatic reduction. (Figs. 10 and 12 in [9])

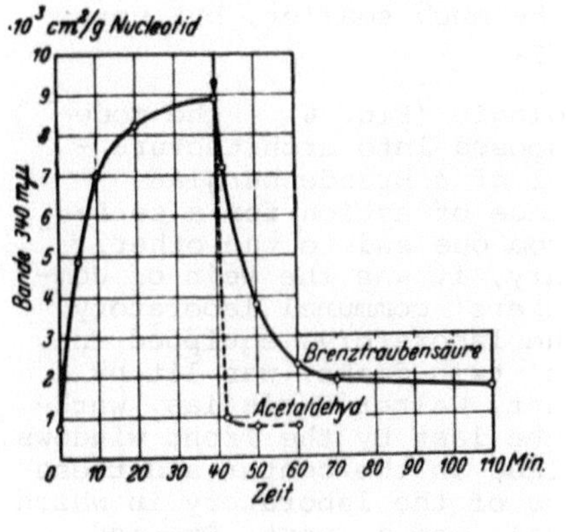

Fig. 4

Fig. 5

Fig. 4. "Ascending curve: hydrogenation of NAD by carbohydrate, descending curves: dehydrogenation by acetaldehyde or pyruvic acid", in the presence of enzyme fractions from yeast juice. (Fig. 15 in [9])

Fig. 5. Crystals of alcohol dehydrogenase from yeast. (Fig. 13 in [9])

But also in the preparative sector, procedures of general importance were conceived, for example phenol extraction and crystallization of metabolic enzymes (Fig. 5).

In the thirties, a popular biography of Pasteur and other "microbe hunters" was given broad distribution. It described how the young Pasteur demonstrated to the venerable Biot the rotation of the plane of polarized light in solution of those tartrate crystals that he had classified into "rights" and "lefts" on the basis of the asymmetry of their hemieders. Biot is supposed to have said: "Mon enfant, cela me fait battre le coeur". At the time, I had most certainly understood very little of the above-remembered achievements. Nonetheless, the precision of style and the clarity of the advance from complexity into the causality of molecular structures made my heart pound: I had met the Pasteur of our century. By the way, Warburg asked for Professor Hermann Leuch's consent himself. In fact, I was allowed to leave the uninviting rooms of the chemical institute and the world of old-fashioned laboratory instruments behind me, in order to embark on the years of apprenticeship as a scientist.

How was it possible that a man who treated an unknown student in such an understanding, even comradelike manner, had the reputation of being arrogant or extravagant?

The Laboratory

From 1914 to 1931 Warburg was a member of the Dahlem Kaiser-Wilhelm-Institute for Biology. Emil Fischer had placed him in this institute, headed from the beginning by Carl Correns. In 1930, following the already mentioned Herter Lecture, the Rockefeller Foundation put an extremely generous sum of money at Warburg's disposal [48], which allowed not only the purchase of a large piece of land, but also the erection of two institutes, the big KWI for Physics along with the director's house and swimming pool as well as the much smaller, but very well equipped Institute for Cell Physiology.

The Kaiser-Wilhelm-Institut für Zellphysiologie (Fig. 6) - the mode of life and working style of Warburg transposed into architecture - had been designed in 1929/30 from the model of a Brandenburgian country manor and had been planned as a place of action for a self-contained group. A hall corridor passed from one end to the other. Large enough to hold the oval of the library, it was the vein of communication between Warburg's and his coworkers' communal laboratory and functional rooms in larger numbers. The laboratory, equipped in classical manner with five or six chemists' twin desks, was lit by seven big windows. Warburg and his assistant, Walter Christian, worked at the first table, Erwin Negelein at the last by the front windows. Fritz Kubowitz (Fig. 7) had the working place in the center and these three positions set the permanent structure of the laboratory in which the younger collaborators, Griese and Brömel, and a guest, Dr. med. Paul Ott, a radiologist, were incorporated. E. Haas had already emigrated to the United States when I joined.

Fig. 6. Kaiser-Wilhelm-Institute for Cell Physiology (Richard Gradewitz Building), Berlin Dahlem [48]. *First Floor* Central laboratory (*left*), Library (*middle*), Microanalytical and optical rooms (*right*). *Attic* Flats of the families Kubowitz and Lüttgens, Secretariat and guest rooms. *Ground floor* "Factory"-laboratories, cold room

"Sich gesund erhalten und stetig arbeiten: Geistesblitze können sich auf einen bis zwei pro Jahr beschränken".[5] According to this pattern set by Otto Warburg, the working atmosphere of the laboratory was free from stress. Each project was a new step forward, and the necessary time could be given to each job. "Laboratory chatting" was uncon-

5 "Keep in good health and work steadily: brain-waves can be limited to one or two per year"

Fig. 7. Collaborators at the Institute's entrance 1937. *Left to right* F. Kubowitz, W. Christian, E. Haas, W. Lüttgens, A. Griese, H. Brömel

strained, all the more so as there was tacit unanimity in the judgement of the political circumstances. New projects were, as a rule, developed from preceding ones; if, not, they were inaugurated in such a way that those who took part had the feeling they had participated in the concept. The picture of Utopia is completed by the fact that - except for the secretariat and the microanalysis - all rooms and equipment were available for common use. Not even Warburg had a separate room. His typewriter stood in the library. Order was maintained according to the simple principle that one leaves an instrument in working order even if it was found in a bad state. Only the bombs succeeded in putting a sudden end to that paradise of a research community led by genuine, generous authority.

Due to the lack of knowledge of the reality, possibly also due to academic conceit, the members of the research group have been called "technicians". It was said: "Warburg is only working with laboratory assistants". In his biography, H.A. Krebs called Fritz Kubowitz, to name an example, with whom he had published his first paper from Warburg's laboratory in 1927, a "technical collaborator". It is certain that in every respect there was a qualitative difference between Warburg and his collaborators. This does not mean, however, that such a difference in knowledge and originality existed between men like Negelein and Kubowitz and other biochemists of their time.

Fritz Kubowitz

Fritz Kubowitz, born in 1902, to whom I was assigned when I entered the institute, had completed an apprenticeship as a precision tool maker and evening classes at the technical school in Berlin, when he applied for a position in Warburg's laboratory in the mid-twenties.

The positive experience made when employing the 5 years older Erwin Negelein who, like Kubowitz, had been trained at Siemens, may have made Otto Warburg decide to repeat the procedure. Till the end of the twenties, Kubowitz assisted Warburg, mainly with manometric measurements. His first more independent development was the refinement, together with E. Haas, of the photochemical action spectrum of the respiratory ferment by observations at 37 wavelengths. Carried out with methodical refinement, the measurements were available in time for the Nobel Lecture (Table 1). By characterizing the long-wave minor bands, they introduced the constitutional classification of the oxygen-binding heme. By the extension of the measurements into the ultraviolet up to the range of the absorption maxima of the aromatic amino acid residues they helped Warburg to gain a late success. In 1945, he made a correct estimate of the molecular weight of cytochrome a_3 [10] (Fig. 8).

Table 1. Absolute extinction coefficients of the carbon monoxide compound of cytochrom oxidase of acetobacteria (*B. pasteurianum*). Irradiation at + 10°C ([6] "Messungen von F. Kubowitz and E. Haas")

Wellen-länge	Lichtquelle	Absoluter Absorptionskoeffizient der Kohlenoxydverbindung des Ferments $\frac{cm^2}{\text{Grammatom Eisen}}$
μμ		Grammatom Eisen
253	Zinkfunke	0.70×10^8
283	Magnesiumfunke	2.00×10^8
309	Effektkohle (Aluminiumsalz)	
313	Quecksilberdampflampe	0.55×10^8
326	Effektkohle (Kupfersalz)	
333	Zinkfunke	0.51×10^8
344	Cadmiumfunke	0.50×10^8
356	Effektkohle (Thalliumsalz)	0.59×10^8
366	Quecksilberdampflampe	0.51×10^8
383	Effektkohle (Magnesiumsalz)	0.35×10^8
405	Quecksilberdampflampe	0.90×10^8
422	Effektkohle (Strontiumsalz)	3.05×10^8
430	Effektkohle (Calciumsalz)	3.70×10^8
436	Quecksilberdampflampe	3.60×10^8
448	Magnesiumfunke	1.30×10^8
460	Effektkohle (Lithiumsalz)	0.40×10^8
494	Effektkohle (Magnesiumsalz)	0.15×10^8
517	Effektkohle (Magnesiumsalz)	0.19×10^8
524	Effektkohle (Strontiumsalz)	0.18×10^8
535	Thalliumdampflampe	0.30×10^8
546	Quecksilberdampflampe	0.30×10^8
553	Effektkohle (Magnesiumsalz)	0.26×10^8
560	Effektkohle (Calciumsalz)	0.23×10^8
578	Quecksilberdampflampe	0.30×10^8
589	Natriumdampflampe	0.54×10^8
596	Effektkohle (Strontiumsalz)	0.38×10^8
603	Effektkohle (Calciumsalz)	0.20×10^8
610	Effektkohle (Lithiumsalz)	0.12×10^8
652	Effektkohle (Strontiumsalz)	0.02×10^8
670	Effektkohle (Lithiumsalz)	0.005×10^8

Warburg let Kubowitz work completely on his own. He was also formally promoted to scientific assistant and produced a sequence of beautiful experimental investigations which I should mention.

After H. Kempner had discovered the CO inhibition of molecular hydrogen evolution in the fermentation of butyric acid, Kubowitz undertook to determine the photochemical action spectrum of the hydrogen-evolving principle of the butyric acid bacteria [11]. We know today that he was at that time quite close to the discovery of ferredoxin, a redox mediator of unusually negative mid-point potential, all the more so as the photochemical splitting of carbon monoxy ferroglutathione had been included in the investigations [12]. A new chapter on heavy metal catalysis was opened with the isolation, copper splitting and resynthesis of phenol oxidase from potatoes [9,13]. Meanwhile, however, the aims of the group had moved into a different direction.

That is to say that Negelein had successfully carried out the crystallization of alcohol dehydrogenase [14]. The isolation of glyceraldehyde phosphate dehydrogenase had progressed in the hands of Warburg and Christian [15]. With this, the entry into the new research topic was indicated to Kubowitz. His project which had, of course, been coordinated with Warburg - the isolation of lactate dehydrogenase - indicated the transition of tissue extracts of warm-blooded animals.

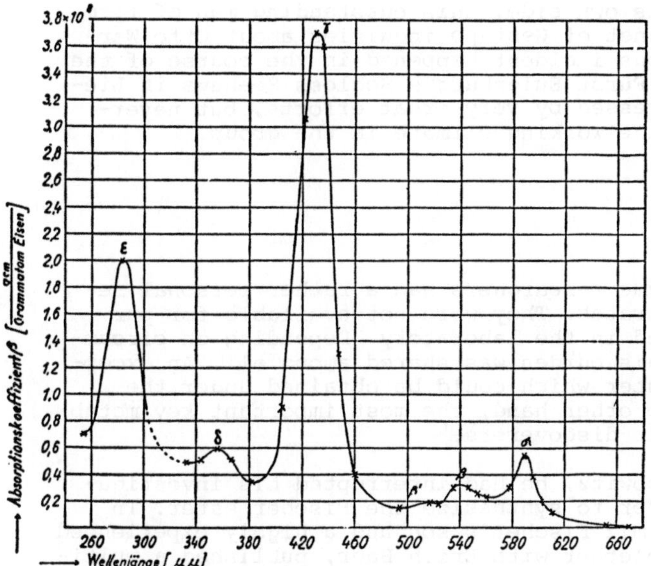

Kohlenoxydverbindung des sauerstoffübertragenden Ferments der Atmung

Fig. 8. Action spectrum [6,49] of the photodissociation of CO-cytochromoxidase as a basis of Warburgs calculation of its molecular weight in 1945 [10]: Down to 313 nm the spectrum coincides essentially with the one of the CO-hemochromogene. But three quarters of the ε-band at 280 nm must be attributed to the protein moiety, the heme absorbance at that wavelength beeing about the same as at 360 nm. In model experiments with CO-myoglobin [26] radiation quanta absorbed in the globin had been found as efficient in the process of photodissociation as those absorbed in the CO-heme moiety. Using the quantum efficiency one, and using the average value of $2 \cdot 10^3$ for the absorption coefficient (ln-based) of protein [cm^2/gram] the molecular weight of cytochrome a_3 (per gram atom of CO binding iron) results as $0.75 \cdot 2 \cdot 10^8 / 2 \cdot 10^3 =$ 75,000 gram. This is about one half of recent estimations of the molecular weight of the isolated cytochrome aa_3 unit

These experiments, the isolation of the enzyme from rat muscle, Jensen sarcoma and human muscle, are outstanding pioneer achievements of comparative enzymology. For the first time, the techniques of the analytical use of the antienzyme and heat denaturation had been demonstrated extensively.

It does not appear in any of Kubowitz' publications that in order to unburden the large spectrophotometer from routine work, and applying experience gained during his photochemical investigations, he had set up a smaller device for photometric enzyme tests. It was further developed later and became the basis of today's spectral line filter photometers.

Kubowitz was probably able to cope with any practical or theoretical task, provided he had time. Among the many things I learned from him was foremost the way in which he followed up failures. When, during the test of a purification step of the lactate dehydrogenase the acetone precipitate and the supernatant proved to be enzymatically inactive, he dried the precipitate by high vacuum and thus eliminated the acetone. He had observed that the remaining activity increased slowly during the test. In fact, he not only regained almost all total activity, but could also register a significant increase in the specific activity.

Not without some fault on his own side, this outstanding man of firm character got caught in the net of Gestapo inquiries about Otto Warburg. This extremely dangerous incident happened in the course of the removal of the Institute to Fürst Eulenburg's Schloss Seehaus in Liebenberg. It was finaly suppressed by very great efforts, but nevertheless weighed heavily on the working climate in the group.

Fischer Ester

Right into the forties, enzyme researchers had a rather personal relationship with enzyme substrates. They were not available for purchase, but had to be produced in the laboratory. Depending on circumstances and capabilities, this burden was shared among all. An exception was the Harden-Young ester which could be obtained under the trade mark Candiolin. On the other hand, the most important key metabolites were named after their discoverers.[6]

When I took up work with Kubowitz, he had interrupted his investigations for a few weeks in order to synthesize the Fischer ester. In 1932 Hermann O.L. Fischer, Emil Fischer's son and a highly experienced carbohydrate chemist had, together with Erich Baer, published a multistep synthesis of racemic glyceraldehyde 3-phosphate [17]. The availability of Fischer's ester had initiated the final phase of the efforts for the mechanism of fermentations which had lasted for decades. It was a matter of the reaction sequence of the anaerobic part of oxidative phosphorylation. In essential parts this was discovered in Warburg's laboratory [18,19]. I could finally contribute to it, too [20]. Larger quantities of the Fischer ester were required for the

6 Harden-Young ester (D-fructose-1,6-biphosphate): Sir Arthur Harden (1859-1944), head of Biochemical Department (1897-1930), Lister Institute, London. Robison ester (D-glucose-6-phosphate): Robert Robison (1883-1941). Harden's successor (1931-1941) at the Lister Institute. Nilsson ester (D-glyceric acid-3-phosphate): Ragnar Nilsson, member of the v. Euler-Myrbäck group, Stockholm

isolation of the "Oxydizing enzyme of fermentation", the glyceraldehyde-phosphate dehydrogenase of today, but especially for the isolation of the back-reacting product, the "R-diphosphoglyceric acid". The synthesis according to Fischer and Baer, starting from acroleindiethylacetal, develops into a 10-step reaction, the last of which, the phosphorylation of the benzylcycloacetal of glyceraldehyde and the reductive splitting of the phenylic ether by palladium and hydrogen, called for great care with material which had already become very precious.

My Own Project

There was hardly anything to assist Kubowitz with. Moreover, I happened to see one day when I worked overtime in the library, how Kubowitz returned to the laboratory from his attic flat after dinner in order to rinse once more the glassware which I had washed for him during the day. Thereby I hardly needed to be offended, even Warburg himself washed his equipment sometimes after Christian had fished it out of the hot chromic sulphuric acid. It was said to be economical to have a personal grip on all possible sources of errors in the experiments.

Several weeks later, tolerated by Kubowitz with extraordinary patience, I met Otto Warburg in the glass store whilst I was searching for a round-bottom flask for my master. Looking with pleasure at the beautiful flask, he spoke to me. In a friendly manner, like a senior officer in the officer's mess would speak to the lieutenant, he told me that I should not have the impression that I had really learnt anything in the course of my studies. It was all half-understood stuff, not applicable. One must always acquire the fundamentals by one's own work. Even as a scientific member of the Kaiser-Wilhelm-Institute he had been fooled, though he should have known better. For when, after the war, he wanted to measure the quantum yield of assimilation of carbonic acid, his father's assistants had focused for him red spectral fields with the pocket spectroscope. However, a large portion of ultra red had been overlooked and the quantum yield had been determined as far too high. My task was now to apply productively the accumulated knowledge I had received. Many failed in this. In fact, even today the scientific prospects of graduate students can be judged during the first weeks by their ability to get away from the material learned and to gather experience. In the language of Warburg's familiar quotations: "Klug ist, wer nicht nur aus eigener Erfahrung lernt". "Mit der geistigen Armut verhält es sich so, wie mit der Armut. Sie ist keine Schande, aber eine Ehre ist sie wiederum auch nicht".[7]

The next morning he brought a piece of paper into the institute. It was headed "Bücher" and contained the following six-point working program (Fig. 9):

1. Is adenine-nucleotide really the coferment of phosphorylation?

Test: Inorganic phosphate + sugar + catalytic quantity of adenine nucleotide (Disappearance of inorganic phosphate in presence of protein).

---

7 "He who learns not only from his own experience is intelligent". "Intellectual poverty can be compared to poverty. It is not a disgrace, but neither is it an honor"

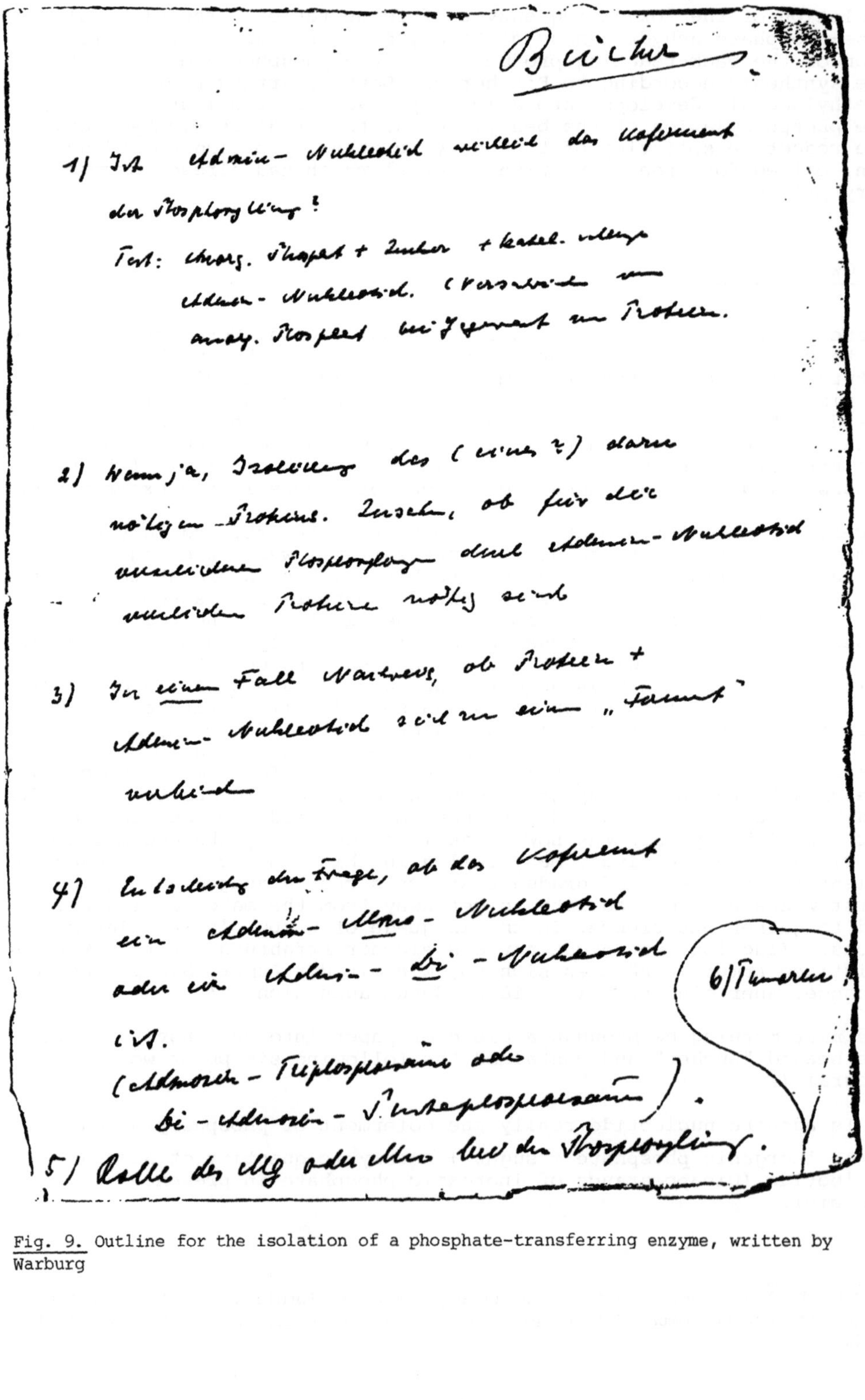

Bücher

1) Ist Adenin-Nukleotid vielleicht das Koferment der Phosphorylierung?
Test: anorg. Phosphat + Zucker + kochb.-Menge Adenin-Nukleotid. (Verschwinden von anorg. Phosphat bei Zugabe von Protein.

2) Wenn ja, Isolierung des (eines ?) dann nötigen Proteins. Zusehen, ob für die verschiedenen Phosphorylierungen durch Adenin-Nukleotid verschiedene Proteine nötig sind

3) In einem Fall Nachweis, ob Protein + Adenin-Nukleotid sich zu einem „Ferment" verbinden

4) Entscheidung der Frage, ob das Koferment ein Adenin-Mono-Nukleotid oder ein Adenin-Di-Nukleotid ist.
(Adenosin-Triphosphorsäure oder Di-Adenosin-Pentaphosphorsäure).

5) Rolle des Mg oder Mn bei der Phosphorylierung.

6) Tumoren

Fig. 9. Outline for the isolation of a phosphate-transferring enzyme, written by Warburg

2. If so, isolation of the (only one?) protein necessary for phosphorylation. Check whether for the different phosphorylations by adenine nucleotide different proteins are necessary.

3. In *one* case give evidence if protein + adenine-nucleotide bind to a "ferment".

4. Decide whether the coferment is an adenine-*mono*-nucleotide or an adenine-*di*-nucleotide (adenosine-triphosphoric acid or di-adenosine-pentaphosphoric acid).

5. Role of Mg or Mn in phosphorylation.

6. Tumors.

Sechse treffen, sieben äffen?[8]

This sheet of paper, in a sense a snapshot of one of the most dynamic phases of energy-metabolic developments, is in many ways interesting: It offers an insight into the structure of Warburg's thinking and working and also sheds some light on some of the pecularities of his polemic.

Adverse to speculation, Warburg takes the theoretical statement only so far as an experimental starting point opens up. Further steps are left to experimental questioning. The investigations of his former pupil Otto Meyerhof in the autumn of 1938 met with reserve on Warburg's part. For one reason, as already in the case of Thunberg, the experimental approaches of the Meyerhof group appeared too complex to him. He mockingly called mixtures of rough-and-ready prepurified components "Wolfsschlucht", after the key scene of the Freischütz[8], and in a more sarcastic frame of mind he would remark "das linke Auge eines Luchses!".

A "Missing Link" in Oxidative Phosphorylation

Warburg's assumption that Meyerhof had been led astray by his thermodynamic speculations carried more weight. Most probably Meyerhof had imagined the linkage of oxidation and phosphorylation in the heart of fermentation, the "energetic coupling" as he called it, to be actually some new kind of physicochemical mechanics. Warburg, however, basing his judgement in principle and from experience on the simplest working hypothesis, declared: "If the reactions work, then the energetics must be correct. Everything depends on the knowledge of a coupling intermediate metabolite!"

About 100 years earlier, Berzelius had coined the term "catalytic power". Liebig did not like that at all. He was convinced, and said

8 "Six hit, seven befool", Warburg - active lay-musician until he got married to science - refers here to the so-called "Wolfsschlucht Szenen" (Wolf's Glen scenes, second act, scenes five and six) of C.M. Webers romantic opera Der Freischütz. Max and Kaspar mold magic bullets from a variety of ingredients, such as "Das rechte Auge eines Wiedehopfes, das linke eines Luchses!" (The right eye of a hoopoe, the left eye of a lynx!). Six of these bullets meet their target unfailingly, but the seventh, the devil's bullet, will later hit his sweetheart Agathe. Warburg uses this motif to warn against over-complex experimental set-ups.

in all frankness that Berzelius - as expressed in Warburg's words - had "given away" a reaction mechanism which was still to be discovered. There was a big fuss. "Don't you admit", writes Liebig on 2.3.1837 to his friend Wöhler, who tried to negotiate, "that, if nitrous oxide did not form red vapors with air and if nitrous acid were unknown, the process of formation of sulfuric acid would have to be considered as a catalytic process; don't you admit that the whole idea of the catalytic "power" is wrong? And why shouldn't I talk, when talking is a duty, and holding back would be a vile action against myself?"

In fact, the pioneering achievements of the Meyerhof group have hardly been taken into consideration in the publications on the *Oxydationsreaktion der Gärung* [18,19] from Warburg's laboratory. This troubled Meyerhof, all the more so as he was soon to find himself in difficulties due to his emigration.

For the task set to me, the neglect of thermodynamic considerations proved to be a handicap. After having shaken manometers and analyzed phosphate for weeks as point one of my program, linking up experimentally with the fermentation tests of 1936 [20], Negelein secretly gave me the tip to use the Harden-Young ester or - when possible - the Fischer ester instead of sugar. For the start he gave me some "alpha" and "gamma". That was the name then given to ADP/ATP and NAD in the laboratory jargon. I soon obtained active, though complex, protein fractions of Lebedev juice, an extract of autolysing baker's yeast [21]. These enabled me to change over to photometric tests. I was, in any case, the only one in the laboratory left to shake manometers.

The Manometric Era

Warburg, coming out of the photometer room with a quartz cuvette in his hands, met me as I was dragging a ready-for-use battery of manometers to the thermostat room. On passing by he hissed: "I can't stand seeing these things anymore!" In fact, the knell of the heroic era of manometry had already sounded several years before I entered the group: Photometry, a labor-saving method which had the same high sensitivity and precision as manometry, was now available.

What we know today as the Warburg Manometer, he called in all fairness the "Haldane-Barcroft blood-gas manometer" or "the beautiful blood gas apparatus of Haldane-Barcroft". In 1910, he had seen this principle during a visit to Cambridge, that permanent source of methodical innovations, in Barcroft's laboratory. Out of a delicate and accurate instrument of quantitative analysis, ("Endwert-Analytik"), he formed the methods adequate for measuring metabolic rates. He conceived refined calculating procedures, developed the "tissue-slice technique" and carried out systematic investigations for their maximum permissible thickness. A variety of forms of recipients, of which we show a selection (Fig. 10), reflect a wealth of manometric projects of great elegance. H.A. Krebs, whose life-work is based to a great extent on the "Warburg technique", has gone into this development with particular expertise [1,2]. As we all know, it concerned the photosynthesis in suspensions of *Chlorella*, the classification of normal and pathological tissues according to their relationship between cell fermentation and cell respiration, as well as the mechanisms of cell respiration itself.

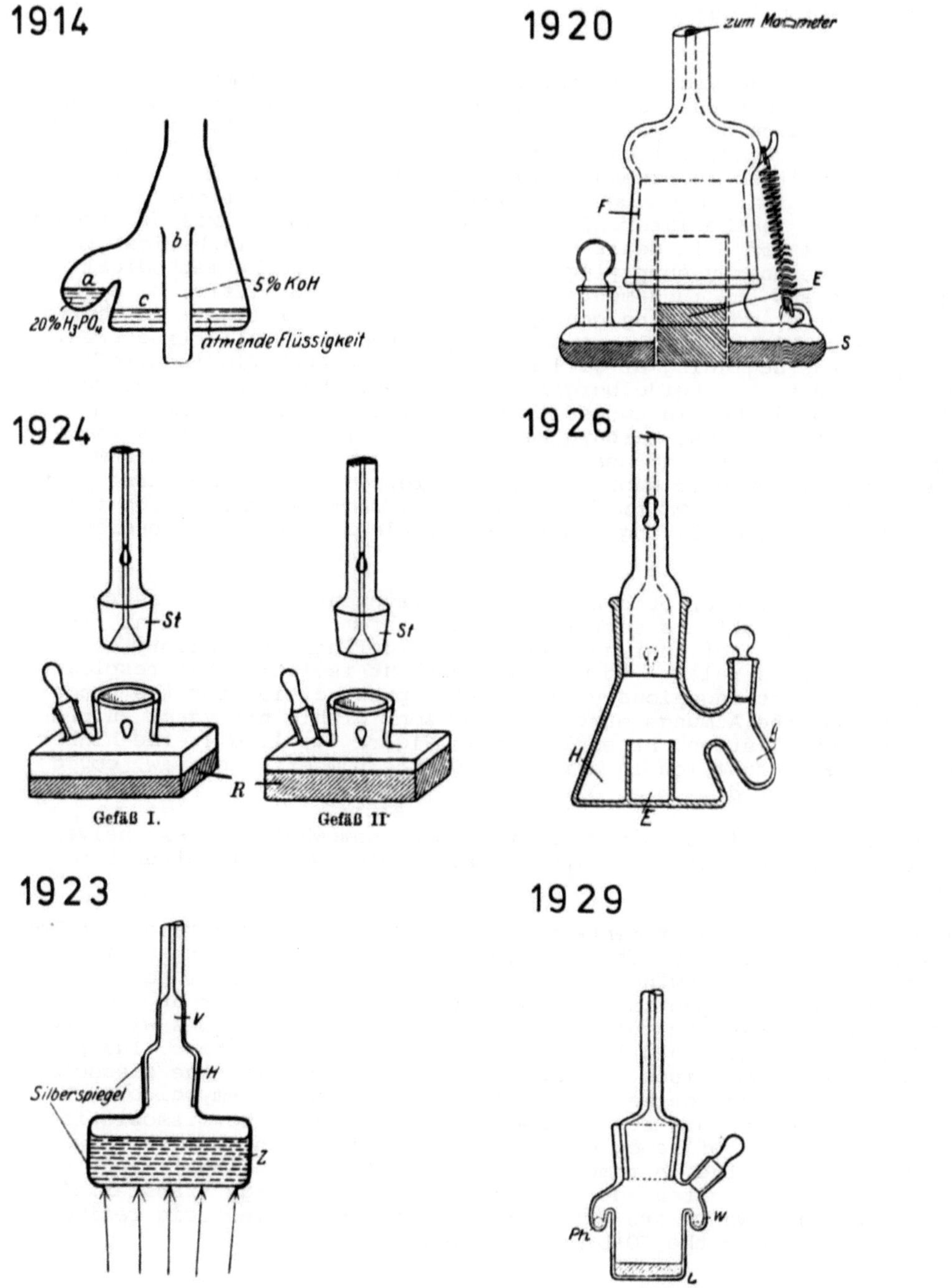

Fig. 10. Manometer vessels. 1914: Respiration of the sea-urchin egg [50]; 1920: Nitrate assimilation in *Chlorella* [27]; 1923: Action spectrum of the carbon dioxide assimilation [51]; 1924: Vessel pair method [52]; 1926: Effect of light on respiration and fermentation of CO-inhibited yeast [53]; 1929: CO-ferrocysteine (*Ph* yellow phosphor) [54]

Cell Respiration 1909-1945

In 1914 - almost 30 years old and already a scientific member of the Kaiser-Wilhelm-Institut für Biologie - before leaving for almost four years' war service as a volunteer with the Potsdam Guards Lancers "in one of the most beautiful uniforms of the old Prussian army", Otto Warburg had written an extensive article *Beiträge zur Physiologie der Zelle* for Volume 14 of the *Ergebnisse der Physiologie* [24]. With rigor of thought, though not yet in the later rigor of style - the latter was acquired in Ludendorff's staff - it deals with thermodynamic and biochemical foundations of the energy-supplying mechanisms. Besides the methodical achievements and the fundamentals of heavy metal catalysis, he also explained the trend-setting concept that cell respiration - in contrast to the glycolytic system - cannot be separated from cell structures. Already at the beginning of 1911 he reported on such findings in a lecture given to the Heidelberger Naturhistorisch-Medizinischer Verein. It has been published in two sequences in the *Münchener Medizinische Wochenschrift* [23]. Thus, Warburg had consequently turned toward those models of structure-bound heavy metal catalysis which, together with the effects of narcotic substances, were reproduced in the first part of his book *Über die katalytischen Wirkungen der lebendigen Substanz* (On the catalytic effects of living substance). Table 2 contains a concise summary.

The most impressive of Otto Warburg's achievements is the measurement of the absolute action spectrum of CO-cytochromoxidase - the at first sight incredible fact that he succeeded in defining the absorption spectrum which is related to molarity without isolating the respiratory pigment. The ingeniously simplifying precondition for the procedure was that the Atmungsferment was present within his yeast suspensions in high dilution. As explained in Figure 11, line 4, the higher members of an e-function in a row, as in the Lambert-Beer law, could have been neglected.

The equations outlined in the figure appear somewhat more extensive when the photochemical splitting is reversible, which is often the case.

All photochemical measurements which augmented Warburg's fame have been measured manometrically. It was therefore understandable that, in 1941, in the face of the growing acceptance of the Optical Test, he was also anxious to register photometrically the light effects for at least one model. I was allowed to collaborate in these investigations with Erwin Negelein [25], whom I had learnt to admire during that time. The problem lay in the comparison of the quantum requirement in the dissociation of carbon monoxide-hemoglobin and carbon monoxide-myoglobin. The manometric measurements required 4 light quanta for the dissociation of a Fe-CO binding of hemoglobin. The question was whether there was a connection with the tetrameric structure. As already mentioned, these studies were later extended to the phenomenon of energy transfer of radiation quanta which are transduced by aromatic amino acid residues of the protein onto the CO-binding heme [26].

Erwin Negelein

Born in 1897 as the son of a carpenter, Erwin Negelein went through elementary school and then absolved a 4-year apprenticeship as a mechanic at Siemens and Halske. For a further 4 years, he continued

Table 2. Chemistry of cell respiration 1908-1945

1908-1914	"Observations of the oxidation processes in the sea-urchin egg" [22]
	Effect of hydrocyanic acid
	Cell structure and oxidation rate
	Inhibition of cell respiration by substances of homologous rows
	"Investigations on oxidation processes in cells" [23]
	Combustions on blood-charcoals
	Manometry
	Role of iron in the cell respiration
	Cell respiration residing in 'Grana' isolated from liver cells
	"Contributions to the physiology of the cell" [24]
1921-1925	Heavy metal analysis and "Autooxydation"
	Vessel pair method
	Tissue-slicing techniques
	"On the metabolism of tumors" (Berlin, 1926) [55]
1926	Respiration inhibition by CO
	Effect of carbon monooxide and light
	Classification of the tissues after metabolism ('Meyerhof Quotient', 'Pasteur Effect')
1927	Relative action spectrum of the "respiration ferment"
	"On the catalytic effects of the living substances" (Berlin, 1928) [56]
1928-1929	Absolute action spectrum of the "respiration ferment". "The oxygen transferring enzyme of respiration" (Stockholm, 1931) [6]
1930-1932	Mixed-coloured hemoglobins
	Isolation and sum formula of cytohemin
1933	CO-inhibition of butyric acid fermentation (ferredoxin)
	CO-ferroglutathione
1945	"Molecular weight of the oxygen-binding ferment" [10]
	Heavy metals as active groups of ferments (Berlin, 1948) [57]

working there as a mechanic. Already a year after beginning his employment in 1919 as a laboratory mechanic in Warburg's department at the Kaiser-Wilhelm-Institute for Biology, he appears as co-author of a paper on the light action on the nitrate assimilation of *Chlorella* [27].

Right up to the early thirties when the reinforcement of "man power" by Kubowitz, Haas and Walter Christian made him less indispensable and he could employ collaborators himself, he stood at Warburg's side in all important projects. The previously mentioned publication contains 36 experimental protocolls.

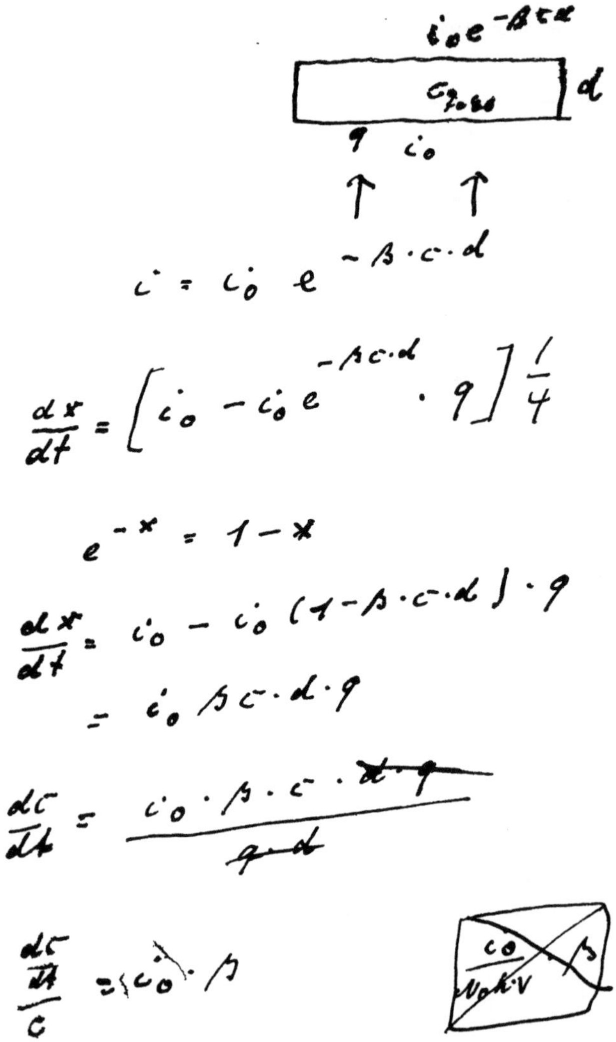

Fig. 11. Note in Warburg's handwriting explaining the derivation of the absorption coefficient β [cm^2/mol] from the kinetics of the photochemical dissociation: A measurement of the "light sensitivity" as per example the half-life time of the photochemical decay at known radiation intensity and known quantum efficiency directly yields β.

Lines *1* and *2* Geometry of an irradiated cube of solution (q = area, d = depth) and the Lambert-Beer law. In correspondence to Einstein's law of photochemical equivalence the dimension of i_o, the flow of radiation, is moles quanta per time- and area unit (energy of a mole quanta = Avogadro number × kν).

Line *3* Photo dissociation dx/dt (x = split moles of sensitive pigment) as function of the parameters in lines 1 and 2, as well as of the quantum efficiency, which is assumed as a quarter in this line (CO-hemoglobin) and is neglected in the following lines.

Line *4* Warburg here reminds me of the mathematical development of an e-function into a row. The higher members of the row can be discarded under conditions x ≪ 1 (in contrast to line 3, x signifies here a dimensionless general variable).

Lines *5* to *8* Derivation of "light sensitivity" $\beta \cdot i_o$, which has *only the dimension of reciprocal time, but not that of concentration*

Not only in the attention to measuring observation, but also in the taste for simple, clear-cut techniques did Negelein agree with Warburg. Skimming through the volumes of collected papers listed in Table 2, work on carbon dioxide assimilation, tumor metabolism and cell respiration, one is impressed by the wealth of constructive solutions found in appliances and instruments. Our photograph (Fig. 12) shows him at the observation telescope of the Askania-Sector of the spectral-photometer, a remarkable installation considering the time of its creation (1929). As one could not trust the linearity of the detector, a combination between an alkaline cell and an electrometer, the measurement itself was transferred into the segments of a rotating disk. During motion, the opening angles of two sectors could be adjusted. When fully opened, the solution to be measured was first placed into the light beam and the response of the electrometer registered. After replacing the measuring cuvette by the control (e.g., water) the sector openings were reduced until the original reading was obtained. The instrument (Fig. 13) was later improved on by Erwin Haas. It has been published in this form [29]. The title reads *On the absorption spectrum of water in the ultraviolet*, as Warburg did not consider the pure publication of a machine on its own a good thing and the dates of the literature diverged: "For UV a fly's blow is enough".

In the course of the years, Negelein became an excellent "bio-organic" chemist. His Ph.D. thesis *On the heme of the oxygen transporting respiratory*

Fig. 12. Erwin Negelein at the observation telescope of the spectrometer (see also Fig. 13). *Left* fluorite monochromators (UV); *right* Glasmonochromators (visible wavelengths)

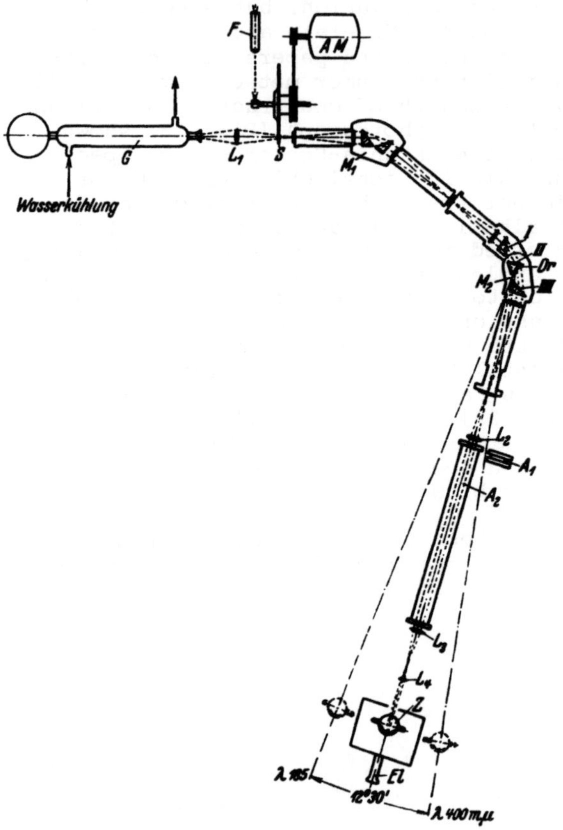

Fig. 13. Photoelectric spectroscopy (Fig. 1 in [9]). *G* hydrogen lamp; *S* rotating sector; *F* telescope observing the degrees of the sector opening during rotation; *M* monochromators with fluorite optics; *A* cuvettes of different lenghts; *Z* photoelectric cell; *El* Lutz-Edelmann electrometer; *L* fluorite lenses

ferment and on some artificial hemoglobins, which he presented to the Philosophical Faculty of the Friedrich-Wilhelms-University in 1932, describes "hemoglobins" in which protoheme had been replaced by "mixed-colored" (red/green) and green hemines derived from chlorophyll and spirographis blood [28]. I have already pointed out the importance of the isolation of crystalline alcohol dehydrogenase for the orientation of the working team. Negelein acquired special esteem by the isolation of the 1.3-diphosphoglyceric acid, a rapidly hydrolyzing substance, as strychnine salt.

In 1927, as an external student, Negelein had graduated from the "gymnasium" and then enrolled at the University of Berlin. He obtained his Ph.D. in 1932. When the Liebenberg Institute closed down, Negelein joined the Department of Biochemistry of the Deutschen Akademie der Wissenschaften in 1946 in Berlin-Buch, GDR. He was appointed professor at the Medical Faculty of the Humboldt University in 1955 and promoted to Director of the Institute of Cell Physiology at the same academy in 1961. Until his retirement in 1964 and also in the following years, he worked on cell physiological problems of tumors and trained a number of collaborators. However, he never again attained the importance of his publications of the twenties and thirties.

Fermentation Enzymes

Before the war really struck home, Warburg and his collaborators had inserted into the framework of glycolysis, which Embden, Meyerhof, Neuberg, Parnas and others had constructed, the active individual enzymes as crystallizable entities. Except for triosephosphate isomerase, which was not crystallized from calf muscle before 1953 [35], and phosphoglycerate mutase, all enzymes of the sectors of the fermentation scheme explained by Meyerhof and Lohmann [36,37] and Meyerhof and Kiessling [38] were represented. The protocol for pyruvate kinase by Negelein, lost in the chaos of the war, is reflected in a publication by Kubowitz [34]. By good fortune, and due to the available test, he (Kubowitz) was able to identify as pyruvate kinase certain cube-shaped protein crystals which were formed in the course of the purification of lactate dehydrogenase. Without crystallization which is also a good purification step for proteins, an isolation project was considered not to have been finished. In the summer of 1942, while Warburg was living in his holiday house at the Northern point of the island of Rügen, I obtained the first crystals of phosphoglycerate kinase and sent him the photograph, together with the birth announcement of our daughter Christiane. By return we received a postcard: "My dear Bücher, success follows success for you at the moment!"

For all enzyme activities, simple or composed photometric tests have been worked out. The role of the phosphate ions in the fluoride inhibition of enolase [30] was likewise discovered, as well as that of zinc ions for the effect of yeast aldolase [31]. The application of antienzymes in comparative enzymological questions [16] has already been mentioned. The discovery of serum aldolase in tumor-bearing rats [39] opened a new era of the serum enzyme diagnosis. Warburg's photometric test is now being used everywhere. Is the idea therefore exaggerated that the aims and methods of the enzymology of metabolism attained a new standard during these few years under menacing surroundings?

Metabolism of Tumors

It was clearly marked above the institute's bell that the general visiting hours to be respected even by prominent guests, were "after 6 o'clock in the afternoon". We were, therefore, very astonished when Professor Schöller (Fig. 14), Director of the laboratories at Schering-Kahlbaum and close friend of Warburg, called one summers's day in 1939 at 10 o'clock in the morning. A few days before that, a series of far-reaching written communications *On the etiology of malignant tumors* by Kögl and Erxleben [40] had been published in Hoppe Seyler's Zeitschrift für Physiologische Chemie. As an "inner" cause of malignant growth it postulated that cancer proteins contained the unphysiological D-glutamic acid. When Warburg was back at his bench I went up to him. We could always approach him quite freely. I asked without ado: "Professor Schöller was here on the matter of the Kögl paper?" Warburg: "Yes". Bücher: "What did you advise him?" Warburg: "That he should keep his fingers away from it!" Bücher: "Is there something fishy about the paper?" Warburg: "Yes!" Bücher: "Kögl is an admired organic chemist. During my studies I heard him lecture at Hans Fischer's in the big lecture

Table 3. Isolation of glycolytic metabolites and enzymes 1936-1944

1936	Diphospho-pyridine-nucleotide [20]
	"Optical test"
	Alcohol dehydrogenase [14]
1938-1939	Glyceraldehyde phosphate dehydrogenase [15]
	1,3-diphosphoglycerate (strychnine salt) [19]
1942	Enolase [30]
	Yeast aldolase [31]
	Phosphoglycerate kinase [32]
	Lactate dehydrogenase of Jensen Sarcoma [16]
	Lactate dehydrogenase of rat muscle [16]
	Muscle aldolase [33]
1943-1944	Pyruvate kinase of rat muscle [34, 34a]
	Pyruvate kinase of human muscle [34]

Fig. 14. Walter Schöller (1880-1965), Professor of medical chemistry (1915), Head of the Scientific Laboratory at Schering (1923-1945), amicably accompanied Warburg's career and gave the hint on nicotine amide

hall on the discovery and constitution of auxin. All of Munich's top people were there. Using refined tests, he isolated a few milligrams of substance out of floods of urine. These were subjected to a complete structural analysis. Everybody was full of admiration." Warburg:

"If so, that's also wrong. I am going to tell you something, Bücher, whoever has worked with human mamma carcinoma knows that they contain only a fraction of genuine cancer tissue. Years ago, a Japanese dragged whole bucketsfull into our laboratory. Most of it was connective tissue and fatty tissue. Kögl once even alluded to his work to me. Why didn't he use experimental tumors as I had already advised him?" We know today how right Warburg was in both statements. Up to the forties, the discussion about D-glutamic acid still went "to and fro". Even Hans Fischer who, on the advice of Butenandt during a Munich conference of the Chemische Gesellschaft had agreed to a repetition of the tests in his laboratory, had been fooled by Kögl's assistant. By the way, our honorary member Ernst Auhagen was orded by his boss to send D-glutamic acid from Elberfeld to Utrecht at that time. Even in our times, the importance of the cancer problem in the eyes of the public as well as ambition sometimes deludes highly intelligent people into swindling [41,42]. I report this primarily in order to illustrate Warburg's unusual consistency in scientific deductions. Also, since his publications on the metabolism of tumors, he had, apart from Kögl, encountered many an eccentric personality.

Throughout most of his career, Warburg resisted the temptation to develop an extensive theory on malignant growth or a therapeutical concept out of his discovery "that cancer under full supply of oxygen ferments like a heavy-working muscle in suffocation".

In 1927, at the end of the Berlin lecture [4], he said: "I see the cause of carcinoma in the anaerobic component of the metabolism of normally growing body cells as well as in the fact that this component is more resistant to damage than respiration is. As a result, all kinds of damage to which the body is subjected cause the anaerobic component to be bred out of the normal" (population) "and thus cells with the characteristics of carcinoma cells".

Only in this sense should he in all fairness be called a witness today. Warburg's investigations on the energy metabolism of normal and malignant tissues have acquired lasting significance by the fact that - in collaboration with Meyerhof, and based on observations on the relationship between glycolysis and respiration going back to Pasteur and Liebig - they made what he called the Pasteur Effect to a cell physiological problem of our times including malignant degeneration.

Liebenberg 1943-1945

Warburg and his team had moved into the institute on Gary Street at the end of 1931. In a world heading for "total war" the building was allowed a time limit of scarcely 12 years when an air mine detonated in the neighborhood and shattered the windows. This was not complete "bombing out" but enough to occupy the staff for weeks with clearing away cement and glass splitters. A few weeks after the new panes had been fixed, a second bomb dropped. Though the putty was still soft and the impediment thus shorter, the warning was taken seriously and the evacuation of the institute to the Uckermark began.

Uckermark is the most Northern and most beautiful part of the Mark Brandenburg, a county of lakes and forests. The idyllically situated "castle" Seehaus (Fig. 15), a dependency built by Liebenberg in 1912, former residence of the princes of Eulenburg, had been abandoned for years and deprived of all its installations. The roof and part of the rotten ceilings had to be renewed. With the help of papers certifying

Fig. 15. Family Lüttgens in front of the Seehaus in Liebenberg 1944. Wilhelm Lüttgens, trained as a microanalysicist at Krupp's, started working with Warburg in 1930. As faithful "governors" Mr. and Mrs. Lüttgens lived through the most difficult days and weeks starting from the arrival of the first Russian soldiers (30.4.45) till the evacuation of the last cases of public and private belongings (1.8.45) from the radically dismantled institute to a never-known destination (Diary of Wilhelm Lüttgen)

the top urgency of the evacuation, the installation of central heating, electricity, even the erection of high-voltage cables and of a transformer station, were carried out. Once the house was warm, lit and painted, Warburg's interest in its further development as an institute faded. Already in 1944 he was expecting the imminent surrender. Could he be blamed, with his beautiful furniture, partly signed by Chippendale, and his "orderly", Jakob Heiss, for wanting to wait for the end? He began to write the book *Heavy metals and enzyme action* [57] a hybrid of science and scientific curriculum vitae which was later to harm his reputation. The rest of us meanwhile had no occupation and pottered around indefatigably completing the institute until, unexpectedly, it proved to be an efficient and comfortable place of work. This was fortunate in the end, for somehow, possibly from Warburg himself, the authorities had heard about his intention not to work any more and had initiated secret investigations against him. The procedure in which Warburg's head might have been at stake could only be suppressed when a high party official from Hitler's immediate surroundings was lured into coming to Liebenberg, where he could satisfy himself as to the fully functioning research activity under Warburg's leadership.

Directly after the departure of this official, Warburg invited me for a stroll across the park. He probably wished to acknowledge my having finished the building against his will. After a few minutes of silence he asked, alluding to his difficulties with his collaborators: "Do you consider me to be anti-social?" My reply, the impertinence of which still makes me feel ashamed today, was: "As far as I am concerned, the problem is settled with the phrase 'noblesse oblige'". Warburg broke off the conversation. Every member of the institute, Otto Warburg as well, had finished a satisfying piece of experimental work in Liebenberg, whilst the bombing squadrons headed roaringly toward Berlin even in broad daylight.

"Down with all the Enemies of Brandenburg!"

Not only I, student that I was, whose application has been so laboriously reported at the beginning, but "all those who came into contact with him whether they shared his views or not" [2] fell under "the spell of Warburg's personality". In face of all others, whoever his opponent might be, he was completely unafraid. In as far as bravery can be acquired by education, the service in the Prussian army, especially after his injury in the high ranks, was an unequivocal root of it. In 1967, the Prussian guards officer still mentions in a letter to Manfred von Ardenne [43]: "The greater the resistance I met with, all the more I attacked and all the better became my arms. 'In Staub mit allen Feinden Brandenburgs' (Down with all the enemies of Brandenburg!)". At the bottom of his heart, Warburg remained a "staff officer in plain clothes" with all reservations against "civilians". Notwithstanding all criticism, he thought highly of the military: "Here a purer air reigns". When, already with a service award, I showed up at the institute during special leave, he came towards me with great strides and said after a glance at the small pin: "Aha, Bücher, now you know that there are also cowards among the German soldiers!" The fact that I was probably the only one of his pupils (see Krebs' description of his separation from Warburg [44]) who had never had any serious quarrel with him, may be attributed to the fact that I, too, had been trained in a Brandenburgian unit. It may, however, also be possible that this makes me overrate this point of view. In any case, the institute in Gary Street had been planned with farsightedness and was run in such a generous and efficient manner that it functioned so to speak "automatically".

A second root which determined everyday laboratory work leads back to Warburg's years of apprenticeship with Emil Fischer. In the beginning, Warburg must have taken things rather easily. As a young man of good family he was, as he described it himself, probably more interested in taking his huge pedigree Great Dane for walks in the Tiergarten. When he delivered a cinchonin salt of L-α-bromopropionic acid with no more than the specific rotation indicated already by previous investigators, Fischer called him to order energetically. Thereupon he recrystallized the preparation 20 times and almost quadrupled the rotation. Fischer extolled this in the introduction to a joint publication [45] and furthered Warburg's career. Not only the best parts of Warburg's work, but also many of the maxims and aphorisms passed on in the laboratory stood in Emil Fischer's tradition.

Though born in Freiburg, Warburg - as far as the surroundings made an impression on his personality - was a son of the Brandenburgian Berlin. His musicality, his wide reading and his wit fit well into this. His genetic roots are more complex.

The family on his father's side had been residents in the Holstein-Danish town of Altona since the seventeenth century [48]. The mother, on the contrary, who probably took the more active part in his education, came from a Wurttemberg civil servant and officer's family. Drive, courage, creative intelligence and noblesse had probably been handed down from both families, the semitic as well as the alemannic; at the same time, however, also hybridic sensitivity which - in the course of the political development of father and mother country - burdened him to breaking-point.

Epilog

Above all, after the declaration of war against the United States, when he had to manage without the protecting hand of the Rockefeller Foundation, he began to make mistakes. With hardly restrained impatience, Warburg waited for the end of the war and the break-down of Hitler's power. Almost nothing hurt him as much as the icy coldness and contempt with which he was treated as a traitor by the emigrant and the old-American Jewish circles at the end of the war. In addition, came the indignation of the Anglo-Saxon colleagues when the book [57] that had been written in Liebenberg became known. One could not help noticing that Warburg's polemics were based in some part on manipulated quotations. Finally, Keilin proved in a review in *Nature* [47] that even the motto of the publication had been placed in a misleading context. What would the colleagues have said knowing Warburg's original motto - borrowed from an English biography? I quote from memory: "One would think that the best way to fight error would be to expose the truth. But in this perverse world, error does not yield to such gentle methods". I still have, from that time, an exchange of letters with Hans Krebs amongst my papers. It reflects with such liveliness the turbulence of those years which now date back more than a generation that I dare to reproduce relevant passages even though the fighting of the pioneers cannot upset us any more, but rather stimulate us. Looked at from this angle, the *Schwermetalle* is really a rather amusing book, the reading of which can be highly recommended.

Krebs to Bücher (Sheffield, 10, 16th June 1947):

"...Warburg sent me recently a copy of his new book. I was most grateful to have this interesting and valuable account of his work and views. I was sorry, however, that he had thought it necessary to intersperse his story with so many unpleasant polemic remarks. Many of these polemics are, in my view, unfair (because they misrepresent the views of the opponents) and they are often discourteous because he suggests that his victims are dishonest people who deliberately tried to confuse the issue. I feel very distressed about this and have written to Warburg about this, conveying to him these my views. I have implored him to eliminate polemics of this kind from his second edition. I hope he will not be angry at my criticisms and will take them seriously, though I have doubts that he will accept them with good grace. I fear (I hope unnecessarily) that his vanity and pride will run away with him. Humility was never his strong point. It is really a great pity that his outstanding brain is not matched by equal qualities of character...".

Bücher to Krebs (Lübeck, 5th July 1947):

"...I have been very upset by what you wrote about Warburg's last book and about himself. You have recalled to me the abundance of contradictory thoughts and sentiments which occupied my mind during my 6 years of working in Warburg's institute. Personally, I owe a lot to Warburg. He is my master. He brought me back from the front and prevented my being called up again without my being obliged to cheat the war machinery by some scientific half-fraud. That means, at my age and with my military training as an assault engineer officer, nothing less than that I owe him my life. From your lines I can see the same worry which weighs on me and on some of Warburg's friends. Your remarks are very much to the point and you characterize the situation with a precision that is probably only possible in the English language.

Warburg's book was written in days of extraordinary agitation in great as well as in small events: at the beginning of the final phase of the war, of the emptying and evacuation of the institute from Dahlem to Liebenberg (summer 1943 - spring 1944), a time which was filled with dreadful, inevitably escalating disputes internally between Warburg and his coworkers, quarrels which came to a horrifying end when Kubowitz and Negelein were called up to join an active Volkssturm squadron and moved to the collapsing front at the Oder.

I know him from many evening talks in Liebenberg in which I, living under the same roof with Warburg, tried with completely insufficient means to calm him down and to fight his bad spirits ... This year's visit was under the same fatal star of condemnation of his oldest collaborators ...

As I see it, light and shadow have the same source in Warburg: the enormous potential that boils up in the depth of a volcano, his extraordinary temperament that forces him to follow up any of his ideas with incredible force and consequence. I have often admired the efforts with which he kept himself in check, for he has a very fine and sensitive conscience ... I believe that one can influence him more by good deeds than by anything else...".

The chapter entitled "Personality" in the 1972 biography [1], conceived with moving maturity, in no way covering up eccentricies and weaknesses, makes us realize that Sir Hans would today hesitate to apply "humility" and "character", as understood in his letter, as a measuring stick. Probably he did not know that Warburg could secretly have feld himself persecuted, even persecuted to death, on rare occasions even by people close to him. This explains perhaps some of his more eccentric actions. They must not confuse us. If you agree to entrust yourself to my recollection and the thought which moved me whilst writing them, then Otto Warburg will continue to live in our memory not only as an outstanding master of our field, but also - in spite of the severity of his demands - as an intrinsically kind man.

I wish to thank G. Fried (München), H. Gibian (Berlin), B. Hess (Dortmund), W. Lüttgens (Mainz), F. Noll (Berlin-Buch) and K. Wallenfels (Freiburg) for information and documentation. I am grateful to H. Drewitz and I. Linke for their help in making this manuscript.

References

1. Krebs HA (1972) Otto Heinrich Warburg 1883-1970. Biographical Memoirs of Fellows of The Royal Society, 18:629-699
2. Krebs HA, Schmid R (1979) Otto Warburg, Zellphysiologe - Biochemiker - Mediziner 1883-1970. Große Naturforscher 41, Stuttgart
3. Krebs HA (1927) Über den Stoffwechsel der Netzhaut. Biochem Z 189:57

4. Warburg O (1927) Über den heutigen Stand des Carzinomproblems. Naturwissenschaften 15, 1. Physiol 8:519
5. Editorial (1928) Prof. Dr. Otto Warburg: Über das Atmungsferment. Die Umschau 32:228
6. Warburg O (1932) Das sauerstoffübertragende Ferment der Atmung. Les Prix Nobel en 1931, Stockholm. Angewandte Chem. 45:1 (1932)
7. Warburg O (1930) Enzyme Problem and biological oxidations (Herter Lecture). Bull Johns Hopkins Hospital 46:341
8. Barron ESG, Harrop GA Jr (1928) Studies on blood cell metabolism II, the effect of methylene blue and other dyes upon the glycolysis and lactic acid formation of mammalian and avian erythrocytes. J Biol Chem 79:65
9. Warburg O (1938) Chemische Konstitution von Fermenten. Ergebnisse der Enzymforschung. Nord FF, Weidenhagen R (eds) Leipzig, 7:210; bes. Seiten 211 und 239
10. Warburg O (1946) Molekulargewicht des sauerstoffübertragenden Ferments. Naturwissenschaften 33:94
11. Kubowitz F (1934) Über die Hemmung der Buttersäuregärung durch Kohlenoxyd. Biochem Z 274:285
12. Kubowitz F (1935) Kohlenoxyd-Ferrogluthation. Biochem Z 282:277
13. Kubowitz F (1938) Re-Synthese der Phenoloxydase aus Protein und Kupfer. Biochem Z 296:443
14. Negelein E, Wulff H-J (1937) Kristallisation des Proteins der Acetaldehydreduktase. Biochem Z 289:436
15. Warburg O, Christian W (1939) Isolierung und Kristallisation des Proteins des oxydierenden Gärungsferments. Biochem Z 303:40
16. Kubowitz F, Ott P (1942) Isolierung und Kristallisation eines Gärungsferments aus Tumoren. Biochem Z 314:94
17. Fischer HOL (1932) Über die 3-Glycerinaldehyd-phosphorsäure. Ber dt chem Ges 65:337
18. Warburg O, Christian W (1939) Isolierung und Kristallisation des Proteins des oxydierenden Gärungsferments. Biochem Z 303:40
19. Negelein E, Brömel H (1939) R-Diphospho-glycerinsäure, ihre Isolierung und Eigenschaften. Biochem Z 303:132
20. Warburg O, Christain W (1936) Pyridin, der wasserstoffübertragende Bestandteil von Gärungsfermenten (Pyridin-Nucleotide). Biochem Z 287:291
21. Lebedev A v (1911) Hoppe-Seyler's Z phys Chem 73:447
22. Warburg O (1908) Beobachtungen über die Oxydationsprozesse im Seeigelei. Hoppe-Seyler's Z phys Chem 57:1
23. Warburg O (1911) Untersuchungen über die Oxydationsprozesse in Zellen. Münch Med Wschr 58:289; 59:2550 (1912)
24. Warburg O (1914) Beiträge zur Physiologie der Zelle, insbesondere über die Oxydationsgeschwindigkeit in Zellen. Ergebn Physiol 14:253
25. Bücher Th, Negelein E (1942) Photochemische Ausbeute bei der Spaltung des Kohlenoxyd-Hämoglobins. Biochem Z 311:163
26. Bücher Th, Kaspers J (1947) Photochemische Spaltung des Kohlenoxydmyoglobins durch ultraviolette Strahlung (Wirksamkeit der durch die Proteinkomponente des Pigments absorbierten Quanten). Biochim Biophys Acta 1:21
27. Warburg O, Negelein E (1919) Über die Reduktion der Salpetersäure in grünen Zellen. Biochem Z 110:66
28. Warburg O, Negelein E (1932) Über das Hämin des sauerstoffübertragenden Ferments der Atmung, über einige künstliche Hämoglobine und über Spirographis-Porphyrin. Biochem Z 244:9
29. Haas E (1935) Über das Absorptionsspectrum des Wassers im Ultraviolett. Biochem Z 282:224
30. Warburg O, Christian W (1942) Isolierung und Kristallisation des Gärungsferments Enolase. Biochem Z 310:384
31. Warburg O, Christian W (1942) Wirkungsgruppe des Gärungsferments Zymohexase. Biochem Z 311:209
32. Bücher Th (1942) Isolierung und Kristallisation eines phosphatübertragenden Gärungsferments. Naturwissenschaften 30:756; Biochim Biophys Acta 1:292 (1947)

33. Warburg O, Christian W (1942) Isolierung und Kristallisation des Gärungsferments Zymohexase. Naturwissenschaften 30:731
34. Kubowitz F, Ott P (1944) Isolierung von Gärungsfermenten aus menschlichen Muskeln. Biochem Z 317:193
34a. Bücher Th, Pfleiderer G (1955) Pyruvate kinase from muscle. Methods in Enzymology, Colowick SP, Kaplan NO (eds) vol I, pp 435-440, New York
35. Meyer-Arendt E, Beisenherz G, Bücher Th (1953) Isolierung und Kristallisation der Triose-Phosphat-Isomerase. Naturwisschenschaften 40:59
36. Meyerhof O, Lohmann K (1934) Über die enzymatische Gleichgewichtsreaktion zwischen Hexosediphosphorsäure und Dioxyacetonphosphorsäure. Biochem Z 271:89
37. Lohmann K, Meyerhof O (1934) Über die enzymatische Umwandlung von Phosphoglycerinsäure in Brenztraubensäure und Phosphorsäure. Biochem Z 273:60
38. Meyerhof O, Kiessling W (1935) Über die Isolierung der isomeren Phosphoglycerinsäuren aus Gäransätzen und ihre enzymatischen Gleichgewichte. Biochem Z 276:239
39. Warburg O, Christian W (1942) Zymohexase im Blutplasma von Tumortieren. Naturwissenschaften 30:731
40. Kögl F, Erxleben H (1939) Zur Äthiologie der malignen Tumoren. 1. Mitteilung über die Chemie der Tumoren. Hoppe-Seyler's Z physiol Chem 158:57
41. Hixon J (1976) The patchwork mouse - Politics and intrigue in the campaign to conquer cancer. Anchor Press, Doubleday, New York
42. Stein MD (1981) Spectacular cancer mechanism doubted - Cornell retracts reports of kinase cascade. Nature 293:93
43. Ardenne M v (1972) Ein glückliches Leben für Technik und Forschung - Autobiographie. Kindler Verlag, Zürich und München, bes S 343 (s. auch S 334)
44. Krebs H (1978) Otto Warburg - Biochemiker, Zellphysiologe, Mediziner. Vortrag am 8. März 1978 im Harnack Haus, Berlin Dahlem. Jahrbuch der Max-Planck-Gesellschaft für 1978, bes S 96
45. Fischer E (1905) Synthese von Polypeptiden XI. Justus Liebigs Annal Chem 340:123, bes S 126
46. Artikel "Warburg" in "Jüdisches Lexikon", Herlitz G, Kirschner B (eds) Jüdischer Verlag, Berlin 1930, Band V
47. Keilin D (1950) Metal catalysis in cellular metabolism. Nature 165:4
48. Die Umschau (1933) 37:807
49. Kubowitz F, Haas E (1932) Ausbau der photochemischen Methoden zur Untersuchung des sauerstoffübertragenden Ferments (Anwendung auf Essigbakterien und Hefezellen). Biochem Z 255:247
50. Warburg O (1914) Über die Rolle des Eisens in der Atmung des Seeigeleis nebst Bemerkungen über einige durch Eisen beschleunigte Oxydationen. Hoppe-Seyler's Z physiol Chem 92:231
51. Warburg O, Negelein E (1923) Über den Einfluß der Wellenlänge auf den Energieumsatz bei der Kohlensäureassimilation. Z Phys Chem 106:191
52. Warburg O (1924) Verbesserte Methode zur Messung der Atmung und Glykolyse. Biochem Z 152:51
53. Warburg O (1926) Über die Wirkung von Kohlenoxyd auf den Stoffwechsel der Hefe. Biochem Z 177:471
54. Warburg O, Cremer W (1929) Reaktionen des Kohlenoxyds mit Metallverbindungen des Cysteins. Biochem Z 206:228

Books

55. Über den Stoffwechsel der Tumoren. Springer, Berlin 1926
56. Über die katalytischen Wirkungen der lebendigen Substanz. Springer, Berlin 1928
57. Schwermetalle als Wirkungsgruppen von Fermenten. 2. Auflage, Arbeitsgemeinschaft medizinischer Verlage GmbH. Saenger, Berlin 1948

NAD-Dependent Dehydrogenases

Structure-Function Relationships of NAD-Dependent Dehydrogenases

M. G. Rossmann[1]

High resolution structural studies of four NAD-dependent dehydrogenases and five NADP-dependent enzymes have now been completed (Table 1). In many cases, extensive results are also available for complexes of these enzymes with substrates, inhibitors, coenzymes, and coenzyme analogs. I shall review here the wealth of information obtained from these studies and show the stimulus derived in some quite unexpected topics.

Domain Structure, Exons, Introns, and Evolution

Structural comparison of LDH, MDH, GAPDH, and LADH showed immediately that there was a common NAD-binding domain (Fig. 1) and a variable catalytic domain (Fig. 2). Although there was no significant amino acid sequence homology within the various NAD-binding domains, yet four residues were completely conserved and other residues showed some degree of conservation (Table 2). Not only did this demonstrate the greater conservation of tertiary structure, but also served as the basis for the prediction of similar folds in other proteins (Wootton 1974; Benyajati et al. 1981). This suggested gene fusion to express enzymes with more sophisticated functions. Indeed, domains appeared to be the basis for convenient evolutionary changes (Rossmann and Liljas 1974). They were stable folding units with simple functions. Although dehydrogenases were the first structures to show clearly the importance of domains to protein architecture, other examples soon came to light.

Then Gilbert (1978) suggested that the newly discovered exons were the basis of gene shuffling and increased evolutionary rates in eukaryotes. Blake (1978) made the link between domains and exons. Soon some good examples of this relationship appeared as in the comparison of phage and hen egg white lysozyme (Rossmann and Argos 1976; Remington and Matthews 1978; Artymiuk et al. 1981) or the central exon of the globin gene which has a heme binding function (Argos and Rossmann 1979; Craik et al. 1981; Eaton 1980).

Unfortunately, the exon-domain hypothesis is not necessarily always valid. Not only are many exons very small, but predicted regions of domain fusion are not necessarily associated with introns in eukaryotes. A typical example is that of *Drosophila melanogaster* ADH. The gene has been isolated and then expressed in bacteriophage λ permitting

Abbreviations: LDH lactate dehydrogenase. MDH malate dehydrogenase. GAPDH glyceraldehyde-3-phosphate dehydrogenase. LADH liver alcohol dehydrogenase. ADH alcohol dehydrogenase. FDP fructose 1,6-diphosphate

1 Department of Biological Sciences, Purdue University, West Lafayette, Indiana 47907, U.S.A.

34. Colloquium - Mosbach 1983
Biological Oxidations

Table 1. X-ray structural investigations of NAD(P)-utilizing enzymes

Enzyme[a]	Source	Complex	Space Group	Resolution (Å)	References
LDH	Dogfish muscle	Apo	*F*422	2.0	Adams et al. (1970a)
		Tetraiodofluorescein		4.0	Wassarman and Lentz (1971)
		Adenosine		2.8	Chandrasekhar et al. (1973)
		AMP		2.8	Chandrasekhar et al. (1973)
		ADP		2.8	Chandrasekhar et al. (1973)
		ADPR		2.8	Chandrasekhar et al. (1973)
		NAD^+		5.0	Adams et al. (1970b)
		Citrate		2.8	Adams et al. (1973)
		NAD-pyruvate	$P42_12$ and $P2_12_12_1$	3.0	Rossmann et al. (1972); White et al. (1976)
		NAD^+/oxalate		3.0	White et al. (1976)
		NADH/oxamate		3.0	White et al. (1976)
	Mouse testes	Apo	*P*1	2.9	Musick and Rossmann (1979b)
	Pig muscle	NADH/oxamate	$P22_12_1$	6.0	Hackert et al. (1973)
	Pig heart	NADH/oxamate	*C*2	6.0	Eventoff et al. (1975)
		S-lac-NAD^+	$P3_221$	2.7	Grau et al. (1981a,b)
	B. stearothermophilus	Apo		2.5	Schär et al. (1982);
		FDP/NADH			unpublished results
	L. casei	FDP/Co^{2+}		4.0	Buehner et al. (1982)
GAPDH	Lobster muscle	NAD^+	$P2_12_12_1$	2.9	Buehner et al. (1974a,b)
					Moras et al. (1975)
		8-Br-NAD^+		2.9	Olsen et al. (1976a)
		NAD^+ (meso)		2.9	Olsen et al. (1976a)
		NAD^+/citrate		2.9	Olsen et al. (1976b)
		NAD^+/trifluoroacetone		2.9	Garavito et al. (1977b)
		Apo	*P*1	2.9	Murthy et al. (1980)
	B. stearothermophilus	NAD^+ (holo)		2.7	Biesecker et al. (1977)
		Apo		6.0	Wonacott and Biesecker (1977)
	Human	NAD^+ (holo)		5.0	Watson et al. (1972)
ADH	Horse liver	Apo	$C222_1$	2.4	Eklund et al. (1976)
		ADPR		2.9	Abdallah et al. (1975)
		8-Br-ADPR		4.5	Abdallah et al. (1975)
		ADP		3.7	Zeppezauer et al. (1975)
		3-I-$PyAD^+$/imidazole[b]		4.5	Samama et al. (1977)

Table 1 (cont.)

Enzyme[a]	Source	Complex	Space Group	Resolution (Å)	References
ADH (cont.)		$PyAD^+$/imidazole[c]		4.5	Samama et al. (1977)
		$PyAD^+$		4.5	Samama et al. (1977)
		5-iodosalicylate		4.5	Einarsson et al. (1974)
		1,10-phenanthroline		4.5	Boiwe and Brändén (1977)
		Imidazole		2.9	Boiwe and Brändén (1977)
		Iodoacetate (mod.)		4.5	Zeppezauer et al. (1975)
		H_2NADH dimethylamino cinnamaldehyde		2.9	Cedergren-Zeppezauer et al. (1982)
		Pyrazole		3.2	Eklund et al. (1982a)
		NADH/DMSO	$P1$	2.9	Eklund and Brändén (1979); Eklund et al. (1981)
		Cibacron blue F3GA		3.4	Biellmann et al. (1979)
		NAD(H)-bromobenzyl		2.9	Eklund et al. (1982b)
		NAD^++trifluoroethanol		4.5	Plapp et al. (1978a)
		NAD(H)+solvent alcohol		4.5	Plapp et al. (1978a)
		H_2NADH dimethylamino cinnamaldehyde		2.9	Cedergren-Zeppezauer et al. (1982)
		NAD^+ pyrazole		2.9	Eklund et al. (1982a)
		NAD^++4-iodopyrazole		2.9	Eklund et al. (1982a)
		Isonicotinamidylated amino groups	$C2$		Plapp et al. (1978b)
s-MDH	Pig heart	NAD^+	$P2_12_12$	2.4	Hill et al. (1972); Banaszak and Bradshaw (1975); Birktoft and Banaszak (1983)
		Apo		5.5	Weininger et al. (1977)
m-MDH	Pig heart	Apo	$P6_2$ or $P6_4$ and $P2_1$		Weininger and Banaszak (1978)
6-PGDH			$C222_1$	6.0	Adams et al. (1977a,b)
		$NADP^+$		6.0	Abdallah et al. (1979)
		NADPH		6.0	Abdallah et al. (1979)
		NADPH+6-phosphogluconate		6.0	Abdallah et al. (1979)

Table 1 (cont.)

Enzyme[a]	Source	Complex	Space Group	Resolution (Å)	References
β-HADH	Pig heart	Apo	$C222_1$ and $P3_121$ or $P3_112$		Weininger et al. (1974)
ICDH	*E. coli*	Apo	$C2$		Hackert et al. (1977)
	A. vinelandii	Apo	$P4_22_12$		Czerwinski et al. (1977)
GluDH	Tuna liver	Apo	$I23$		Birktoft et al. (1979)
	Rat liver	Apo and PLP modified	$P6_222$		Birktoft et al. (1980)
GDH	Chicken muscle	Apo	$P1$		McPherson and White (1980)
FH_2R	*E. coli*	Methotrexate	$P6_1$	1.7	Matthews et al. (1977); Bolin et al. (1982)
	L. casei	Methotrexate/NADPH	$P6_1$	1.7	Matthews et al. (1978); Bolin et al. (1982)
	Chicken liver	NADPH+phenyltrizine		2.9	Volz et al. (1982)
GTR	Human	FAD	$B2$	3.0	Thieme et al. (1981)
Ferredoxin R	Spinach	Apo	$C2$	3.7	Sheriff and Herriott (1981)
p-OHBz-hydroxylase	*P. fluorescens*	*p*-OHBz/FAD	$C222_1$	2.9	Hofsteenge et al. (1980)

[a]Enzymes: LDH, lactate dehydrogenase; GAPDH, glyceraldehyde-3-phosphate dehydrogenase; ADH, alcohol dehydrogenase; MDH malate dehydrogenase; 6-PGDH, 6-phosphogluconate dehydrogenase; β-HADH, L-3-hydroxyacyl coenzyme A dehydrogenase; ICDH, isocitrate dehydrogenase; GluDH, glutamate dehydrogenase; GDH, glycerol-3-phosphate dehydrogenase; FH_2R, dihydrofolate reductase; GTR, glutathione reductase; *p*-OHBz-hydroxylase, *p*-hydroxybenzoate hydroxylase

[b]3-I-PyAD, 3-iodopyridine adenine dinucleotide

[c]PyAD, pyridine adenine dinucleotide

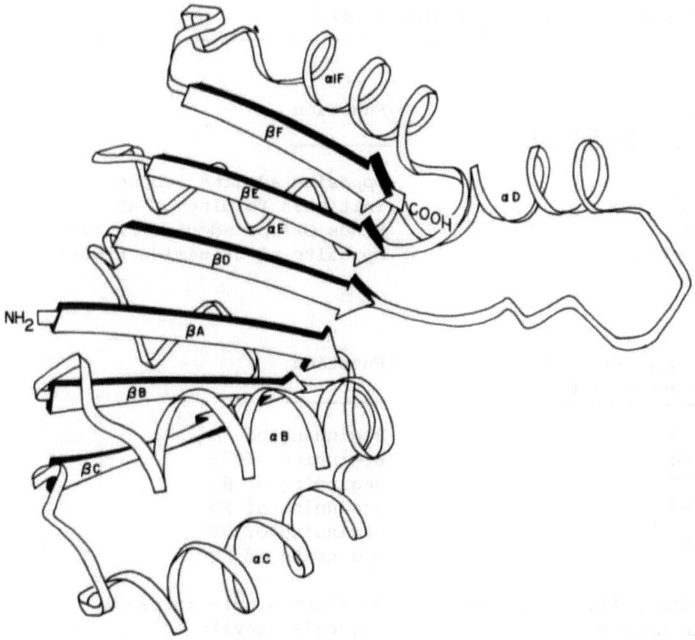

Fig. 1. The NAD-binding domain common to NAD-dependent dehydrogenases. There are six parallel strands in the β-sheet. The first three (βA, βB, βC) with their connecting α-helices (αB and αC) bind AMP at their carboxy-terminal end. The last three strands (βD, βE, βF) bind NMN in a twofold-related equivalent position

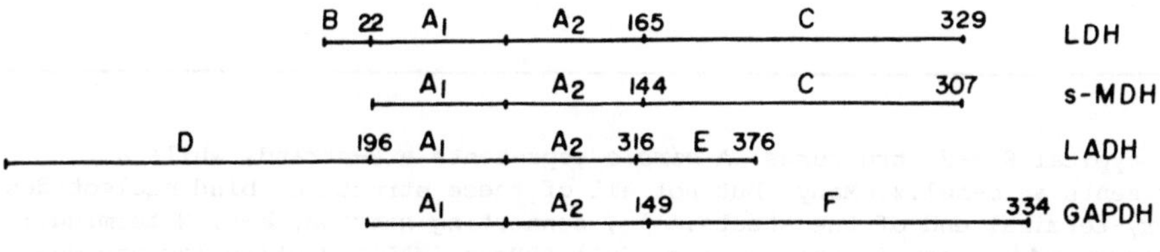

Fig. 2. LDH, s-MDH, LADH, and GAPDH all contain a common NAD-binding domain A_1A_2 and different catalytic domains (C, D, and F). The NAD-binding domain consists of two similar halves for binding AMP and NMN

sequencing of the DNA (Benyajati et al. 1981). Secondary structural predictions together with the known characteristics of NAD-binding domains (Rossmann et al. 1974, 1975) helped to identify the domain boundary. However, no intron was found corresponding to the junction with the catalytic domain. On the other hand, one intron occurred at the αB-βB corner of the NAD-binding domain and another occurred in the catalytic domain. These results are, however, consistent with the observation of Craik et al. (1982), who showed that introns nearly always relate to corners at the molecular surface in the protein structure.

Table 2. Conservation of residues in the NAD-binding domain

1) *Invariant*

Residue	LDH Sequence Number	Function
Gly	28	Opposite adenine ribose
Gly	33	Initiates folding of αB
Asp[a]	53	Binds O2' of adenine ribose
Gly	99	Opposite nicotinamide ribose

2) *Conservative Changes*

Preferred Residue	Equivalent LDH Sequence Number	Function
Lys	23	Beginning of βA
Lys or Arg	48	Beginning of βB
Lys	77	Beginning of βC
Lys	92	Beginning of βD
Lys or Arg	134	Beginning of βE
Lys or Arg	159	Beginning of βF
Hydrophobic	Alternative residues in sheets	Residues within internal molecular cavity

[a] Spatially equivalent position is arginine 43 for NADP-binding *L. casei* dihydrofolate reductase (Matthews et al. 1978)

Fig. 3. Some typical β-α-β structures. A *circle* represents a β-strand, while a *hexagon* represents an α-helix. Many, but not all of these structures bind nucleotides at the carboxy-terminal end of the sheets. ——, connecting section; ▶—, N terminus; —▶, C terminus; —◀ ⊢—, gap in the section. (*a*1) LDHase (NAD), dogfish lactate dehydrogenase. Part of the NAD binding domain (22 to 163); (*a*2) MDHase (NAD), beef heart cytoplasmic malate dehydrogenase. Part of the NAD binding domain; (*b*) ADHase (NAD), horse liver alcohol dehydrogenase. The NAD binding domain (168 to 318); (*c*) GPDHase (NAD), lobster glyceraldehyde-3-phosphate dehydrogenase. The NAD binding (1 to 147); (*d*) PGKase (ATP), horse phosphoglycerate kinase. Part of the ATP-binding domain; (*e*) Rhodanese, bovine liver rhodanese (23 to 113). This figure also represents part of the second domain (160 to 249), which has a similar fold; (*f*) Subtilisin, subtilisin BPN' (26 to 153); (*g*) ADHase (CAT), horse liver alcohol dehydrogenase. Part of the catalytic domain (348 to 374); (*h*) GPDHase (CAT), lobster glyceraldehyde-3-phosphate dehydrogenase. Part of the catalytic domain (236 to 312); (*i*) PGKase (N), horse phosphoglycerate kinase. Part of the N-terminal domain; (*j*) Papain, papaya latex papain (111 to 131 and 200 to 210); (*k*) TIMase, chicken triose phosphate isomerase (1 to 248); (*l*) Flavodoxin, Clostridial flavodoxin (1 to 138); (*m*) PGMase, yeast phosphoglycerate mutase (1 to 215); (*n*) Cyt b_5, calf liver cytochrome b_5 (1 to 80); (*o*) Hexokinase, yeast hexokinase (86 to 161 and 321 to 450); (*p*) CAase, human carbonic anhydrase C (188 to 258); (*q*) AKase, porcine adenyl kinase (10 to 194); (*r*) CPase, bovine carboxypeptidase A (60 to 306); (*s*) Thioredoxin, *E. coli* thioredoxin S_2 (1 to 58). [Reprinted with permission from Sternberg and Thornton (1976). Copyright by Academic Press Inc. (London) Ltd.]

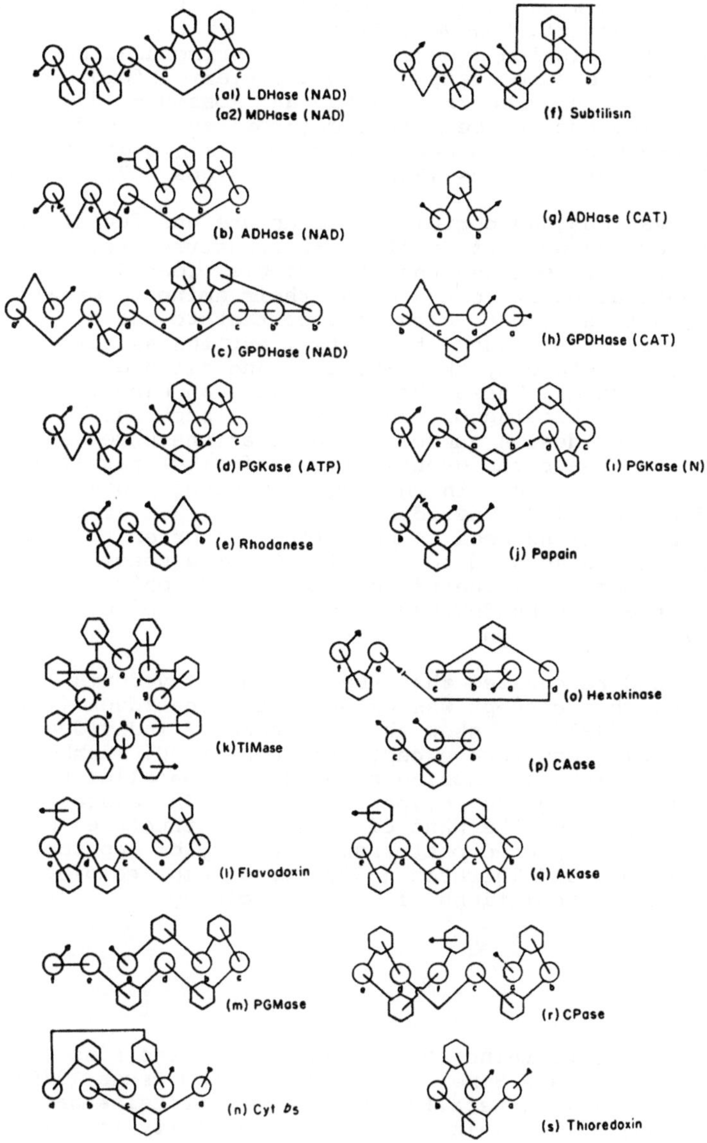

Fig. 3 (Figure caption see opposite side)

The Nucleotide Binding Fold and Supersecondary Structures

Rao and Rossmann (1973) have shown that the structure of flavodoxin and the NAD-binding domain of LDH were remarkably similar and that the FMN and NAD coenzymes bound in a similar manner to the carboxyl end of the parallel β-sheet of these structures. Rossmann et al. (1974) and Ohlsson et al. (1974) suggested further that these structures might represent generalized nucleotide binding folds. Indeed, it soon became apparent that nearly all structures which required the binding of mono-

or dinucleotides had a domain with a β-α-β structure (Levitt and Chothia 1976; Fig. 3). It was not, however, clear whether these similarities were due to divergent evolution from some common primordial nucleotide binding domain or due to convergence to a preferred folding motif (supersecondary structure as defined by Rao and Rossmann 1973) with the function of binding nucleotides. This controversy raged for some considerable time and generated a search for criteria to differentiate these alternative possibilities (Rossmann and Argos 1976, 1977).

Although the above observations had correlated structure and function, yet the physical basis for the properties of β-α-β structures was not made until Hol et al. (1978) recognized that the dipole created by α-helices liberated a partial positive charge at their amino-terminal ends. It is, therefore, not the sheet which is important to the β-α-β structures (although the sheet provides stability), but the associated partial positive charges, provided by the α-helices and aligned near the carboxyl ends of the strands. These charges provide an electrostatic field suitable for the binding of negative ions, such as the phosphate moieties in nucleotides. A number of calculations have been performed in an attempt to assess the quantitative affect of parallel helices or their affect on each other in the requirement of protein folding (Warwicker and Watson 1982; Sheridan et al. 1982; Hol et al. 1981). However, these calculations need to be refined in order to show how a particular β-α-β fold is adapted to the binding of a specific nucleotide. Such calculations must, therefore, include not only the electrostatic forces, but also consider hydrophobic clefts and crevices into which the nucleotide is inserted.

Rao and Rossmann (1973) had recognized that the LDH and flavodoxin folds had a similar hand. This concept was extended by Richardson (1976) and Sternberg and Thornton (1976) to show that the hand of individual β-α-β segments is, with only very rare exceptions, conserved. Furthermore, there is a preference for neighboring strands in a parallel β-sheet to be near each other along the polypeptide chain. The discovery of the explicit folding rules for the hand of crossover structures has led to a search for other unsuspected patterns. A variety of apparent preferences has led Sternberg and Thornton (1978) to attempt complete three-dimensional structural predictions of protein folding.

Thermal Stability

Organisms tend to adapt their proteins to function within their normal environmental temperature (Low and Somero 1976; Feeney and Osuga 1976; Hochachka and Somero 1973). This generally implies that proteins have a limited temperature range within which structural integrity is maintained. Outside this thermal span, denaturation occurs with corresponding loss of function, such as enzymic activity. The thermal stability of a protein can be changed intrinsically by alteration of amino acids or extrinsically by addition of suitable stabilizing effectors (e.g., cations, coenzymes, membranes, and peptides). Yet the structure of the three-dimensional polypeptide backbone of proteins is highly conserved (Rossmann et al. 1975; Kretsinger 1975), despite alterations in the primary sequence. Thus, with the use of the known tertiary structure as well as the amino acid sequences of thermophilic and mesophilic variants, it should be possible to determine some of the principles with which a protein achieves structural tolerance within defined temperature limits.

Only a few tertiary structures of proteins have been determined where amino acid sequences are known for species of widely different thermal environments. Ferredoxin is one such protein, although the variation of thermal stability is not extreme (Perutz and Raidt 1975). By far the best examples are GAPDH and LDH. The structures of the mesophilic lobster (Buehner et al. 1974a; Moras et al. 1975) and thermophilic *Bacillus stearothermophilus* (Biesecker et al. 1977) proteins have been independently determined for GAPDH. Although Wonacott and Biesecker (1977) suggest a change in hydrogen bonding in the vicinity of the active center, this structural alteration does not involve an amino acid change and remains to be validated by a careful comparison of the electron densities. In addition, they also note the possibility of additional salt bridges for stabilizing the thermophile. More strikingly, the amino acid sequence differences for these two GAPDH molecules involve 130 substitutions within 331 positions. It would thus seem possible that many small alterations would in net effect produce the desired thermal stability for this 140000 molecular weight tetramer. A more careful study of GAPDH is possible with the sequence of an extreme thermophile, namely *Thermus aquaticus* (Harris and Walker 1977). The variation of *B. stearothermophilus* LDH (H. Zuber, unpublished results, but see Table 6) to dogfish LDH is similar to that of *B. stearothermophilus* GAPDH and lobster GAPDH.

In Table 3 references to the protein sequences are given, and their associated "optimal" temperatures of stability, used to determine the statistical effect of specific amino acid changes by Argos et al. (1979). The optimal temperatures are assumed to correspond to normal environmental temperatures for bacteria (Buchanan and Gibbons 1974) or as body temperatures for pig and chicken. While the exact mean thermal stable range for a given enzyme is debatable, the general temperature trend is clearly reasonable.

Table 3. Amino acid sequences used in study on thermal stability

Protein	Species	References	Assumed Midpoint of Temperature Stability (°C)
Ferredoxin	*C. thermosaccharolyticum*	Tanaka et al. (1973)	55
	C. tartarivorum	Tanaka et al. (1971)	46
	P. elsdenii	Azari et al. (1973)	37
	C. acidi-urici	Rall et al. (1969)	41
	C. pasteurianum	Tanaka et al. (1966)	37
	M. aerogenes	Tsunoda et al. (1968)	37
GAPDH	*T. aquaticus*	Harris and Walker (1977)	71
	B. stearothermophilus	Harris and Walker (1977)	65
	Yeast	Jones and Harris (1972); Holland and Holland (1979)	37
	Pig	Harris and Perham (1968)	37
	Lobster	Davidson et al. (1967)	20
LDH	*B. stearothermophilus*	Zuber (1983)	65
	Chicken M_4	Eventoff et al. (1977)	37
	Pig M_4	Eventoff et al. (1977); Kiltz et al. (1977)	37
	Dogfish M_4	Eventoff et al. (1977); Taylor (1977)	20

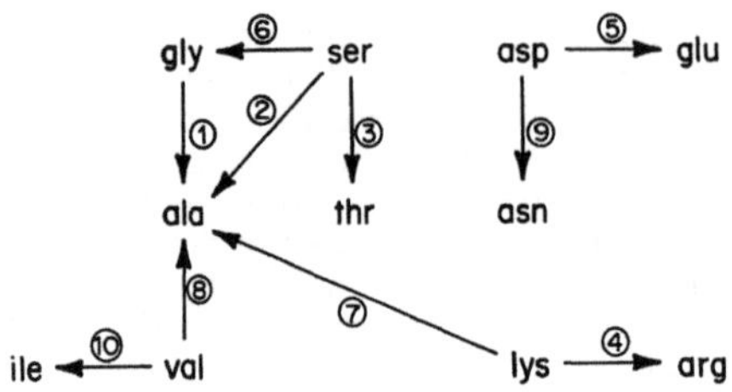

Fig. 4. Direction of preferred exchanges for producing a thermally stable enzyme. *Arrows* point from the mesophilic to the thermophilic protein. *Numbers* indicate the ranking of the significance for a given exchange

Table 4. Benefit of exchange to stability

Exchange Cold→Hot	Helical Region	Sheet Region	Internal Hydrophobicity	External Polarity	Internal Packing
Gly→Ala	++		++		++
Ser→Ala	++		+		+
Ser→Thr		+			-
Lys→Arg	-	+	++	++	+
Asp→Glu	+				-
Ser→Gly	-			+	-
Lys→Ala		+	++	--	++
Val→Ala	++	-	--	+	--
Asp→Asn	--		+	--	-
Val→Ile		--	+		+

The precise analysis of these data was given by Argos et al. (1979). Suffice it here to state that there are some obvious trends. For instance, many lysine residues are replaced by arginines in those proteins stable at higher temperatures. Such observations are also consistent with the results of lysine modifications of pig H_4 and M_4 LDH (Tuengler and Pfleiderer 1977; Shibuya et al. 1982). The trends were essentially independent of the enzyme studied and are shown diagrammatically in Fig. 4. These were then analyzed in terms of various physical properties of the amino acids and their position within the known structure (Table 4) and shown to be beneficial to stability.

The increase or decrease in hydrophobicity is extremely marked whenever the exchange is primarily internal or external, respectively. This property is the most important stabilizing property. Improvement in the internal packing arrangement of residues is less marked while increase of external polarity exhibited little consistency. The helical regions have a strong trend for stabilization. However, all these trends are superimposed on a random background of changes. Thermal stability is largely achieved by an additive series of very small improvements at many locations within the molecule.

Amino Acid Sequences and the Active Center of 2-Hydroxy Acid Dehydrogenases

A large amount of amino acid sequence information is now available for the 2-hydroxy acid dehydrogenases (Table 5). Unfortunately, quite a substantial part of this data base has not yet been published. Nevertheless, the available published data are collected in Table 6. Some of these data have been analyzed by Li et al. (1983) for determining a phylogenetic tree (Fig. 5). They suggest, based on this analysis, that the gene duplication of the LDH-C gene is older than 750×10^6 years.

Table 5. Amino acid sequences of 2-hydroxy acid dehydrogenases

Dogfish muscle LDH	Taylor (1977)
Pig muscle } LDH Pig heart }	Kiltz et al. (1977)
Chicken muscle } LDH Chicken heart }	Eventoff et al. (1977)
Rat } testes LDH Mouse }	Pan et al. (1983)
Bacterial LDH: 7 different bacilli	Zuber (1983)
Bacterial LDH: *Lactobacillus casei*	Hensel (1983)
Pig heart mitochondrial MDH	Birktoft et al. (1982)

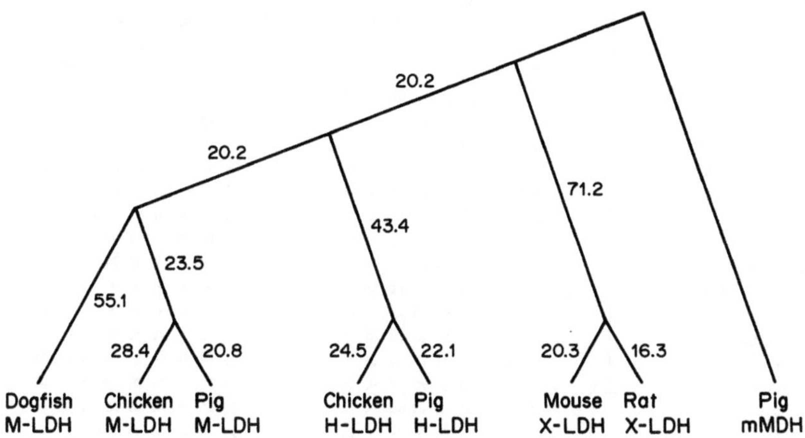

Fig. 5. Phylogenetic tree derived from homologous 2-hydroxy acid dehydrogenases amino acid sequences. (Adapted from Li et al. 1983)

The tree also rules out that the LDH-C gene was derived from LDH-B as had been suggested (Markert et al. 1975). The greater similarity of LDHs from the same gene in different species is consistent with immunological and hybridization experiments (Eventoff et al. 1977).

The rate of evolution for different parts of the amino acid sequence is substantially different. The largest concentration of changes occurs in the amino-terminal arm and in the region 294-310. These two regions complement each other in that they form the principal *R*-axis-generated subunit contacts. For MDH (which is a *Q*-axis dimer), the arm is entirely missing and for the tetrameric bacterial LDHs, the arm starts at around residue 15 and is, therefore, only about 6 residues long. The "loop" region and the βJ-α1G corner are also particularly variable.

In both LDH (Table 6) and GAPDH (Buehner et al. 1974a), the coenzyme binding domain is more conserved than the catalytic domain. Residues used for coenzyme binding and substrate binding are the most conserved. Residues in the *Q*-axis subunit contact surfaces (relating NAD-binding domains) are more conserved than those in the *P* or *R* - axis contact surfaces.

Table 6. Amino acid homology for the 2-hydroxy acid dehydrogenases

```
                                                  |<- βA ->|        |<---- αB ----
                      10                  20 22           30                 40
      DM      A T L K D K L I G H L A T S Q E P R S Y N K I T V V G V G A V G M A C A I S I L
      CM      · · L K D H L I H N V H K E E H A H A H N K X X X X X X X X X X X M A C A I S I L
      PM      A T L K D Q L I H N L L K E E H · V P H N K I T V V G V G A V G M A C A I S I L
      CH      A T L K E K L I T P V A A G S T · V P S N K X X X X X X X X X X X M A C A I S I L
LDH   PH      A T L K E K L I A P V A Q Q E T T I P N N K I T V V G V G Q V G M A C A I S I L
      MX      S T V K E E L I Q N L V P D K L · · S R C K I T V V G V G D V G M A C A I S I L
      RX      S T V K E E L I Q N L A P D K Q · · S R C K I T V V G V G D V G M A C A I S I L
      BS      · · · · · · · · · · · · · · M K N N G G A R V V V I G A G F V G A S Y V F A L M
   m-MDH      · · · · · · · · · · · · · · · · · · · · A K V A V L G · A S G G I G Q P L S L L

              αB→|      |<---- βB ---->| |<---------- αC ---------->|    |<- βC ----
                        50                   60                   70
      DM      M K · D · L A D E V A L V D V M E D K L K G E M M D L Q H G S L F L H T A K I V
      CM      M K · B · L A B Z L T L V B V V Z B K L K G E M M D L Q H G S L F L K T P K I T
      PM      M K · E · L A D E I A L V D V M E D K L K G E M M D L Q H G S L F L R T P K I V
      CH      G K · G · L C D E L A L V D V L E D K L K G E M M D L Q H G S L F L Q T H K I V
LDH   PH      G K · S · L T D E L A L V D V L E D K L K G E M M D L Q H G S L F L Q T P K I V
      MX      L K · G · L A D E L A L V D A D T D K L R G E A L D L L H G S L F L S T P K I V
      RX      L K · G · L A D E L A L V D A D E D K L K G E A L D L L H G S L F L S T P K I V
      BS      N Q · G · I A D E I V L I D A N E S K A I G D A M D F N H G K V F A P K P V D I
   m-MDH      L K N S P L V S R L T L Y D I A · · H T P G V A A D L S H I E T R A · T V K G Y

              βC→|              |<--- βD --->| |<-------- Loop -------->|<-- αD ----
              80             90                 100*            *110
      DM      S G K D Y S V S A · · G S K L V V I T A G A R Q Q E G E S R L N L V Q R N V N I
      CM      S G K D Y S V T A · · N S K L V I V T A G A R Q Q E G E S R L N L V Q R N V N I
      PM      S G K D Y N V T A · · N S R L V V I T A G A R Q Q E G E S R L N L V Q R N V N I
      CH      A B K B Y A V T A · · N S K I V V V T A G V R Q Q E G E S R L N L V Q R N V N V
LDH   PH      A N K D Y S V T A · · N S K I V V V T A G V R Q Q E G E S R L N L V Q R N V N V
      MX      F G K D Y N V S A · · N S K L V I I T A G A R M V S G Q T R L D L L Q R N V A I
      RX      F G K D Y S V S A · · N S K L V I I T A G A R M V S G Q S R L A L L Q R N V T I
      BS      W H G D Y D D C R · · D A D L V V I C A G A N Q K P G E T R L D L V D K N I A I
   m-MDH      L G P E Q L P D C L K G C D V V V I P A G V P·R K P G M T R D D L F N T N A T I
```

Table 6 (cont.)

```
              ——αD——→|←——αE——→| |←————βE————→| |←————————α1F————————→|
              120                130              140              150
       DM     F K F I I P N I V K H S P D C I I L V V S N P V D V L T Y V A W K L S G . . . .
       CM     F K F I I P N V V K Y S P D C K X X X X X X X X X X X X V A W K I S G . . . .
       PM     F K F I I P N I V K Y S P N C K L L V V S N P V D I L T Y V A W K I S G . . . .
LDH    CH     F K F I I P Z I V X X X X X X X X V V S N P V D I L T Y V T W K L S G . . . .
       PH     F K F I I P Q I V K Y S P N C I I I V V S N P V D I L T Y V T W K L S G . . . .
       MX     M K A I V P G V I Q N S P D C K I I V V T N P V D I L T Y V V W K I S G . . . .
       RX     M K A I V P G V I Q N S P D C K I M I V T N P V D I L T Y V V W K I S G . . . .
       BS     F R S I V S E V M A H G F Q G L F L V A T N P V D I L T Y A T W K F S G . . . .
    m-MDH     V A T L T A A C A Q H C P D A M I C I I S N P V N S T I P I T A E V F K K H G V

                  |←—βF—→|   |←————————α2F————————→|                  |←—βG—→|
                  160             *  170*              180                  190
       DM     L P M H R I I G S G C N L D S A R F R Y L M G E R L G V H S C S C H G W V I G E
       CM     F P K H R V I G S G C N L D S A R F R H L M G E R L G I H P L S C H G W I V G Q
       PM     F P K N R V I G S G C N L D S A R F R Y L M G E R L G V H P L S C H G W I L G E
LDH    CH     L P K H R V I G S G C N L D T A R F R Y L M A E R L G I H P T S C H G W I L G E
       PH     L P K H R V I G S G C N L D S A R F R Y L M A E K L G V H P S S C H G W I L G E
       MX     F P V G R V I G S G C N L D S A R F R Y L I G E K L G V N P T S C H G W V L G E
       RX     L P V S S V I G S G C N L D S A R F R Y L I G E K L G V N P T S C H G W V L G E
       BS     L P H E R V I G S G T I L D T A R F R F L L G E Y F S V A P Q N V H A Y I I G E
    m-MDH     Y N P N K I F G V T . T L D I V R A N A F V A E I K G L D P A R V S V P V I G G

                |←————————βH————————→|      |←βJ→|                |←——α1G————
              *     200                  209    210 211              220
       DM     H G D S V P S V W S G M W N A . . . . L K E L H P E L G T N K D K Q D W K K L H
       CM     H G D S S V P V W X X X X X X X X X . X X N L H P D M G T B A B K Z X X K E V H
       PM     H G D S S V P V W S G V N V A G V S . L K N L H P E L G T D A D K E H W K A V H
LDH    CH     H G D S S V A V W S G V N V A G V S . L Q Q L N P A M G T B K B S E N W K E V H
       PH     H G D S S V A V W S G V N V A G V S . L Q Q L N P E M G T D N D S E N W K E V H
       MX     H G D S S V P I W S G V N V A G V T . L K S L N P A I G T D S N K Q H W K N V H
       RX     H G D S S V P V W S G V N V A G V T . L K S L N P A I G S D S N K Q E W K T V H
       BS     H G D T E L P V W S Q A Y I . G V M . P I R K L V E S K G E E A Q K D L E R I F
    m-MDH     H A G K T I I P L I S Q C T P K D F P Q D Q L S T . L . T . . . . . . . . . . .
```

Table 6 (cont.)

α1G→|←α2G→| |←α3G→|

		230 240 250 260
	DM	K D V V D S A Y E V I K L K · G Y T S W A I G L S V A D L A E T I M K N L C R V
	CM	K Q V V D S A Y E V I K L L · G Y T S W A I G L S V A D L A E T I M K N L R R V
	PM	K E V V D S A Y E V I K L K · G Y T S W A I G L S V A D L A E S I M K N L R R V
	CH	K Q V V E S A Y E V I R L K · G Y T N W A I G L S V A Z L C Z T M L K N L Y R V
LDH	PH	K M V V E S A Y E V I K L K · G Y T N W A I G L S V A D L I E S M L K N L S R I
	MX	K Q V V E G G Y E V L D M K · G Y T S W A I G L S V T D L A R S I L K N L K R V
	RX	K Q V V D G G Y E V L D L K · G Y T S W A I G L S V T D L A E S I L K N L K R V
	BS	V N V B D A A Y Q I I E K K · G A T Y Y G I A M G L A R V T R A I L H N E N A I
	m-MDH	G R I Q E A G T E V V K A K A G A G S · · A T L S M A Y A G A R F V F S L V D A

|←βK→| |←βL→| |←βM→|

		270 280 290 301
	DM	H P V S T M V K D F · · · · Y G I K D N V F L S L P C V L N D H G I S N · I V K
	CM	H P I S T A V K G M · · · · H G I K D D V F L S V P C V L G X X X X X X · X X X
	PM	H P I S T M I K G L · · · · Y G I K E N V F L S V P C I L G Q M G I S D · V V K
	CH	H S V S T L V K G T · · · · Y G I Q D D V F L S L P C V L S A S G L T S · V I N
LDH	PH	H P V S T M V Q G M . . . . Y G I E N E V F L S L P C V L N A R G L T S · V I N
	MX	H P V T T L V K G F · · · · H G I K E E V F L S I P C V L G E S G I T D · F V K
	RX	H A V T T L V K G L · · · · Y G I K E E I F L S I P C V L G E S G I T D · L V K
	BS	L T V S A Y L D G L · · · · Y G E R D · V Y I G V P A V I N R N G I R E · V I E
	m-MDH	M N G K E G V V E C S F V K S Q E T D C P Y F S T P L L L G K K G I E K N L G I

|←αH→|

		310 320 330
	DM	M K L K P N E E Q Q L Q K S A T T L W D I Q K D L K · F · · · · ·
	CM	X I L K P D E E E Q I K K S A D T L W G I Q K E L G · F · · · · ·
	PM	V T L T P E E E A H L K K S A D T L W G I Q K E L Q · F · · · · ·
	CH	Q K L K D D E V A K L K K S A D T L W S I Q K D L K D L · · · · ·
LDH	PH	Q K L K D D E V A Q L K N S A D T L W G I Q K D L K D L · · · · ·
	MX	V N M T A E E E G L L K K S A D T L W N M Q K N L E · L · · · · ·
	RX	V N M N T E E E A L F K K S C D I L W N I Q K N L E · L · · · · ·
	BS	I E L N D D E K N R F H H S A A T L K S V L A R A F T R · · · · ·
	m-MDH	G K I S P F E E K M I A E A I P E L K A S I K K G E E F V K N M K

LDH from: DM, dogfish muscle (LDH-A); CM, chicken muscle (LDH-A); PM, pig muscle (LDH-A); CH, chicken heart (LDH-B); PH, pig heart (LDH-B); MX, mouse testes (LDH-C); RX, rat testes (LDH-C); BS, *B. stearothermophilus*. m-MDH, mitochondrial malate dehydrogenase from pig heart. Secondary structural elements are shown. Residues essential for catalysis are marked with an *

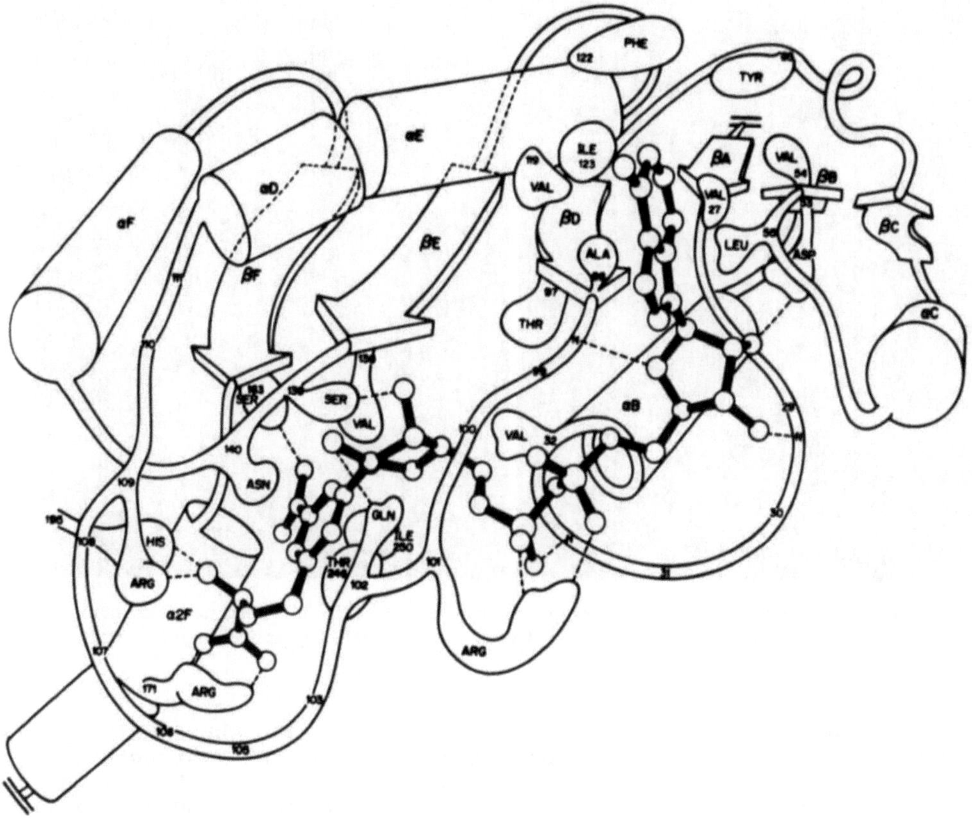

Fig. 6. Nucleotide binding fold and active site in pig H_4 LDH with bound *S*-lac-NAD^+. (Grau et al. 1981a)

The relationship between sequence and antigenicity had been briefly discussed by Eventoff et al. (1977). Recently, Goldberg and co-workers (Gonzales-Prevatt et al. 1982) have been able to isolate a peptide (152-159) which reacts specifically with rabbit anti-mouse LDH-X. Immunization of rabbits with this peptide conjugated to bovine serum albumin induces an immune response which is specific for the peptide. It is situated on the very outside of the molecule and is at the carboxyl end of α1F.

The essential residues at the active center of LDH are His 195 (the proton sink and source), Arg 171 (substrate binding; Holbrook et al. 1975), and Arg 109 (controlled by the loop movement). A more general view of the NAD-substrate site is shown in Fig. 6. Recently, Banaszak and his colleagues (Birktoft and Banaszak 1983; Birktoft et al. 1982) have recognized that s-MDH Asp 152 forms a hydrogen bond with the essential histidine residue in the holoenzyme structure. They suggest that this is important for catalysis by forming an electron relay system similar to that found in serine protease (Blow et al. 1969; Kraut 1977). This aspartate is entirely conserved being residue no. 168 in the LDH numbering system. The refined structure of dogfish M_4 apo LDH

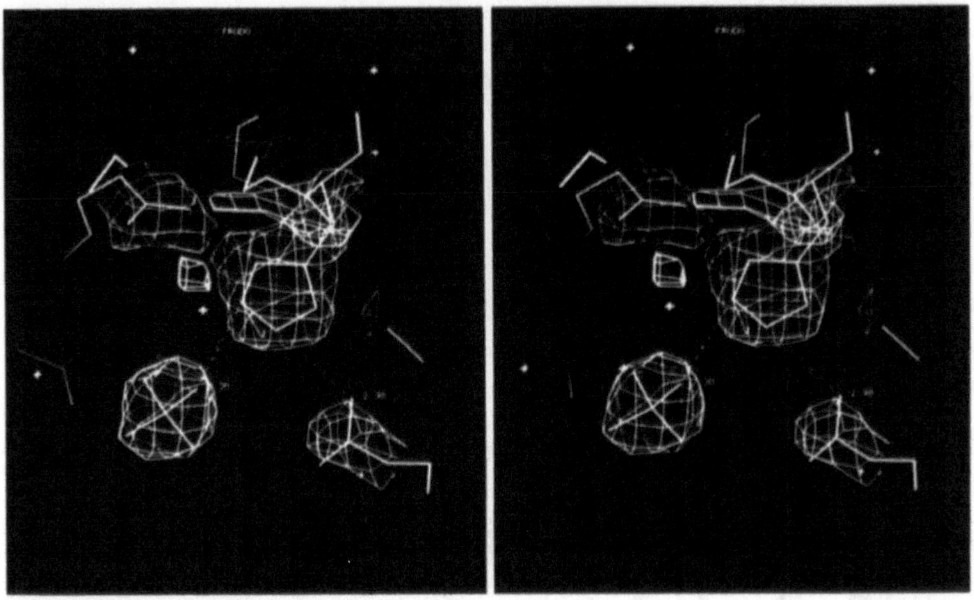

Fig. 7. The Asp 168-His 195 charge relay system as seen in the refined structure of dogfish M_4 apo LDH. The stereo pair was taken on the MMS-X graphics system using Alwyn Jones' FRODO program

(Abad-Zapatero and Sussman, unpublished results) also shows this hydrogen bonding system (Fig. 7), as does the apo mouse LDH-X structure (Musick and Rossmann 1979a). The refined dogfish M_4 apo LDH shows an intricate hydrogen bonding network around His 195 involving Asp 168, Asn 140, and SO_4^{2-} at the substrate binding site (Fig. 7). This network does not seem to exist in the ternary complex (White et al. 1976; Grau et al. 1981a) due to a conformational change of His 195, although these structures must still be refined to ascertain the accuracy of this assertion. Previous postulates on the mechanism have not taken into account such a "charge relay" system. Figure 8 shows a possible mechanism which includes the effect of Asp 168, the insertion of Arg 109 into the active center during catalysis, and the oil-water histidine proposal of Parker and Holbrook (1977) as modified by Grau et al. (1981a) and Birktoft and Banaszak (1983).

In Fig. 8a, the apoenzyme with the NAD in the process of binding is depicted. As the adenine finds its hydrophobic pocket, so Arg 101 is attracted to the pyrophosphate group which is the trigger for closing down the loop. However, before this event is complete, the substrate enters and binds into the substrate pocket (Fig. 8b). The positive charge on Arg 109 induces a negative charge on C2 of the substrate thus pushing a proton toward His 195, which in turn permits cancellation of the charge on Asp 168 (Fig. 8c). Indeed, the reduced solvent accessibility would cause the electroneutral form of the His-Asp pair to be favored, resulting in an increase of pK of His 195. Thus, Arg 109 transiently stabilizes the negative charge on the substrate and, therefore, permits the transfer of a hydride ion to the nicotinamide (Fig. 8d). Removal of the hydride now causes the substrate C2 atom to alter its geometry to a trigonal keto form. His 195 is thus again available to hydrogen bond with Asp 168. The additional proton on His 195 will repel Arg 109 causing the loop to open and to discharge of product into water.

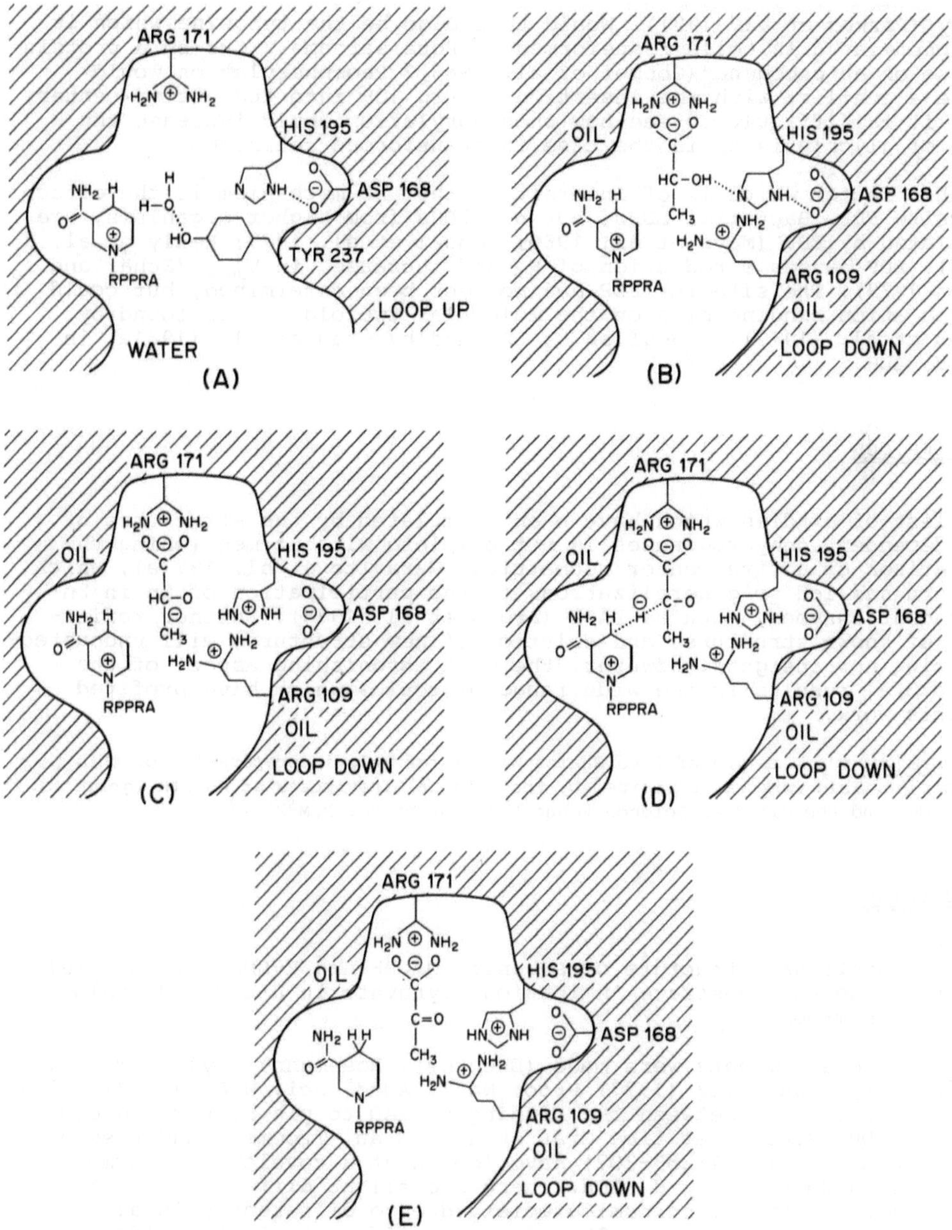

Fig. 8. The possible participation of various essential residues in the catalysis of LDH

When the loop comes down to make the ternary complex, Arg 109 replaces a water molecule held in position by Tyr 237 on helix α2G. This tyrosine is moved away in the ternary position. Tyrosine 237 has been identified as having an altered pK in the apo and ternary structure (Parker et al. 1982). Chemical modification of this residue affects the motional stability of this residue as determined by NMR spectroscopy (Parker et al. 1981). Would this also be the tyrosine which is phosphorylated in cells transformed by Rous sarcoma virus by a protein kinase of an oncogene (Cooper et al. 1983)? Phosphorylation would cause a conflict with the insertion of Arg 109 into the active center, thus it is difficult to see how phosphorylation would increase the rate of glycolysis as is the case in transformed cells.

Another factor which is of interest in the LDH mechanism is the effector site FDP. Bacterial LDHs, but not LDHs from higher organisms, are regulated by FDP (Mayr et al. 1980; Hensel et al. 1977; Kelly et al. 1978). FDP causes a reduction of K_m and increase in V_{max} (Schär and Zuber 1979). The site for FDP has not yet been determined, but could be the anion binding site on the molecular twofold *P*-axis found by Adams et al. (1973) in dogfish M_4 LDH and by Grau et al. (1981a) in pig H_4 LDH.

Conclusions

The list of studies which have been stimulated by the structures of NAD-dependent dehydrogenases is not complete without mentioning the comparison of active center geometries (Garavito et al. 1977a), which in turn has led to generalizations to the participation of Zn in Zn-dependent enzymes, such as LADH (Argos et al. 1978). Further refinement of known structures and solution of new structures will undoubtedly bring new insights. However, the most astonishing aspect of the structural results is the wide range of topics which have profited by this work.

Acknowledgments. I am grateful to Sharon Wilder for help in preparation of this manuscript. This work was supported by the National Institutes of Health (grant no. GM 10704) and the National Science Foundation (grant no. PCM82-07747).

Discussion

Bücher: Functional structure of subunits M-LDH and H-LDH with special reference to the substrate inhibition (pyruvate in H-LDH and oxaloacetate in m-MDH).

Rossmann: Some comments were made (Eventoff, Rossmann, Taylor, Torff, Meyer, Keil, and Kiltz (1977) Proc. Natl. Acad. Sci. USA 74:2677-2681) on the differences between the binding of NAD to pig H_4-LDH and dogfish M_4-LDH. Since that time Grau et al. (Grau, Trommer, and Rossmann (1981) J Mol Biol 151:289-307) have looked at a covalent coenzyme-substrate adduct in pig H_4-LDH. There are slight differences in the environment of the nicotinamide moiety due to differences in amino acids in the active center. The slightly different nicotinamide position and orientation will no doubt be responsible, at least in part, for differences in properties of the H and M isozymes.

Buehner: You indicated in your lecture that Len Banaszak's proposal of a hydrogen bond-proton shuttle between His-195 and Asp-168 might not bear on the enzymatic activity of LDH, because this hydrogen bond is not visible in the structures of ternary complexes. How large is the difference between apo- and ternary LDH, in other words do you think it possible that thorough refinement of ternary structures might bring the His-Asp H-bond back to life?

Rossmann: I had not wished to say that the His-195-Asp-168 did not bear on the enzymatic activity. On the contrary, it certainly will have considerable influence. However, I had hoped I had said that the hydrogen bond is broken in the ternary complex. Hence, it is the formation and breakage of this bond which is important. You are right in asking how well the ternary structure is refined, for in fact we have not done any serious refinement. Nevertheless, ternary complexes of the dogfish M_4 and pig H_4 structures both show the same results.

Schulz: The role of the C-terminal position of βB for discriminating against the binding of NADPH is also discussed in our contribution.

Rossmann: Thank you. I gather that glutathione reductase solves this problem in a manner equivalent to dihydrofolate reductase. This is indeed most interesting.

Schulz: Do you know the dihedral angles of the conserved glycines? Are they side-chain-forbidden or not?

Rossmann: I do not have that information here. I will look up the information when I return to Purdue. You are obviously wondering whether this is a standard β turn where glycine is (almost) invariant. This is a good point and I gather that in glutathione reductase the dihedral angle in fact conforms to the requirements of a glycine in this position.

Trommer: Michael, you have mentioned that Manfred Bühner has solved the structure of *Lactobacillus casei* LDH. This enzyme differs considerably from all the other LDHs studied so far in as much as it is allosterically regulated by manganese and FDP. Have the binding sites of these effectors already been identified?

Rossmann: Reinhardt Hensel has shown (manuscript in preparation) that modification of His-188 inhibits the binding of FDP. This residue controls the binding of anions on the *P*-axis (Adams, Liljas, and Rossmann (1973), J Mol Biol 76:519-531). Clearly then the two FDP molecules are positioned similarly to the four citrate molecules which we studied with respect to dogfish M_4 LDH. This site also involves Arg-173 which is close (but pointing in the opposite direction) to Arg-171, the substrate binding residue. Hence, the *P*-axis anion binding site is well-suited for communication between *P*-axis-related subunits.

Hensel: Metal ion-binding of the allosteric L-LDH from *Lactobacillus casei*: We think that the Me^{2+} ions bind close to the FDP binding region. Possibly the Me^{2+} ions compensate the negative charges of FDP bound in the subunit contacts across the molecular *P*-axis.

Wallenfels: 30 years ago Horst Sund discoverd a linear relationship between the number of free -SH groups in crystalline YADH preparations of different age and the enzyme activity. Can you explain that?

Rossmann: No, I am afraid not. Some light on this property may be shed when the structure of yeast ADH has been determined.

References

Abdallah MA, Biellmann JF, Nordström B, Brändén CI (1975) Eur J Biochem 50:475-481
Abdallah MA, Adams MJ, Archibald IG, Biellmann JF, Helliwell JR, Jenkins SE (1979) Eur J Biochem 98:121-130

Adams MJ, Ford GC, Koekoek R, Lentz PJ Jr, McPherson A Jr, Rossmann MG, Smiley IE, Schevitz RW, Wonacott AJ (1970a) Nature (London) 227:1098-1103
Adams MJ, McPherson A Jr, Rossmann MG, Schevitz RW, Wonacott AJ (1970b) J Mol Biol 51:31-38
Adams MJ, Liljas A, Rossmann MG (1973) J Mol Biol 76:519-531
Adams MJ, Helliwell JR, Bugg CE (1977a) J Mol Biol 112:183-197
Adams MJ, Archibald IG, Helliwell JR (1977b) In: Sund H (ed) Pyridine nucleotide-dependent dehydrogenases. Walter de Gruyter, Berlin pp 72-84
Argos P, Rossmann MG (1979) Biochemistry 18:4951-4960
Argos P, Garavito RM, Eventoff W, Rossmann MG, Brändén CI (1978) J Mol Biol 126: 141-158
Argos P, Rossmann MG, Grau UM, Zuber H, Frank G, Tratschin JD (1979) Biochemistry 18:5698-5703
Artymiuk PJ, Blake CCF, Sippel AE (1981) Nature (London) 290:287-288
Azari P, Glantz M, Tsunoda J, Yasunobu KT (1973) unpublished results cited in Yasunobu KT, Tanaka M (1973) Syst Zool 22:570-589
Banaszak LJ, Bradshaw RA (1975) In: Boyer PD (ed) The enzymes, 3rd edn, Vol XI. Academic, New York pp 369-396
Benyajati C, Place AR, Powers DA, Sofer W (1981) Proc Natl Acad Sci US 78:2717-2721
Biellmann JF, Samama JP, Brändén CI, Eklund H (1979) Eur J Biochem 102:107-110
Biesecker G, Harris JI, Thierry JC, Walker JE, Wonacott AJ (1977) Nature (London) 266:328-333
Birktoft JJ, Banaszak LJ (1983) J Biol Chem 258:472-482
Birktoft JJ, Miake F, Banaszak LJ, Frieden C (1979) J Biol Chem 254:4915-4918
Birktoft JJ, Miake F, Frieden C, Banaszak LJ (1980) J Mol Biol 138:145-148
Birktoft JJ, Fernley RT, Bradshaw RA, Banaszak LJ (1982) Proc Natl Acad Sci US 79:6166-6170
Blake CCF (1978) Nature (London) 273:267
Blow DM, Birktoft JJ, Hartley BS (1969) Nature (London) 221:337-340
Boiwe T, Brändén CI (1977) Eur J Biochem 77:173-179
Bolin JT, Filman DJ, Matthews DA, Hamlin RC, Kraut J (1982) J Biol Chem 257:13650-13662
Buchanan RE, Gibbons NE (eds) (1974) Bergey's manual of determinative bacteriology, 8th edn. Williams and Wilkins, Baltimore
Buehner M, Ford GC, Moras D, Olsen KW, Rossmann MG (1974a) J Mol Biol 90:25-49
Buehner M, Ford GC, Moras D, Olsen KW, Rossmann MG (1974b) J Mol Biol 82:563-585
Buehner M, Hecht HJ, Hensel R, Mayr U (1982) J Mol Biol 162:819-838
Cedergren-Zeppezauer E, Samama JP, Eklund H (1982) Biochemistry 21:4895-4908
Chandrasekhar K, McPherson A Jr, Adams MJ, Rossmann MG (1973) J Mol Biol 76:503-518
Cooper JA, Reiss NA, Schwartz RJ, Hunter T (1983) Nature (London) 302:218-223
Craik CS, Buchman SR, Beychok S (1981) Nature (London) 291:87-90
Craik CS, Sprang S, Fletterick R, Rutter WJ (1982) Nature (London) 299:180-182
Czerwinski EW, Bethge PH, Mathews FS, Chung AE (1977) J Mol Biol 116:181-187
Davidson BE, Sajgó M, Noller HF, Harris JI (1967) Nature (London) 216:1181-1185
Eaton WA (1980) Nature (London) 284:183-185
Einarsson R, Eklund H, Zeppezauer E, Boiwe T, Brändén CI (1974) Eur J Biochem 49: 41-47
Eklund H, Brändén CI (1979) J Biol Chem 254:3458-3461
Eklund H, Nordström B, Zeppezauer E, Söderlund G, Ohlsson I, Boiwe T, Söderberg BO, Tapia O, Brändén CI, Åkeson Å (1976) J Mol Biol 102:27-59
Eklund H, Samama JP, Wallén L, Brändén CI, Åkeson Å, Jones TA (1981) J Mol Biol 146: 561-587
Eklund H, Samama JP, Wallén L (1982a) Biochemistry 21:4858-4866
Eklund H, Plapp BV, Samama JP, Brändén CI (1982b) J Biol Chem 257:14349-14358
Eventoff W, Hackert ML, Rossmann MG (1975) J Mol Biol 98:249-258
Eventoff W, Rossmann MG, Taylor SS, Torff HJ, Meyer H, Keil W, Kiltz HH (1977) Proc Natl Acad Sci US 74:2677-2681
Feeney RE, Osuga DT (1976) Comp Biochem Physiol 54A:281-286
Garavito RM, Rossmann MG, Argos P, Eventoff W (1977a) Biochemistry 16:5065-5071
Garavito RM, Berger D, Rossmann MG (1977b) Biochemistry 16:4393-4398
Gilbert W (1978) Nature (London) 271:501
Gonzales-Prevatt V, Wheat TE, Goldberg E (1982) Mol Immunol 9:1579-1585

Grau UM, Trommer WE, Rossmann MG (1981a) J Mol Biol 151:289-307
Grau UM, Rossmann MG, Trommer WE (1981b) Acta Crystallogr B37:2019-2026
Hackert ML, Ford GC, Rossmann MG (1973) J Mol Biol 78:665-673
Hackert ML, Harris BA, Poulsen LL (1977) Biochim Biophys Acta 481:340-347
Harris JI, Perham RN (1968) Nature (London) 219:1025-1028
Harris JI, Walker JE (1977) In: Sund H (ed) Pyridine nucleotide-dependent dehydrogenases. Walter de Gruyter, Berlin pp 43-61
Hensel R (1983) unpublished results
Hensel R, Mayr U, Stetter KO, Kandler O (1977) Arch Microbiol 112:81-93
Hill E, Tsernoglou D, Webb L, Banaszak LJ (1972) J Mol Biol 72:577-591
Hochachka PW, Somero GN (1973) Strategies of biochemical adaptation. Saunders WB, Philadelphia
Hofsteenge J, Vereijken JM, Weijer WJ, Beintema JJ, Wierenga RK, Drenth J (1980) Eur J Biochem 113:141-150
Hol WGJ, van Duijnen PT, Berendsen HJC (1978) Nature (London) 273:443-446
Hol WGJ, Halie LM, Sander C (1981) Nature (London) 294:532-536
Holbrook JJ, Liljas A, Steindel SJ, Rossmann MG (1975) In: Boyer PD (ed) The enzymes, 3rd edn, Vol XI. Academic, New York pp 191-292
Holland JP, Holland MJ (1979) J Biol Chem 254:9839-9845
Jones GMT, Harris JI (1972) FEBS Lett 22:185-189
Kelly N, Delaney M, O'Carra P (1978) Biochem J 171:543-547
Kiltz HH, Keil W, Griesbach M, Petry K, Meyer H (1977) Hoppe-Seyler's Z Physiol Chem 358:123-127
Kraut J (1977) Ann Rev Biochem 46:331-358
Kretsinger RH (1975) In: Carafoli E, Clementi F, Drabikowski W and Margreth A (eds) Calcium transport in secretion and contraction. North-Holland, Amsterdam pp 469-478
Levitt M, Chothia C (1976) Nature (London) 261:552-558
Li SSL, Fitch WM, Pan YCE, Sharief FS (1983) J Biol Chem 258:7029-7032
Low PS, Somero GN (1976) J Exp Zool 198:1-12
Markert CL, Shaklee JB, Whitt GS (1975) Science 189:102-114
Matthews DA, Alden RA, Bolin JT, Freer ST, Hamlin R, Xuong N, Kraut J, Poe M, Williams M, Hoogsteen K (1977) Science 197:452-455
Matthews DA, Alden RA, Bolin JT, Filman DJ, Freer ST, Hamlin R, Hol WGJ, Kisliuk RL, Pastore EJ, Plante LT, Xuong N, Kraut J (1978) J Biol Chem 253:6946-6954
Mayr U, Hensel R, Kandler O (1980) Eur J Biochem 110:527-538
McPherson A, White H (1980) Biochem Biophys Res Commun 93:607-610
Moras D, Olsen KW, Sabesan MN, Buehner M, Ford GC, Rossmann MG (1975) J Biol Chem 250:9137-9162
Murthy MRN, Garavito RM, Johnson JE, Rossmann MG (1980) J Mol Biol 138:859-872
Musick WDL, Rossmann MG (1979a) J Biol Chem 254:7621-7623
Musick WDL, Rossmann MG (1979b) J Biol Chem 254:7611-7620
Ohlsson I, Nordström B, Brändén CI (1974) J Mol Biol 89:339-354
Olsen KW, Garavito RM, Sabesan MN, Rossmann MG (1976a) J Mol Biol 107:577-584
Olsen KW, Garavito RM, Sabesan MN, Rossmann MG (1976b) J Mol Biol 107:571-576
Pan YCE, Sharief FS, Okabe M, Huang S, Li SSL (1983) J Biol Chem 258:7005-7016
Parker DM, Holbrook JJ (1977) In: Sund H (ed) Pyridine nucleotide-dependent dehydrogenases. Walter de Gruyter, Berlin pp 485-501
Parker DM, Holbrook JJ, Birdsall B, Roberts GCK (1981) FEBS Lett 129:33-35
Parker DM, Jeckel D, Holbrook JJ (1982) Biochem J 201:465-471
Perutz MF, Raidt H (1975) Nature (London) 255:256-259
Plapp BV, Zeppezauer E, Brändén CI (1978a) J Mol Biol 119:451-453
Plapp BV, Eklund H, Brändén CI (1978b) J Mol Biol 122:23-32
Rall SC, Bolinger RE, Cole RD (1969) Biochemistry 8:2486-2496
Rao ST, Rossmann MG (1973) J Mol Biol 76:241-256
Remington SJ, Matthews BW (1978) Proc Natl Acad Sci US 75:2180-2184
Richardson JS (1976) Proc Natl Acad Sci US 73:2619-2623
Rossmann MG, Argos P (1976) J Mol Biol 105:75-95
Rossmann MG, Argos P (1977) J Mol Biol 109:99-129
Rossmann MG, Liljas A (1974) J Mol Biol 85:177-181
Rossmann MG, Adams MJ, Buehner M, Ford GC, Hackert ML, Lentz PJ Jr, McPherson A Jr, Schevitz RW, Smiley IE (1972) Cold Spring Harbor Symp Quant Biol 36:179-191
Rossmann MG, Moras D, Olsen KW (1974) Nature (London) 250:194-199

Rossmann MG, Liljas A, Brändén CI, Banaszak LJ (1975) In: Boyer PD (ed) The enzymes, 3rd edn, Vol XI. Academic, New York pp 61-102
Samama JP, Zeppezauer E, Biellmann JF, Brändén CI (1977) Eur J Biochem 81:403-409
Schär HP, Zuber H (1979) Hoppe-Seyler's Z Physiol Chem 360:795-807
Schär HP, Zuber H, Rossmann MG (1982) J Mol Biol 154:349-353
Sheridan RP, Levy RM, Salemme FR (1982) Proc Natl Acad Sci US 79:4545-4549
Sheriff S, Herriott JR (1981) J Mol Biol 145:441-451
Shibuya H, Abe M, Sekiguchi T, Nosoh Y (1982) Biochim Biophys Acta 708:300-304
Sternberg MJE, Thornton JM (1976) J Mol Biol 105:367-382
Sternberg MJE, Thornton JM (1978) Nature (London) 271:15-20
Tanaka M, Nakashima T, Benson A, Mower H, Yasunobu KT (1966) Biochemistry 5:1666-1681
Tanaka M, Haniu M, Matsueda G, Yasunobu KT, Himes RH, Akagi JM, Barnes EM, Devanathan T (1971) J Biol Chem 246:3953-3960
Tanaka M, Haniu M, Yasunobu KT, Himes RH, Akagi JM (1973) J Biol Chem 248:5215-5217
Taylor SS (1977) J Biol Chem 252:1799-1806
Thieme R, Pai EF, Schirmer RH, Schulz GE (1981) J Mol Biol 152:763-782
Tsunoda JN, Yasunobu KT, Whiteley HR (1968) J Biol Chem 243:6262-6272
Tuengler P, Pfleiderer G (1977) Biochim Biophys Acta 484:1-8
Volz KW, Matthews DA, Alden RA, Freer ST, Hansch C, Kaufman BT, Kraut J (1982) J Biol Chem 257:2528-2536
Warwicker J, Watson HC (1982) J Mol Biol 157:671-679
Wassarman PM, Lentz PJ Jr (1971) J Mol Biol 60:509-522
Watson HC, Duée E, Mercer WD (1972) Nat New Biol 240:130-133
Weininger MS, Banaszak LJ (1978) J Mol Biol 119:443-449
Weininger M, Noyes BE, Bradshaw RA, Banaszak LJ (1974) J Mol Biol 90:409-413
Weininger M, Birktoft JJ, Banaszak LJ (1977) In: Sund H (ed) Pyridine nucleotide-dependent dehydrogenases. Walter de Gruyter, Berlin pp 87-100
White JL, Hackert ML, Buehner M, Adams MJ, Ford GC, Lentz PJ Jr, Smiley IE, Steindel SJ, Rossmann MG (1976) J Mol Biol 102:759-779
Wonacott AJ, Biesecker G (1977) In: Sund H (ed) Pyridine nucleotide-dependent dehydrogenases. Walter de Gruyter, Berlin pp 140-156
Wootton JC (1974) Nature (London) 252:542-546
Zeppezauer E, Jörnvall H, Ohlsson I (1975) Eur J Biochem 58:95-104
Zuber H (1983) unpublished results

Binding of Coenzyme Analogs to NAD-Dependent Dehydrogenases

J. F. Biellmann[1]

The study of alcohol dehydrogenase from horse liver started in our laboratory with the preparation and the enzymatic studies of coenzyme analogs and received great impetus from the structural analysis of this enzyme by the cristallographic group of Brändén (Uppsala). The collaboration of both groups has led us to draw a number of conclusions of general interest to biochemistry.

For the preparation of coenzyme analogs, nature has provided an enzyme: NAD^+ glycohydrolase, which catalyses the exchange of the nicotinamide ring with substituted pyridines (Fig. 1). This enzyme renders feasible the preparation of a number of analogs. Beside this enzymatic method, a number of chemical transformations may be carried out on the coenzyme and its analogs (Anderson 1982). The most illustrative example is given by the Hofmann degradation of the amid group of NAD^+ (Fisher et al. 1973).

3-Benzoylpyridine adenine dinucleotide (Fig. 2) has been prepared by enzymic transglycosidation and shown to be active only with horse liver alcohol dehydrogenase among the tested dehydrogenases (Anderson et al. 1959a; Anderson and Kaplan 1959b). The replacement of the NH_2 group of the amide by a large phenyl group is indeed a big change. The model study of horse liver alcohol dehydrogenase shows that NAD^+ binds to the enzyme in the binary and ternary complexes with the NH_2 group of the amide of the coenzyme close to the substrate binding site. 3-Benzoylpyridine adenine dinucleotide would then bind to the enzyme in a similar manner to that of NAD^+ with the phenyl group in the substrate

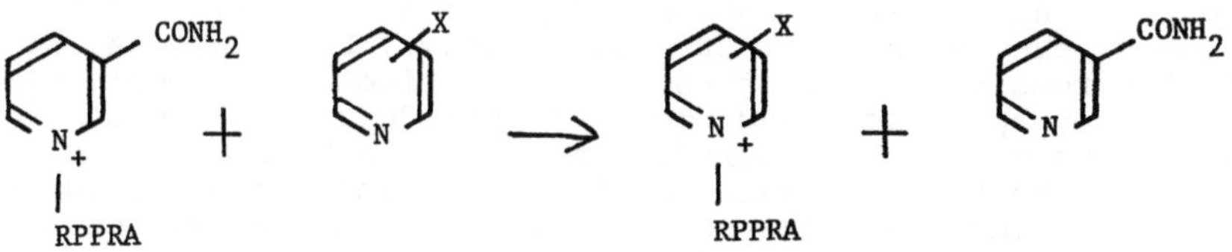

Fig. 1. Exchange catalyzed by NAD^+ glycohydrolase

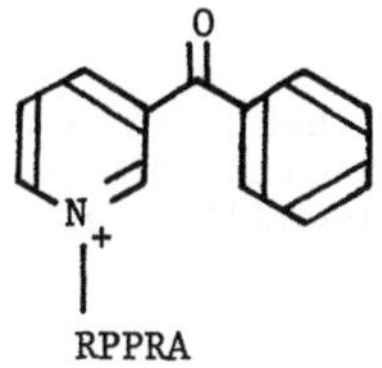

Fig. 2. 3-benzoyl pyridine adenine dinucleotide

1 Laboratoire de Chimie Organique Biologique, Institut de Chimie, Université Louis Pasteur, 1 rue Blaise Pascal, 67008 Strasbourg, France

34. Colloquium - Mosbach 1983
Biological Oxidations

binding site. This should have two consequences: a higher substrate specificity and a different kinetic mechanism. A number of alcohols should be substrate with NAD^+ as coenzyme and should not be substrate with 3-benzoylpyridine adenine dinucleotide. The mechanism should change from Theorell chance with NAD^+ to a random bi-bi mechanism with 3-benzoylpyridine adenine dinucleotide. Indeed a number of primary and secondary alcohols are not substrate on the presence of 3-benzoylpyridine adenine dinucleotide. With ethanol as substrate, the mechanism is random bi-bi (Samama et al. 1983).

Taking advantage of the structure of the substrate binding site, we designed the structures of molecules which should bind to the alcohol dehydrogenase-NADH complex. The polar part, which acts as ligand to the zinc ion, is an amide or formamide function. A hydrophobic skeleton with proper orientation enforces the binding. Through this rational approach using molecular models, diamond lattice (Dutler and Brändén 1981) and the display, some 15 competitive inhibitors to the substrate with a $K_i < 10^{-6}$ *M* were prepared and found to be effective in vitro and in vivo (Freudenreich et al. 1983). This intellectually more satisfying approach rather than the structure-activity relationship will be used more commonly when more receptor structures are available.

In order to provide the protein cristallographs with heavy atom derivatives of the coenzyme, we prepared the 3-halopyridine adenine dinucleotides (Abdallah et al. 1976; Fig. 3). The corresponding models, the 3-halo-pyridine salts were reduced with sodium dithionite. 3-Halo, 1,4-dihydropyridines have similar absorption spectra to that of the corresponding pyridinium salt. In order to prove that these analogs were active as hydride acceptors, we used a substrate, such as cinnamic alcohol giving rise to cinnamaldehyde, whose absorption at 290 nm was used to follow the reaction with alcohol dehydrogenase or an oxidizing system with dyes (Abdallah and Biellmann 1980). 3-Halopyridine adenine dinucleotides were found to be active with liver alcohol dehydrogenase as hydride acceptors (Abdallah et al. 1976). The binding of 3-iodopyridine adenine dinucleotide to the apoenzyme of liver alcohol dehydrogenase was studied in the cristalline state. It occurs in a different mode to that of NAD^+ and related compounds (Samama et al. 1977). Adenine and adenine ribose are bound in the same manner, but starting from pyrophosphate, the remaining part of the molecule binds in a completely different way: the pyrophosphate and the nicotinamide ribose are on the surface of the enzyme and the pyridinium ring in a hydrophilic environment, close to the Lys-228, whose modification increases the turnover rate by accelerating the coenzyme dissociation (Zoltobrocki et al. 1974; Sogin and Plapp 1975). The pyridinium ring is at a distance of 15 Å from active site zinc ion. A similar binding was found for pyridine adenine dinucleotide and 5-methyl nicotinamide adenine dinucleotide (Samama et al. 1981). So 3-iodopyridine adenine dinucleotide binds to the enzyme in two ways: one close to Lys-228 and one in the active site.

A wrong binding or a multiple mode of binding occurs with these derivatives. Some attention to these modes were paid, but the possibility to detect them lacked. We shall now discuss these binding modes with respect to the affinity labeling.

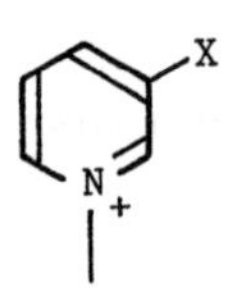

Fig. 3. X = Cl, Br, I 3-Halopyridine adenine dinucleotides

$$E + I \underset{K'}{\overset{k}{\rightleftharpoons}} EI \xrightarrow{k} E_{inact.}$$

$$E + I \rightleftharpoons EI'$$

$$\text{Log}\ \frac{A_o}{A} = \frac{kt}{1 + \frac{K}{K'} + \frac{K}{|I|}}$$

A_o initial activity
A activity at time t
$|I|$ inactivator concentration

$$\frac{1}{K_{app}} = \frac{1}{K} + \frac{1}{K'}$$

$$k_{app} = \frac{k}{1 + \frac{K}{K'}}$$

if $K' \ll K$: $K_{app} = K'$ and $k_{app} = k\,\frac{K'}{K}$

Fig. 4. Criteria for affinity labeling

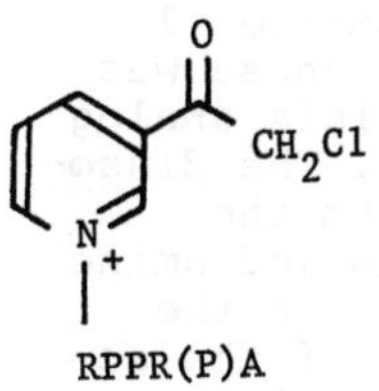

Fig. 5. 3-Chloroacetylpyridine adenine dinucleotide (3-CAPAD$^+$) and 3-chloroacetylpyridine adenine dinucleotide phosphate (3-CAPADP$^+$)

In case of the multiple binding mode of an affinity label, the kinetics of the labeling may not give information about the reversible complex giving rise to the covalent bond. What is actually detected is the complex of highest affinity. One of the criteria for affinity labeling is the identity of the binding constant determined by inactivation kinetics to that found by physical methods or activity determination with the affinity label or an isosteric compound. This complex may not lead to inactivation. So this criteria is only of little value (Fig. 4).

As an affinity label of dehydrogenase, we prepared the 3-chloroacetylpyridine adenine dinucleotide: 3-CAPAD$^+$ (Biellmann et al. 1974) and 3-chloroacetylpyridine adenine dinucleotide phosphate: 3-CAPADP$^+$ (Biellmann et al. 1978; Fig. 5). The labeling of some dehydrogenases was studied. For the present purpose, only the affinity labeling of 17β-hydroxysteroid dehydrogenase will be reported; this enzyme uses NAD$^+$ and NADP$^+$ as coenzyme, NADP$^+$ showing stronger binding than NAD$^+$. So we can compare the behavior of the two affinity labels (Biellmann et al. 1976; Biellmann et al. 1979), 3-CAPADP$^+$ binds more strongly and forms more slowly the covalent bond with the enzyme compared to 3-CAPAD$^+$. It can be explained in the following manner: the complex of high affinity is not responsible for the formation of the covalent bond. Other complexes of lower affinity and not detectable by the usual methods, where the pyridinium rings are differently positioned, give rise to the labeling. Since with 3-CAPADP$^+$, the binding tends to be more in favor of the nonactive complexes as with 3-CAPAD$^+$, the labeling is slower with CAPADP$^+$ than with 3-CAPAD$^+$.

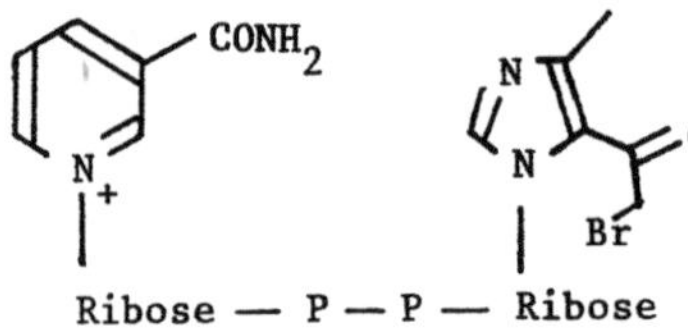

Fig. 6. Nicotinamide 4-methyl-5-bromoacetylimidazole dinucleotide

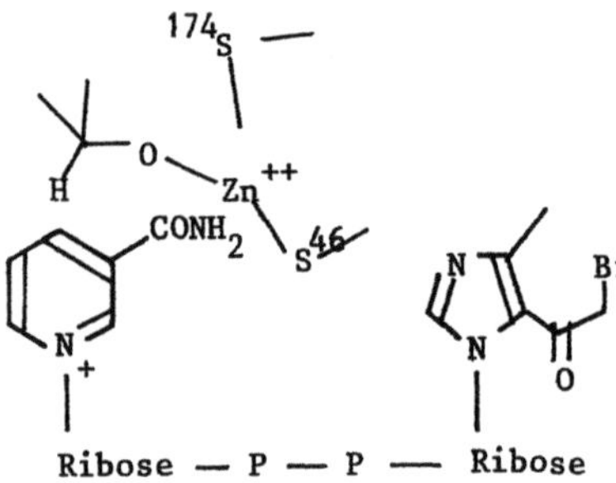

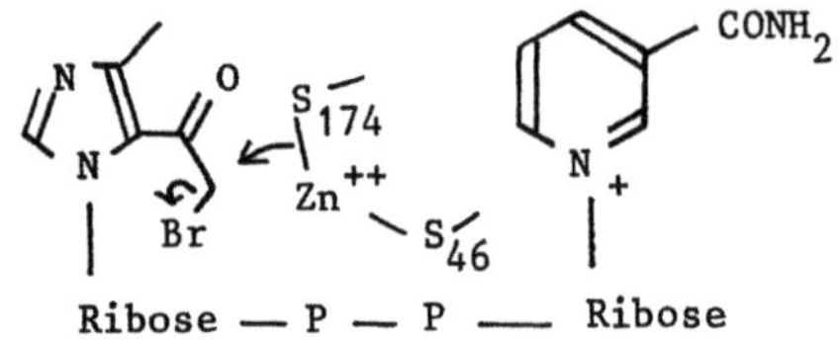

△

Fig. 7b. Complex leading to labeling of Cys-174

◁ Fig. 7a. Hydride transfer complex

A very illustrative example is given by the NAD^+ analog nicotinamide 4-methyl-5-bromoacetylimidazole dinucleotide (Woenckhaus et al. 1970; Fig. 6), where the adenine is replaced by a 4-methyl 5-bromoacetyl imidazole. The reactivity with horse liver alcohol dehydrogenase was studied (Woenckhaus and Jeck 1971; Jörnvall et al. 1975). This analog is active as hydrid acceptor and acts as an affinity label. The dissociation constant determined by labeling kinetics is close to the Michaelis constant determined by activity kinetics. The labeled amino acid is Cys-174, ligand of the active site zinc ion and not in the adenine part binding site of the coenzyme. Although Cys-46 of the active site zinc ion reacts with the other electrophilic reagents, no reaction of Cys-174 had been detected in the native state of the enzyme. This excludes a quasi-affinity where besides formation of a reversible binary complex labeling by a second order process occurs. In a quasi-affinity, labeling of Cys-46 is indeed expected to react and not Cys-174. We propose that for this affinity label two complexes are formed: the first (Fig. 7a) in which the nicotinamide binds to the active site and where the enzymic redox reaction occurs and the second one (Fig. 7b) with reverse binding where the nicotinamide is bound to the adenine site of NAD^+ and the substituted imidazole is close to the active zinc ion and which gives rise to the alkylation of Cys-174. The binding of the pyridinium ring to the adenine binding part is reasonable in view of the fact that the bis-nicotinamide dinucleotide binds to the alcohol dehydrogenase and acts as hydride acceptor.

A related situation has been recently described for the suicide inhibitor: (Likos et al. 1982; Ueno et al. 1982); the reaction of serine O-sulfate with glutamate decarboxylase and aspartate aminotransferase. The electrophilic species does not react with a nucleophile when bound to the coenzyme, but has to dissociate from the coenzyme and to undergo some movements in order to react with the coenzyme, while the carboxylate remains in the anion binding site.

The multiple modes of binding must have consequences in the drug design, where it could be detected in abnormality of the structure-activity relationships. For the design of affinity label, one has to compromise with two requirements: the first one is that the molecule will act as an affinity label and the second one is that the labeled residue has a mechanistic significance. The first requirement implies that the reactive group in the complexes with the enzyme is free to explore a large area in order to hit a reactive group. For the suicide inhibitors, it is required that the inhibitor be activated during the enzymic reaction and that the activated group reaches a reactive group. More flexible groups for suicide inhibitors should be of great utility.

The work presented here has been made possible because we could see what we were doing. The molecular models and the interactive display of the tridimensional structure of enzymes have been innovative tools in this work. Quite clearly the easier access through the interactive display of the tridimensional protein structures has started to change bioorganic chemistry.

The present work has greatly benefited from the very strong interaction with the groups of Brändén (Uppsala). The work was supported by the Université Louis Pasteur of Strasbourg, by the Centre National de la Recherche Scientifique, and by the Institut de Recherches Scientifiques, Economiques et Sociales sur les Boissons for the work on the alcohol dehydrogenase inhibitors.

Discussion

Ghisla: Most of your coenzyme analogs appear to have a weaker binding to the target enzyme. Does this result from an increased k_{off} of the analog as compared to the normal coenzyme?

Biellmann: We have not determined the modifications of the kinetics for the coenzyme analogs. It should be mentioned that careful kinetic studies have been done with horse liver alcohol dehydrogenase with 3-acetylpyridine adenine dinucleotide and thio NAD^+ by Shore and Luisi in their groups.

Massey: In the reaction with the chloroacetyl NAD where three different residues were derived, was the stoichiometry determined? I ask this question because, especially when the "affinity reagent" binds fairly weakly and one has to use high concentrations, there is always the danger of obtaining nonspecific reactions. But in this case of course, one may obtain more than one mole residue of reagent derived per mole protein.

Biellmann: For both reagents 3-chloroacetylpyridine adenine dinucleotide and its phosphate, the stoichiometry was found one per subunit. The mononucleotide 3-chloroacetylpyridine ribose phosphate does not inactivate the enzyme at high concentration.

Jeck: I want to call your attention to a group of NAD analogs recently synthesized by Woenckhaus' group in Frankfurt, in which diazonium residues act as reactive group and I will do this for three reasons:

First of all, under appropriate conditions, these diazonium compounds proved to be rather specific affinity labeling reagents and this fact is still far from being common knowledge.

Secondly, the azomodified amino acid side chains were sufficiently stable to allow enzymatic degradation of the labelled dehydrogenases and last, but not least, the azomodified amino acids show clearly distinct absorption spectra in the long wavelengths UV and short wavelengths visible region.

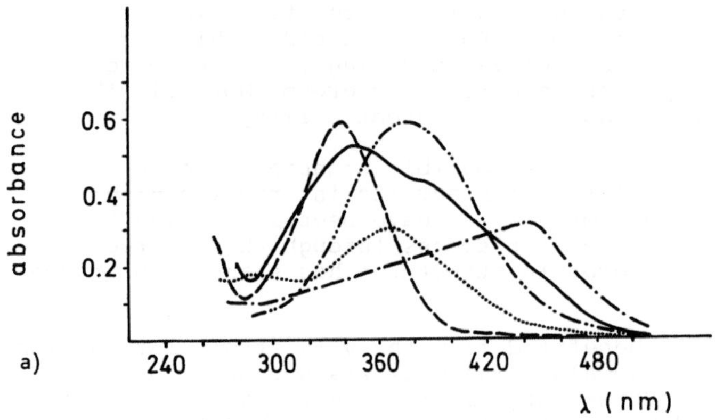

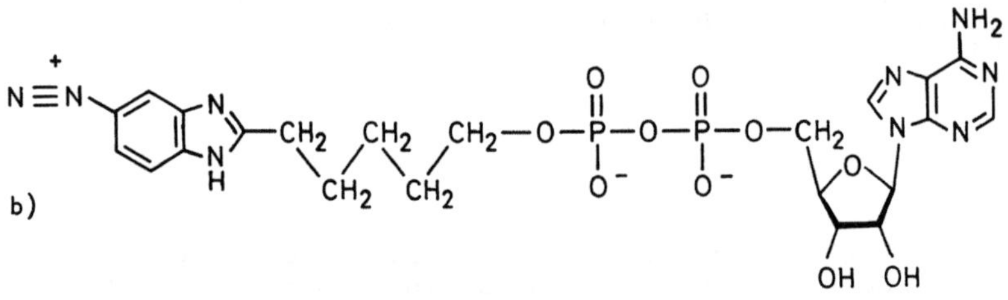

Fig. 8a,b. Absorption spectra (a) of N_α-acetyl amino acids modified with 5-diazotized benzimidazole-2-pentyl ester of adenosine diphosphate (b). Cys ---; His (compound 1) -·-··-; His (compound 2) -·-·-; Lys ·····; Tyr ———

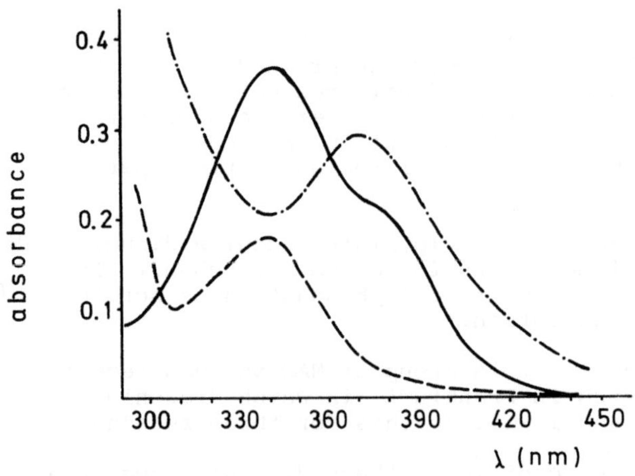

Fig. 9. Absorption spectra of dehydrogenases modified with 5-diazotized benzimidazole-2-pentyl ester of adenosine diphosphate. LADH — — —; GAPDH ———; GlDH -.-.-

Figure 8 shows the absorption spectra of amino acid model compounds prepared by the reaction of N-acetyl-amino acids with a diazotized benzimidazole derivative of adenosinediphosphate.

Figure 9 shows the absorption spectra of three dehydrogenases —halophilic glutamate dehydrogenase (GlDH), glyderaldehyde-3-phosphate dehydrogenase (GAPDH) from rabbit muscle and horse liver alcohol dehydrogenase (LADH) —modified by the same nucleotide derivative.

Comparing both figures, one would guess easily that cysteine is modified in the case of horse liver alcohol dehydrogenase and glyceraldehyde-3-phosphate dehydrogenase from rabbit muscle, while lysine is involved in the modification of halophilic glutamate dehydrogenase, assumptions which were confirmed by successful isolation of the modified amino acids. In any case, the absorption spectra of azomodified proteins yield reliable hints in the identification of active site residues.

References

Abdallah MA, Biellmann JF, Samama JP, Wrixon AD (1976) Eur J Biochem 64:351-360
Abdallah MA, Biellmann JF (1980) Eur J Biochem 112:331-333
Anderson BM, Ciotti CJ, Kaplan NO (1959a) J Biol Chem 234:1219-1225
Anderson BM, Kaplan NO (1959b) J Biol Chem 234:1226-1232
Anderson BM (1982) Everse J, Anderson B and Sa You K (eds) The pyridine nucleotide coenzymes. Academic, New York pp 91-133
Biellmann JF, Branlant G, Foucaud BY, Jung M (1974) FEBS Letters 40:29-32
Biellmann JF, Branlant G, Nicolas JC, Pons M, Descomps B, Crastes de Paulet A (1976) 63:477-481
Biellmann JF, Eid P, Goulas PR, Hirth C (1978) Biochimie 60:417-419
Biellmann JF, Goulas PR, Nicolas JC, Descomps B, Crastes de Paulet A (1979) Eur J Biochem 99:81-84
Dutler H, Brändén CI (1981) Bioorganic Chem 10:1-13
Fisher TL, Vercellotti SV, Anderson BM (1973) J Biol Chem 248, 4293-4299
Freudenreich C, Samama JP, Eklund H, Biellmann JF (1983) unpublished
Jörnvall H, Woenckhaus C, Johnscher G (1975) Eur J Biochem 53:71-81
Likos JJ, Ueno H, Feldhaus RW, Metzler DE (1982) Biochem 21:4377-4386
Samama JP, Zeppezauer E, Biellmann JF, Brändén CI (1977) Eur J Biochem 81:403-409
Samama JP, Wrixon AD, Biellmann JF (1981) Eur J Biochem 118:479-486
Samama JP, Hirsch D, Goulas PG, Biellmann JF (1983) unpublished
Sogin DC, Plapp BV (1975) J Biol Chem 250:205-210
Ueno H, Likos JJ, Metzler DE (1982) Biochem 21:4387-4393
Woenckhaus C, Zoltobroki M, Berghäuser J (1970) Hoppe-Seyler's Z Physiol Chem 351: 1441-1448
Woenckhaus C, Jeck R (1971) Hoppe-Seyler's Z Physiol Chem 352:1477-1423
Zoltobrocki M, Kim JC, Plapp BV (1974) Biochem 13:899-903

What NAD-Dependent Dehydrogenases Teach Us About the Folding and Association of Oligomeric Proteins[1]

R. Jaenicke and R. Rudolph[2]

Introduction

Folding in vivo vs Folding in vitro

The term "protein folding" is commonly used to refer both to the description of the spatial arrangement of the amino acid residues in a functional protein and to the (kinetic) mechanism by which the covalent polypeptide chain achieves its three-dimensional structure (Jaenicke 1980). In this article, we refer to the second alternative which may be considered the postlude of translation. The experimental approach makes use of the reversibility of protein denaturation, focusing on refolding in vitro rather than in vivo folding. The reason is that no direct approach is available to study the acquisition of the native structure of the nascent molecule either in vivo or in the cell-free system (Wetlaufer and Ristow 1973). Evidently the in vitro process must be different from the vectorial in vivo process because refolding starts from the denatured complete chain, while folding of the nascent chain may occur as a co-translational event (Hamlin and Zabin 1972; Bergman and Kuehl 1979). The question of whether refolding indeed reflects the in vivo process cannot be answered unambiguously. However, two criteria suggest that the two reactions follow similar pathways: (1) under certain experimental conditions the kinetics of reconstitution are in the same time range as the folding of nascent polypeptide chains in vivo; (2) the final product of reconstitution after denaturation-renaturation is found to be indistinguishable from the initial native state.

Monomeric vs Oligomeric Proteins

In going from the folding of single polypeptide chains to the formation of the stoichiometrically and geometrically well-defined quaternary structure of oligomeric proteins, association has to be considered in

---

Abbreviations: ADH Alcohol dehydrogenase, L-ADH and Y-ADH for the enzymes from horse liver and yeast, respectively (EC 1.1.1.1). GAPDH Glyceraldehyde-3-phosphate dehydrogenase (EC 1.1.1.12). LDH Lactate dehydrogenase, H_4 and M_4 for the isoenzymes from heart and skeletal muscle, respectively (EC 1.1.1.27). MDH Malate dehydrogenase, m-MDH and s-MDH for the enzymes from mitochondria and cytoplasm, respectively (EC 1.1.1.37). ODH Octopine dehydrogenase (EC 1.5.1.11). PDH Pyruvate dehydrogenase, containing pyruvate decarboxylase, E1 (EC 1.2.3.1), lipoate acyl transferase, E2 (EC 2.3.1.12), and lipoamide dehydrogenase, E3 (EC 1.6.4.3), respectively. (*M*) Unfolded monomers (parentheses indicate unstable intermediates which are not populated). M, M^* Folded monomers. D,T Dimers and tetramers, respectively. k', k" First- and second-order rate constants

1 Dedicated to Lothar Jaenicke on the occasion of his sixtieth birthday

2 Institut für Biophysik and Physikalische Biochemie der Universität Regensburg, 8400 Regensburg, FRG

34. Colloquium - Mosbach 1983
Biological Oxidations

addition to folding. Both reactions must be properly coordinated because the formation of the native assembly requires the surface of the structured monomers to be preformed in the correct way such that specific recognition can occur. As demonstrated by the inherent tendency of self-organization, both the structure formation of the monomer and its assembly to oligomeric or multimeric proteins are determined solely by the specific aminoacid sequence on the one hand, and its aqueous or nonpolar environment on the other.

Thermodynamics vs Kinetics

Considering the "anatomy and taxonomy of protein structure" (Richardson 1981) that have been deduced from well over 100 distinct high resolution crystal structures, a number of topological features and energetic principles of protein folding have been derived. From the thermodynamic point of view, the particular topology of folding and association is determined by the free energy decrease obtained by minimizing the hydrophobic surface area and by optimizing the formation of intramolecular interactions, such as ion pairs and hydrogen bonds. Owing to this principle, the "hydrophobic core" in the interior of a protein molecule and nonpolar subunit contact areas on its surface are sequestered away from the aqueous phase, whereas polar groups are exposed to it. The packing in the interior of a protein molecule resembles a crystal lattice of small organic molecules rather than a low molecular weight alkane in its pure liquid state (Jaenicke 1981).

In the case of small single chain proteins, the two-state model of protein denaturation/renaturation comprising only native or denatured molecules at equilibrium, can be regarded as valid to a first approximation (Privalov 1979, 1982). However, intermediates of folding, which are only marginally populated at equilibrium, may appear in the process of unfolding or refolding of small protein molecules (Kim and Baldwin 1982). In the case of more complex proteins "folding by parts" has been observed as a consequence of "domains" with separable compact local regions of supersecondary structure. If they are clearly separated from vicinal structures, isolated domains produced by limited proteolysis may fold independently to functional entities (Kirschner et al. 1980; Zetina and Goldberg 1980; Wetlaufer 1981; Dautry-Varsat and Garel 1981; Zetina and Goldberg 1982).

General energetic aspects from a great number of determinations of protein structure and stability may be summarized as follows (Pfeil 1981): The Gibbs free energy of unfolding is small ($\Delta G_{unf.} \cong 40$ kJ/Mol) despite the great number of noncovalent contacts maintaining the native structure of proteins. Thus, protein stability reflects a delicate balance of stabilizing and destabilizing contributions required to allow conformational flexibility and order-disorder phenomena underlying protein functioning as well as protein turnover (Huber and Bennet 1983). Since size has no significant effect on protein stability, small substructures, such as domains or subunits must have provided a decisive evolutionary advantage by facilitating folding, flexibility, and transport.

Another essential element in the process of evolution is the degeneracy of the "code" that relates sequence and structure. As shown for a number of proteins, a given protein structure may tolerate a wide variety of alterations of amino acids without loss of biological function. On the other hand, there seems to exist only a limited number of amino acid sequences that can provide a unique viable structure in a given environment. Both observations may be responsible for the apparently small number of architectural groups emerging from the diversity of presently solved structures (for review see Rossmann and Argos 1981; Richardson 1981).

It is obvious that the foregoing considerations do not provide any information on the kinetic pathway since they were concerned with energy. As pointed out by Ptitsyn (Ptitsyn and Finkelstein 1980) and Rossmann (Rao and Rossmann 1973; Rossmann and Argos 1981), topological features are intermingled with the kinetics of protein folding since both involve sequentially adjacent stretches of polypeptide chain which may be assumed to represent "folding units" stabilized by hydrophobic interactions. Such structural elements suggest hypothetical pathways of protein folding. However, they do not provide a definite answer to the question of how the nascent polypeptide chain assumes its three-dimensional conformation.

Based on the successful attempt to regain catalytically active species after denaturation/renaturation of enzymes, biologically significant folding of proteins has been assumed to stabilize the global free energy minimum (Anfinsen and Scheraga 1975). A purely random search mechanism to find the most stable state was dismissed by Levinthal (1968) because of statistical considerations. The alternative kinetic hypothesis which allows fast structure formation involves a nucleation-controlled vectorial folding pathway leading to the "kinetically accessible free-energy minimum" (Wetlaufer and Ristow 1973). Neither the "thermodynamic" nor the "kinetic" hypothesis has been proved in an unambiguous way.

Based on the assumption that folding and refolding follow the same pathway, but in the opposite direction, equilibrium studies in vitro were used to examine the mechanism of the in vivo folding of proteins (Burgess and Scheraga 1975). In this context, the occurrence of stable intermediates in the reversible denaturation has been used to characterize the folding pathway. Obviously, this approach is inadequate because thermodynamic considerations of protein stability do not necessarily reflect the kinetic pathway along which polypeptide chains adopt their native structure.

Scope of Folding Studies on Dehydrogenases

To solve the problem of protein folding, small molecules such as ribonuclease, bovine pancreatic trypsin inhibitor, or antibody fragments, which can be completely and reversibly denatured under a variety of conditions, are far better suited than large oligomeric enzymes.

On the other hand, reconstitution experiments with dehydrogenases, which cannot be as easily renatured as, e.g., ribonuclease, should provide a good test for the generalization of folding data obtained with small proteins. In addition, NAD-dependent dehydrogenases may be taken as a model in the attempt to obtain optimum experimental conditions for protein reconstitution from a practical point of view. What NAD-dependent dehydrogenases may tell us is: (1) what the rate of refolding in the case of large multi-domain proteins, such as monomeric dehydrogenase is and what the intermediates under varying solvent conditions are; (2) how folding and subunit assembly depend on each other in the case of the oligomeric dehydrogenases; (3) what the sequence of events in the quaternary structure formation of dimeric, tetrameric, and multimeric systems is; (4) to what extent the coalescence of domains and subunits is specific; and (5) what the mechanisms competing with renaturation are, which are responsible for the decrease in the yield of reactivation

NAD-Dependent Dehydrogenases

Structure and Evolution

NAD-dependent dehydrogenases have been the subject of extensive structural and functional investigations during the last two decades (Sund 1970; Sund 1977). With respect to their quaternary structure, they cover the whole range from single chain enzymes to multimeric complexes of high molecular weight. In fact, the pyruvate dehydrogenase multienzyme complex from *Bac. stearothermophilus* exceeds the size of cell organelles, such as the ribosome, representing the largest functional entity known so far.

High resolution X-ray data have been reported for L-ADH, GAPDH, LDH, and s-MDH. As pointed out by Rossmann et al. (1975), there exists a remarkable homology with respect to the characteristic domain structure of these enzymes: Those parts of the molecules whose function is to bind NAD^+ are closely similar, showing two roughly identical units, each containing a mononucleotide binding area. On the other hand, the overall subunit conformation of the different dehydrogenases varies to a considerable extent. This becomes clear if one compares the backbone structure of the catalytic domains or the widely differing positions of the NAD-binding domain within the polypeptide chains (Fig. 1). The combination of structural elements in the latter scheme clearly reflects the functional similarity with regard to coenzyme binding and the need for diversity required for substrate binding, catalysis, specificity, and formation of various oligomeric structures. The underlying mechanism of evolution may involve gene fusion of proteins with

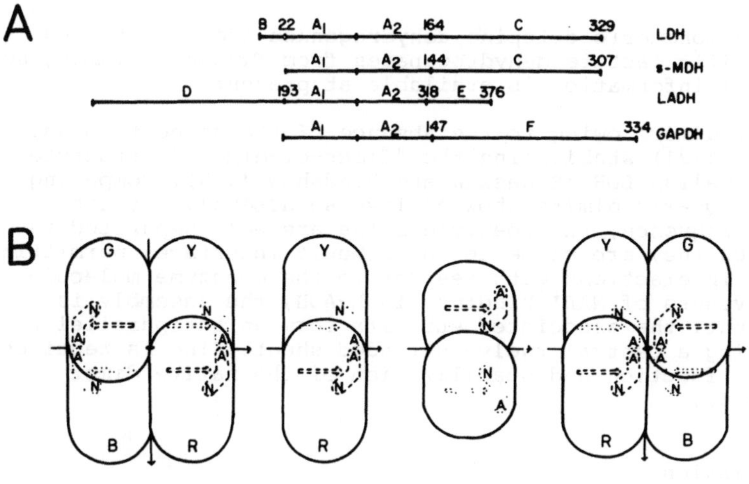

Fig. 1. Diagrammatic comparison of various dehydrogenases. A Positions of the two mononucleotide binding domains, A_1 and A_2, comprising together the NAD-binding site in LDH, s-MDH, L-ADH, and GAPDH. The catalytic domains, C, of LDH and s-MDH are similar in structure, but different to the catalytic domains D and E in L-ADH and F in GAPDH. The N-terminal "arm", B, stabilizes LDH as "dimer-of-dimers". B Association of subunits in LDH, s-MDH, L-ADH, and GAPDH. LDH, s-MDH, and GAPDH form Q-axis dimers. In GAPDH, association of the Q-axis dimers occurs in the opposite manner to LDH, bringing the NAD sites sufficiently close together to generate cooperative effects. (Courtesy of Holbrook et al. 1975)

simpler functions. Comparing the known tertiary structures, evolutionary divergence seems most suggestive although there is no easily recognizable relationship left at the level of the primary structure. In the case of the closely related isoenzymes of LDH and MDH, amino acid homology has been recently demonstrated (Birktoft et al. 1982).[3] The topological similarity may be unequivocal to such an extent that in cases where a complete tertiary structure determination is lacking, the coenzyme binding domains may be recognized by searching for the essential pattern of secondary structure elements predicted from the amino acid sequence (Wootton 1974; Rossmann et al. 1975).

Specific Systems

In the context of the present study, the similarities in the overall folding pattern are just as important as the previously mentioned differences in the quaternary structure.

As indicated, the structural specificity of the catalytic domains is also reflected in differences in the respective subunit interactions. For LDH, s-MDH, and GAPDH, these have been expressed in terms of an orthogonal, right-handed coordinate system P-Q-R (Rossmann et al. 1974). In LDH, the symmetry elements form the three mutually perpendicular molecular twofold axes which may be defined with respect to the direction of the strands in the β-pleated sheet structure. The Q-axis associates subunits by contacts between the NAD-binding domain; it is maintained in going from apo-LDH to holo-LDH, or from LDH to s-MDH (Rossmann ėt al. 1975; Holbrook et al. 1975). By comparing sequence data, Birktoft et al. (1982) concluded that m-MDH forms Q-axis dimers as well. Except for L-ADH, all dehydrogenases, whose structures have been determined, represent Q-axis dimers.

In the case of the monomeric octopine dehydrogenase from *Pecten* and the dimeric D-specific lactate dehydrogenases from *Cardium* or *Limulus*, no detailed structural information is available at present.

s-MDH forms stable dimers owing to the absence of the amino terminal sequence (residues 1-21) stabilizing the "dimer-of-dimers" structure in the case of mammalian LDH (Banaszak and Bradshaw 1975). Comparing GAPDH and LDH, the Q-axis dimers show an inverse assembly pattern (Fig. 1). As a consequence, the coenzyme sites are well separated in LDH, while in GAPDH they are close to the subunit interface, permitting (1) inter-subunit interactions with respect to the coenzyme molecule and (2) cooperativeness of NAD^+ binding. In L-ADH, the assembly is completely different: the associated subunits form an antiparallel β structure generating a central twelve-stranded sheet. Zinc is required for both enzymatic function and stabilization of the native dimer (Eklund et al. 1976).

Folding and Association

In summary, the following points make NAD-dependent dehydrogenases especially interesting objects in the elucidation of protein folding and association.

3 As shown by Rossmann et al. (1974), minimum base changes per codon can be used to measure evolutionary distances when sequence comparison with respect to the primary structure of proteins alone would fail

The different state of association provides us with the possibility to proceed from mere folding, in the case of monomeric ODH, to increasingly complex mechanisms of folding *and* association, comparing dimeric, tetrameric, and multimeric systems.

Specific solvent conditions may be applied to generate "structured monomers" on the pathway of subunit dissociation, thus providing a means of analyzing reassociation in the absence of significant folding.

"Folding by parts", which is suggested by the domain structure of the dehydrogenases, may be investigated by comparing the reconstitution properties of unmodified subunits with isolated domains produced by limited proteolysis. In this context, fragments may be applied to define the minimum size of a folding unit still able to acquire the native structure.

Ligand effects on the processes of both folding and association may be investigated with respect to cofactors, such as zinc in the case of L-ADH and the coenzyme NAD^+/NADH. Reconstitution in the presence of other enzymes (especially in the case of multienzyme complexes) may be used to demonstrate the specificity of subunit recognition.

The mechanism of unfolding and dissociation may be applied to resolve the intermolecular interactions governing the stability of the native quaternary structure (e.g., cold inactivation, pressure dissociation, structured intermediates).

Varying degrees of randomization of the polypeptide chain and varying enzyme concentrations may allow us to characterize the mechanism determining the yield of reconstitution.

Incomplete reconstitution deserves a closer look because it may reflect kinetic competition reactions as well as folding processes involving "wrong" local minima of potential energy.

To attack the given problems, we shall proceed from the simplest case, i.e., monomeric ODH, to oligomeric systems with increasing complexity, ending with the multimeric PDH from *Bac. stearothermophilus*.

Folding of Monomers: Octopine Dehydrogenase From *Pecten Jacobaeus*

ODH is a single-chain enzyme catalyzing the reaction pyruvate + L-arginine + NADH + $H^+ \rightleftharpoons$ D-octopine + NAD^+ + H_2O. As indicated by the close relationship to lactate formation, the enzyme may be considered a substitute for LDH in certain invertebrates (particularly mollusks) containing large amounts of arginine-phosphate in their muscles.

ODH from *Pecten jacobaeus* exists in two electrophoretically distinct forms (A and B), which do not differ significantly with respect to their enzymatic and physical properties. The spectral characteristics of the apo and holo enzymes from various marine invertebrates and their kinetic properties have been the subject of previous studies (Luisi et al. 1977; Monneuse-Doublet et al. 1980; Storey and Dando 1982; Zettlmeißl 1983; Teschner 1983).

Unfolding of the enzyme in 6 *M* guanidine·HCl leads to a decrease in helicity from ≅10% to ≅0% (random coil), as indicated by circular dichroism. Fluorescence emission shows a parallel decrease, accompanied by a significant red shift.

Table 1. Comparison of native and reactivated dehydrogenases[a]

	ODH[b]		s-MDH		m-MDH		LDH-M_4		LDH-H_4		GAPDH	
	N	N_r	N	N_r	N	N_r[c]	N	N_r[c]	N	N_r[c]	N	N_r
$s_{20,w}$ (S)	3.17	-	3.73	-	3.48	-	7.60	7.60	7.38	7.38	7.50	7.50
$M_r \cdot 10^{-3}$	44.4	44.2	67.0	-	63.0	69.7	140	140	142	142	143	144
sp. activity (IU/mg)	750	-	550	540	1040	1130	639	655	405	388	176	177
$K_{M,NADH}$ (m*M*)	0.05	-	0.03	0.03	0.08	0.08	-	-	-	-	-	-
$K_{M,substrate}$[d] (m*M*)	1.2	1.2	0.05	0.05	0.04	0.04	0.24	0.24	0.09	0.09	0.11	0.13
λ_{max} (nm)	327	-	332	332	301	301	339	339	339	339	337	337
$\theta_{207} \cdot 10^{-3}$ ($degr \cdot cm^2 \cdot dmol^{-1}$)	- 6.35	-	- 9.8	-	-	-	-11.4	-12.3	-10.8	-11.5	- 6.2	- 6.8
$\theta_{222} \cdot 10^{-3}$ ($degr \cdot cm^2 \cdot dmol^{-1}$)	- 9.6	-	-11.4	-	-	-	-15.4	-16.0	-12.7	-12.9	- 9.6	-10.7

[a] N, N_r refer to native and reconstituted enzyme. M_r, molecular weight from sedimentation equilibrium and gel filtration; λ_{max}, maximum of fluorescence emission ($\lambda_{exc} \cong 280$ nm); θ_λ, molar ellipticity at given wavelength (mrw = 113)

[b] The given data were determined for both forms, A and B, of ODH. Identical results were obtained within the ranges of error

[c] Values determined after separation of inactive "wrong aggregates"

[d] Michaelis constants for pyruvate (ODH and LDH), oxaloacetate (MDH), and GAP (GAPDH)

In the transition range from 0.4-1.8 *M* guanidine·HCl, unfolding is accompanied by aggregation which is easily detected by light scattering. Under optimum conditions (0.1 *M* phosphate buffer pH 7.6 in the presence of 1 m*M* EDTA and 1 m*M* dithioerythritol), the yield of reactivation is of the order of 70%. The reactivated portion of the enzyme turns out to be indistinguishable from the native ODH, considering all relevant enzymatic properties, heat stability, and susceptibility towards proteolytic attack (cf. Table 1).

The yield of reactivation does not depend on the time of incubation in 6 *M* guanidine·HCl (5 s - 24 h, 0 - 20°C). From this we may conclude that the proportion of the enzyme molecules which can be reactivated is not determined by cis-trans isomerization of X-pro bonds in the denatured state.[4]

If this mechanism would govern the yield, reactivation after short-term denaturation should start from polypeptide chains with their X-pro bonds in the correct, i.e., native isomeric state, thus yielding complete reactivation.

As shown in Fig. 2, the time course of reactivation obeys first-order kinetics with a rate constant $k' = 6.6 \cdot 10^{-4}\ s^{-1}$ (20°C). From the temperature dependence of the reactivation reaction at 5° - 20°C, an activation energy of ≅110 kJ/mol was determined.

Im summarizing the results (which hold equally for ODH from *nemertines*), reactivation of monomeric ODH after "complete denaturation" in 6 *M* guanidine·HCl is characterized by a slow transconformation reaction with a high energy of activation. One rate limiting first-order process is sufficient to characterize the kinetics of reactivation in a quantitative way (Fig. 2, insert). As suggested by the rapid regain of native fluorescence, fast folding steps precede the reactivation reaction, which must be a late event on the folding pathway. The overall reaction may be described by the following sequential model

$$(M) \xrightarrow{\text{fast}} M^* \xrightarrow{k'} M \qquad (1)$$

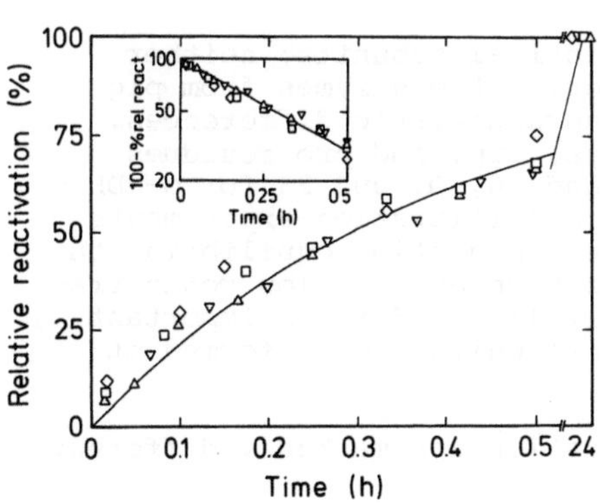

Fig. 2. Kinetics of reactivation of ODH from *Pecten jacobaeus* after 10 min denaturation at 20°C in 2 *M* (◇,▽) or 6 *M* (□,△) guanidine·HCl. Reactivation at 20°C in 0.1 *M* sodium phosphate, pH 7.6, containing 1 m*M* EDTA, 1m*M* dithioerythritol, and 0.1 or 0.15 *M* residual guanidine·HCl. Enzyme concentrations varied between 0.9 and 1.4 µg/ml: (▽,△) ODH form A; (◇,□) ODH form B. Reactivation is calculated relative to the final yield of 70% of the initial activity, as determined after 72 h of reconstitution. The *solid line* is calculated according to a first-order reaction with $k' = 6.6 \cdot 10^{-4}\ s^{-1}$. *Insert*: Determination of the first-order rate constant

4 It has previously been shown that slow kinetic phases in range of seconds (at 20°C) observed in the refolding of numerous small proteins are determined by isomerization reactions around X-pro peptide bonds (Brandts et al. 1975; Kim and Baldwin 1982)

with M as final product of renaturation, indistinguishable from the initial native enzyme.

Folding and Association of Dimers: Malate Dehydrogenase

As indicated by the previously mentioned reversibility of denaturation, folding to the correct native state is an intrinsic property of a given polypeptide chain. In the case of a dimeric enzyme, it is of interest to know whether the isolated chains are able to attain the catalytically active tertiary structure, i.e., whether or not intersubunit interactions are required to maintain the native configuration of the enzyme.

For a number of oligomeric enzymes, dissociation to monomers has been reported at low enzyme concentrations. However, most oligomers are highly stable and do not dissociate unless strong denaturants are applied. Under this condition, the isolated subunits are catalytically inactive, because dissociation is accompanied by a more or less pronounced loss of secondary and tertiary structure. In general, all three processes, dissociation, denaturation, and deactivation, are found to be reversible. Under certain conditions, therefore, structured monomers are expected to be accessible as intermediates of reconstitution. To characterize their properties regarding the correlation of folding, association, and catalytic function, we shall consider the equilibrium properties and the kinetics of reactivation, renaturation, and reassociation of malate dehydrogenase.

The NAD-dependent "simple" malate dehydrogenases occur in virtually all eukaryotic cells in at least two unique forms identified as mitochondrial m-MDH and cytoplasmic s-MDH, according to their cellular location (Banaszak and Bradshaw 1975).[5] The two enzymes are synthesized in the cytoplasm as products of different nuclear genes; s-MDH remains in the cytosol after synthesis, whereas m-MDH is translocated into the mitochondrial matrix (cf. Birktoft et al. 1982). Both are involved in the compartmentation of the malate shuttle providing a means of transporting NADH across the mitochondrial membrane; the mitochondrial enzyme in addition takes part in the TCA cycle.

Each of the two enzymes is composed of identical subunits; neither varies markedly in size (Table 1). Considering the enzymes from pig heart, the amino acid composition shows characteristic differences, especially with regard to the number of lys, trp, and pro residues which amount to 31, 5, and 12 for s-MDH, and 26, 0, and 23 for m-MDH (Banaszak and Bradshaw 1975). As shown by gel filtration experiments as well as meniscus depletion high-speed sedimentation equilibria, the two enzymes retain their dimeric structure even at very low concentrations ($c \geq 0.2$ μg/ml ≙ 3 n*M*) (Jaenicke et al. 1979). This is important in connection with reassociation experiments which are performed at low enzyme concentration.

The kinetic patterns of folding and association are markedly different for the two enzymes.

5 As indicated by different electrophoretic properties, both enzymes can occur in multiple subforms which may result from partial deamination, differences in the amount of covalently bound phosphate, or from differential lipid binding (Glatthaar et al. 1974; Cassman and Vetterlein 1974; Kuan et al. 1979)

Cytoplasmic Malate Dehydrogenase (s-MDH)

Among the various modes of denaturation, 6 *M* guanidine·HCl has been chosen for the following experiments, for several reasons: (1) starting reconstitution from the randomly coiled polypeptide chains may be considered a well-defined process; (2) the kinetics of reactivation do not change markedly upon variation of the time of denaturation between 30 s and 72 h. In contrast to this situation, the reconstitution characteristics after dissociation at acid pH depend on the time of denaturation. With increasing time of dissociation, the rate and yield of reactivation decrease and the kinetic traces become markedly sigmoidal. The dependence of the reactivation behavior on the time of acid denaturation may be caused by rearrangements within the structured monomer, generating a wrong acid-specific structure, the reshuffling of which becomes rate-limiting upon refolding. A similar case has been reported for porcine skeletal muscle lactate dehydrogenase (see below, cf. Zettlmeißl et al. 1981). Unfolding of the acid specific structure by additional guanidine treatment leads back to the usual reactivation pattern. After denaturation by 6 *M* guanidine·HCl, the reactivated enzyme (which amounts to 50% - 70% of the starting material) is indistinguishable from native s-MDH (Table 1).

The kinetics of reactivation illustrated in Fig. 3A are characterized by two distinct slow phases. The fact that no concentration dependence is detected clearly shows that subunit assembly is not rate-limiting in the process of reactivation. Analysis of the reactivation kinetics as two parallel first-order reactions (which is justified by the subsequent cross-linking data), yields the rate constants and the relative amplitudes of the two kinetic phases. At 20°C, the faster phase, comprising about 68% (at 20°C) of the total reactivation reaction, is characterized by a rate constant $k_f' = 13 \cdot 10^{-4}\ s^{-1}$, the slower phase by $k_s' = 7.0 \cdot 10^{-5}\ s^{-1}$. At 0°C, the relative amplitude of the faster phase is slightly decreased and the rate constants of the faster and slower phases are $k_f' = 3.6 \cdot 10^{-4}\ s^{-1}$ and $k_s' = 3.5\ 10^{-5}\ s^{-1}$, respectively. The activation energy of the faster phase is ≅38 kJ/mol; the activation energy of the slower phase, which is low, cannot be determined due to the scatter of the data.

Both the rate constants and the relative amplitudes of the two phases are not affected by the presence of up to 0.42 *M* guanidine during reconstitution. This proves that intermediates are not destabilized by low concentrations of the denaturant, which means that their stability does not differ significantly from the stability of the native enzyme.

From the reactivation data it can neither be decided whether the observed kinetic phases belong to the folding of the monomers or to the reshuffling of dimeric intermediates, nor whether enzymatically active folded monomeric intermediates exist.

Considerably more information about the mechanism of reconstitution was obtained by chemical cross-linking experiments during reconstitution.

Applying covalent cross-linking by glutaraldehyde and subsequent SDS polyacrylamide gel electrophoresis to the reconstituting enzyme (after complete denaturation in 6 *M* guanidine·HCl) yields a series of "snapshots" of the distribution of intermediates of reassociation and reconstituted enzyme which can be directly compared with the kinetics of reactivation (Hermann et al. 1979, 1981). The data show that part of the monomers form dimers with a rate constant and relative proportions similar to the faster phase observed in the reactivation kinetics (Fig. 3B).

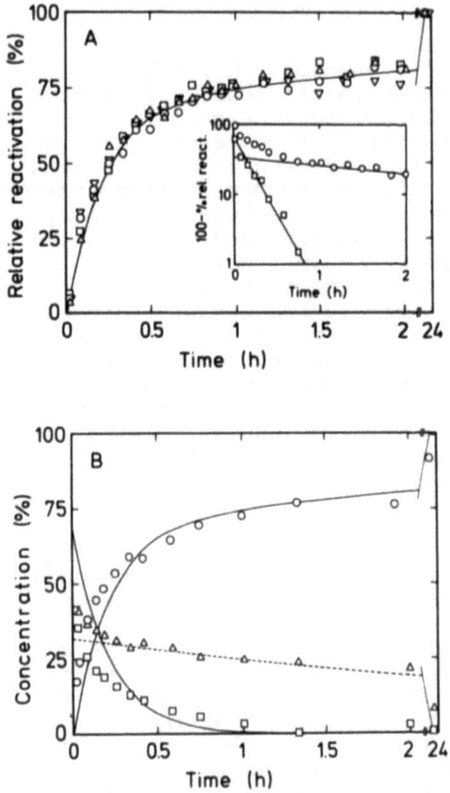

Fig. 3. Kinetics of the reactivation and reassociation of porcine s-MDH after 1 h denaturation at 20°C in 6 *M* guanidine · HCl. Reactivation at 20°C in 0.1 *M* sodium phosphate, pH 7.6, containing 10 m*M* EDTA and 0.012-0.12 *M* residual guanidine HCl. *A* Kinetics of reactivation at varying enzyme concentration (μg/ml): 1.3 (○), 3.7 (△), 6.4 (□), 12.9 (▽). Reactivation is calculated relative to the final yield of ≅50% of the initial activity, as determined after up to 172 h of reconstitution. The *solid line* is calculated according to two parallel first-order reactions with $k_f' = 13 \cdot 10^{-4}\ s^{-1}$ for the faster phase (amplitude 68%) and $k_s' = 7 \cdot 10^{-5}\ s^{-1}$ for the slower phase (amplitude 32%). *Insert*: Determination of the rate constants resolved as the sum of two exponentials for a representative reactivation experiment ($c_{ODH} = 1.3$ μg/ml). *B* Kinetics of reassociation as determined by chemical cross-linking with glutaraldehyde ($c_{MDH} = 6.5$ μg/ml; for experimental details see Hermann et al. 1981 and Zettlmeißl et al. 1982a); the relative concentrations of monomers (□), "dimeric" intermediates D^* (△), and native dimers (○) were determined by densitometry after electrophoresis. The *lines* are calculated for two parallel first-order reactions with the rate constants and amplitudes derived from the reactivation kinetics as shown in Fig. 3A, insert

The remaining monomers rapidly form an intermediate, most probably an incorrectly folded dimer D^*. After cross-linking, this intermediate is characterized by a slightly higher mobility than the native dimers on SDS polyacrylamide gels. This intermediate is also transformed to native dimers, but with the rate constant of the slow phase observed in the reactivation kinetics. Both phases contribute to native dimer formation which in turn parallels reactivation. The nature of the differences of the faster and slower reactivating forms (M_f and M_s) needs further analysis.

Circular dichroism measurements show that most of the backbone structure of the molecule is regained in a fast precursor reaction leading to structured monomers and (to a minor extent) incorrectly structured dimers. The slow regain of the final value of helicity roughly parallels reactivation and the formation of the native dimers. Due to the parallelism of reactivation and dimerization, the folded intermediates of association cannot possess any considerable activity.

The mechanism of reconstitution may be described by the following parallel reactions:

$$(2M_f) \xrightarrow{\text{fast}} 2M^* \xrightarrow{k_f'} (2M) \xrightarrow{\text{fast}} D$$

$$(2M_s) \xrightarrow{\text{fast}} D^* \xrightarrow{k_s'} D \qquad (2)$$

Model calculations with a consecutive first-order/second-order reaction mechanism have shown that the rate constant of the fast association reaction following k_f' must be $\geq 10^6\ M^{-1} \cdot s^{-1}$, i.e., dimer formation must be nearly diffusion controlled.

Mitochondrial Malate Dehydrogenase (m-MDH)

As mentioned earlier, mitochondrial MDH shows reconstitution properties differing from those described for s-MDH. The quaternary structure of the enzyme, the question of subunit activity, and in vitro association experiments after denaturation in various denaturants have been the subject of an earlier report (Jaenicke et al. 1979). Removal of the denaturant and separation of inactive "wrong aggregates" lead back to a final yield of ≅60% of renatured enzyme indistinguishable from native m-MDH (Table 1).

Reactivation as well as the regain of native fluorescence have been shown to be concentration dependent, exhibiting sigmoidal kinetics (Fig. 4). The rapid initial change in fluorescence may indicate that the rate-limiting steps of reactivation are preceded by a fast folding reaction at the monomer level. The reactivation data may be adequately described by a simple consecutive first-order/second-order mechanism:

$$(2M) \xrightarrow{\text{fast}} 2M^* \xrightarrow{k'} 2M \xrightarrow{k''} D \qquad (3)$$

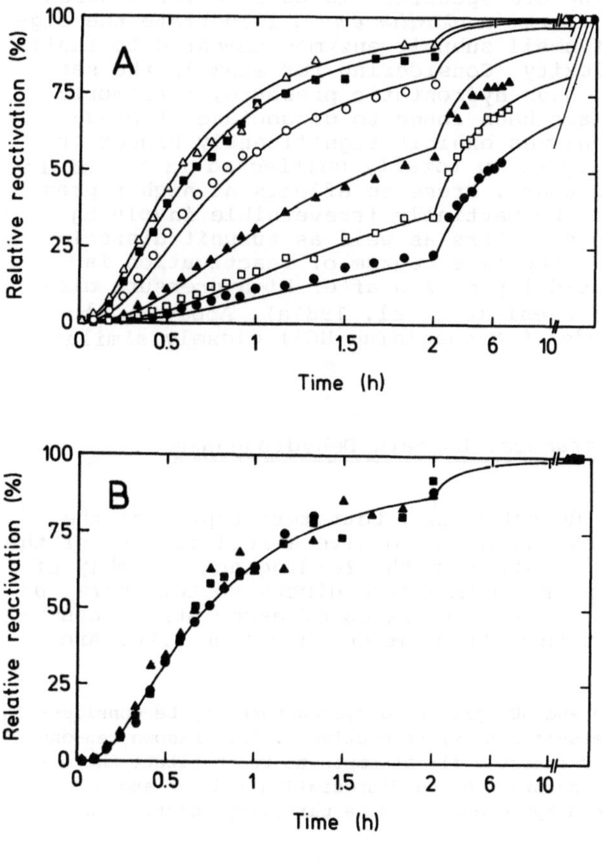

Fig. 4. Kinetics of reactivation of porcine m-MDH. *A* Concentration dependence of the kinetics of reactivation after 5 min denaturation at 20°C in 1 *M* glycine/H_3PO_4, pH 2.3. Reactivation at 20°C in 0.2 *M* potassium phosphate, pH 7.6, containing 10 m*M* EDTA and 10 m*M* dithioerythritol, at the following enzyme concentrations (µg/ml): 0.07 (●), 0.14 (□), 0.35 (▲), 1.2 (○), 3.1 (■), and 5.0 (△). Reactivation is calculated relative to the final yield of ≅60% of the initial activity, as determined after up to 290 h reconstitution. *Solid lines* are calculated according to a sequential first-order/second order mechanism with $k' = 6.5 \cdot 10^{-4}\ s^{-1}$ and $k'' = 3 \cdot 10^4\ M^{-1} \cdot s^{-1}$. *B* Kinetics of reactivation at an enzyme concentration of 2.1 µg/ml after denaturation by 5 min incubation in 1 *M* glycine/H_3PO_4 pH 2.3 (●) or in 6 *M* guanidine·HCl (■), or by 50 min incubation in 6 *M* urea (▲). Reactivation and calculation of the *solid line* as in Fig. 4A

The rate constants (20°C) for the consecutive slow folding/association steps are $k' = 6.5 \cdot 10^{-4}\ s^{-1}$ and $k'' = 3 \cdot 10^{4}\ M^{-1} \cdot s^{-1}$ (Fig. 4A). While k' is of the same order of magnitude as k_f^1 in case of s-MDH, k" must be at least two orders of magnitude slower than the diffusion controlled association of s-MDH. The kinetics of reactivation do not depend on the mode of denaturation, e.g., by acid, urea, or guanidine·HCl (Fig. 4B). Since these denaturants lead to different conformational states of the dissociated enzyme, it is obvious that M^* must be a common intermediate at a relatively highly structured level on the pathway of reconstitution.

An alternative bi-unimolecular kinetic model cannot be excluded on the basis of the given reactivation experiments. Unfortunately, the cross-linking technique cannot be applied in this case, because only incomplete fixation (≅30%) has been achieved under any condition (Brückl 1980). As mentioned, m-MDH has a significantly lower lysine content than s-MDH. The different cross-linking behavior may therefore be explained by differences in the distribution and reactivity of specific lysine residues involved in intersubunit cross-bridges.

Comparison with Other Dimeric Dehydrogenases

Among the other known dimeric NAD-dependent dehydrogenases, there are two which deserve special consideration. Liver alcohol dehydrogenase as a zinc enzyme poses the question of how binding of the metal ion is correlated with folding and dimerization. This problem will be discussed as an example for ligand effects on protein structure formation (see below). Dimeric (D-lactate specific) lactate dehydrogenases, which have been isolated from *Limulus* or *Cardium edule*, illustrate the significantly higher stability of small subunit enzymes compared to their counterparts with higher complexity. Considering for example the sensitively towards the action of high hydrostatic pressure, tetrameric or higher polymeric proteins have been found to dissociate at pressures ≦1 kbar, while dimeric enzymes exhibit significantly higher stability, their enzymatic activity being totally unaffected in the biologically relevant range (p ≦ 1.2 kbar). Pressure effects at higher pressure (up to 4 kbar) turn out to be partially irreversible involving changes in the native backbone structure as well as subunit dissociation. As in the case of m-MDH, the time course of reactivation is characterized by strongly sigmoidal profiles after high pressure dissociation (Müller et al. 1981; Jaenicke et al. 1981a). Applying the traditional denaturation methods (6 *M* guanidine·HCl) closely similar kinetics are observed.

Folding and Association of Tetramers: Lactate Dehydrogenase

Since trimeric dehydrogenases do not exist, tetramers represent the next level of complexity, if one attempts to give a full account of the post-translational mechanisms involved in the folding and assembly of oligomeric dehydrogenases.[6] In proceeding from dimers to tetramers, a number of new questions arise: Do intermediates of association accumulate during reconstitution? If they do, what do they look like? Are

6 Hexameric glutamate dehydrogenase and UDP-glucose dehydrogenase may be considered as dimers-of-trimers and trimers-of-dimers, respectively. For unknown reasons glutamate dehydrogenase has so far resisted all attempts to reconstitute the native quaternary structure after denaturation and dissociation (Müller and Jaenicke 1980), while UDP-glucose dehydrogenase may be partially reactivated (Jaenicke et al. unpublished)

they catalytically active and resistant towards proteolysis? Are there trimeric intermediates? If not, which one of the obligatory consecutive association steps, monomer → dimer or dimer → tetramer, is rate-limiting? Is the reactivated enzyme identical with the native enzyme? Is the given mechanism compulsory or are there side reactions? Do other proteins interfere with the given association steps? To answer these questions, lactate dehydrogenase from skeletal muscle (LDH-M_4) may be used as an example, because its three-dimensional structure is known and its reconstitution has been studied in detail.

Lactate Dehydrogenase from Skeletal Muscle (M_4)

As in the case of the monomeric or dimeric dehydrogenases, the final product of the in vitro reconstitution of LDH-M_4 is indistinguishable from the native enzyme by all physicochemical and enzymatic criteria tested (Table 1).

However, this result only holds true if inactive material (consisting of high aggregates) is separated from the reactivated material (Jaenicke and Rudolph 1977). The mechanism of aggregation, which has been analyzed in detail in the case of LDH-M_4, will be discussed later.

Kinetic Mechanism of Folding and Association

After denaturation of LDH-M_4 by 6 *M* guanidine·HCl, reactivation is determined by sigmoidal kinetics, similar to the reactivation of m-MDH. From the concentration dependence of the rate of reactivation, it is obvious that again consecutive steps of folding and association are involved (Fig. 5A). Since LDH-M_4 can be completely cross-linked by glutaraldehyde, the individual rate-limiting steps were unambiguously identified (Zettlmeißl et al. 1982a). As shown in Fig. 5, during the early phase of reassociation, dimer formation is the only relevant reaction accompanying the decrease in monomers. The latter reaction obeys first-order kinetics with a rate constant $k' = 8.0 \cdot 10^{-4}\ s^{-1}$. The formation of the dimeric intermediate is therefore governed by a first-order folding reaction of the monomers. Similar to the result obtained for s-MDH, the actual dimerization step must be fast, again determined by a rate constant close to that of diffusion controlled reactions.

As in the case of the monomeric and dimeric dehydrogenases, the slow folding reaction of the monomers is preceded by a fast folding step, as judged from circular dichroism measurements.

The second rate-limiting step on the pathway of reconstitution consists of the slow association reaction of dimers to tetramers, as shown by the concentration dependence of the rate of tetramer formation. The rate constant of this association step may be directly determined by reconstitution studies starting from "structured monomers", thus eliminating the slow refolding step. These structured monomers may be produced by acid dissociation in the presence of a stabilizing salt, e.g., 1 *M* Na_2SO_4. Second-order linearization of the respective reactivation and tetramerization kinetics, yields a second-order rate constant[7] (Footnote 7 see the following page) $k'' = 3.0 \cdot 10^4\ M^{-1} \cdot s^{-1}$. On the basis of the individually determined rate constants k' and k" both reassociation and reactivation of LDH-M_4 can be quantitatively described by a consecutive first-order/second-order mechanism.

Deviations in the monomer/dimer ratio at the final stage of reassociation from the curves calculated according to a simple consecutive mechanism may be explained by the fact that the dimeric intermediates are in a fast dissociation equilibrium with monomers (Hermann et al.

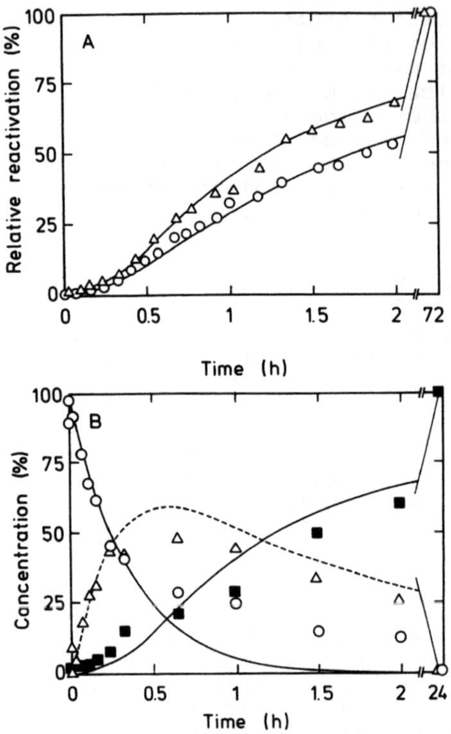

Fig. 5. Kinetics of reactivation and reassociation of porcine LDH-M_4 after 5 min denaturation at 20°C in 6 *M* guanidine·HCl. Reactivation at 20°C in 0.1 *M* sodium phosphate, pH 7.6, containing 1 m*M* EDTA and 0.012 *M* residual guanidine·HCl. *A* Kinetics of reactivation at 0.55 µg/ml (o) and 1 µg/ml (Δ). Reactivation is calculated relative to the final yield of ≅50% of the initial activity determined after 72 h of reconstitution. The *solid lines* are calculated according to a sequential first-order/second-order mechanism with $k' = 8 \cdot 10^{-4}\ s^{-1}$ and $k'' = 3 \cdot 10^{4}\ M^{-1} \cdot s^{-1}$. *B* Kinetics of reconstitution at an enzyme concentration of 0.55 µg/ml, as determined by chemical cross-linking with glutaraldehyde (cf. Zettlmeißl et al. 1982a). The relative amounts of monomers (o), dimers (Δ), and tetramers (■) were determined by densitometry after electrophoresis. The data are corrected for approximately 8% monomers and 22% dimers still present after 24 h of reassociation. These particles most probably stem from the dissociation of partially cross-linked higher aggregates. The *lines* are calculated according to the sequential tetramer formation via dimeric intermediates with the rate constants given in Fig. 5A

1981). The complete mechanism for the reconstitution of LDH-M_4 would therefore be:

$$(4M) \xrightarrow{\text{fast}} 4M^* \xrightarrow{k'} 4M \underset{}{\overset{\text{fast}}{\rightleftharpoons}} 2D \xrightarrow{k''} T \qquad (4)$$

The energy of activation of the dimer → tetramer association after short-term acid dissociation is ≅240 kJ/mol (Rudolph et al. 1977a). A temperature dependent shift of the dissociation equilibrium of the dimers, which was not considered in determining the second-order rate constants, may eventually account for this rather high value.

Since reactivation strictly parallels tetramer formation, it is obvious that the monomeric and dimeric intermediates cannot possess substantial enzymatic activity under standard conditions of the enzymatic assay.

The proposed mechanism for the reconstitution of LDH-M_4, as determined by reactivation and cross-linking, is confirmed by several independent lines of evidence.

7 The rate of the second-order dimer → tetramer association is strongly affected by low concentrations of guanidine·HCl present during reconstitution: guanidine binding to the intermediates of reconstitution yields a significant decrease of the rate of association (Zettmeißl et al. 1979a). The amount of guanidine (0.012M) present during reconstitution was therefore kept constant after both acid and guanidine·HCl denaturation. Considering a fast dissociation equilibrium of the dimers (see below), the rate constant for the reactivation in the absence of guanidine was $k'' = 3.15 \cdot 10^{4}\ M^{-1} \cdot s^{-1}$ (Hermann et al. 1981)

1. Appling medium concentrations of urea or guanidine·HCl, lactate dehydrogenase (M_4 and H_4) has been shown to be partially dissociated to the dimer. Starting reconstitution experiments from this intermediate, the association kinetics have been shown to be indistinguishable from the kinetics describing the monomer → tetramer transition after short-term acid dissociation. Since in the latter case, the tetramer must be formed via the dimer, the dimer → tetramer transition must be the rate determining step in the overall reactivation of the enzyme (Jaenicke et al. 1981c).

2. Thermolysin treatment during reconstitution, which eventually cleaves off the N-terminal "arm", stabilizing the tetrameric state, generates artificial stable "dimers" (Girg 1983). This approach has been used to analyze the kinetics of reassociation to the native tetramer. Applying the proteolytic reaction at various times during reconstitution and subsequently analyzing the reaction products by gel filtration, again proved the dimer → tetramer transition to be rate-limiting for both reassociation and reactivation (Girg et al. 1981).

3. Another method to confirm the proposed kinetic mechanism is based on hybrid formation obtained by mixing reconstituting LDH from skeletal muscle with subunits of the enzyme from heart muscle. The mathematical analysis of the data allows us to calculate the concentrations of all species, monomers, dimers, and tetramers, as well as the equilibrium constant of the monomer-dimer equilibrium preceding the rate-limiting tetramer formation (Hermann et al. 1982).

Thus, LDH-M_4 has provided us with a clear-cut answer regarding the mechanism of the in vitro reconstitution of a tetrameric enzyme, starting from the the unfolded or "structured" monomers, as well as the dimeric intermediate.

Conformational Rearrangements of the Acid Dissociated Monomers

Reactivation after short acid dissociation obeys simple second-order kinetics with high yield, similar to the previously described reconstitution after acid incubation in the presence of 1 *M* Na_2SO_4 (Rudolph and Jaenicke 1976). However, increasing the incubation time at low pH in the absence of the stabilizing salt leads to a marked decrease of the rate and yield of reactivation, caused by conformational changes within the partially unfolded monomers. These slow rearrangements of the acid dissociated monomers are reflected by slight changes of the far UV circular dichroism, indicating a decrease in helix content from 30% immediately after acid dissociation to 25% after completion of the acid transition (Zettlmeißl et al. 1981). The decrease in helicity is parallelled by changes in the susceptibility of certain fragmentation sites towards pepsin digestion (Zettmeißl et al. 1983). Upon incubation at 20°C in 0.1 *M* sodium phosphate pH 2.3, the rate constants for the decrease in either the yield of reactivation, or the helicity, or the increase in the accessibility towards proteolysis are of the order of $3\text{-}8 \cdot 10^{-4}\ s^{-1}$.

Apart from the decrease in the yield, the kinetics of reactivation become increasingly sigmoidal upon long-term acid incubation. Obviously, the monomer is more and more trapped in a "wrong" structure, from which refolding to M proceeds with a low rate. Elimination of the unfavorable configuration, e.g., by 6 *M* guanidine·HCl, restores the normal reactivation properties observed after guanidine unfolding (Fig. 6).

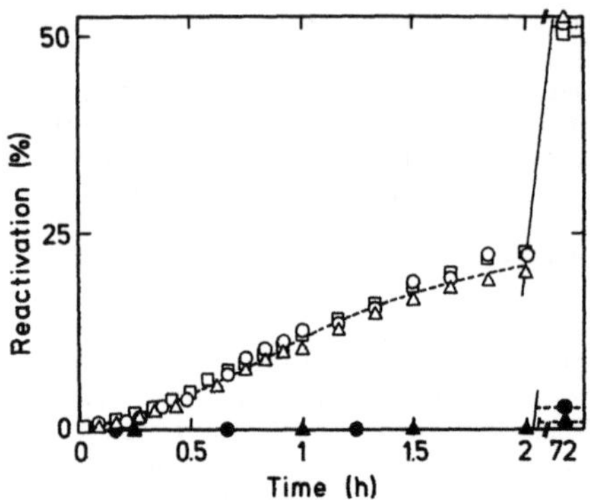

Fig. 6. Kinetics of reactivation of porcine LDH-M_4: Influence of additional guanidine·HCl denaturation after long-term acid dissociation on the kinetics and the yield of reactivation of LDH-M_4 at an enzyme concentration of 1.2 μg/ml. Acid dissociation by 24 h incubation at 20°C in 1 *M* glycine/H_3PO_4, pH 2.3(Δ,▲), or in 0.1 *M* sodium phosphate, pH 2.3 (o,●). Reconstitution in 0.1 *M* sodium phosphate, pH 7.6, containing 1m*M* EDTA, 1 m*M* dithioerythritol, and 0.03 *M* guanidine·HCl after 5 min incubation of the acid dissociated monomers in 6 *M* guanidine·HCl prior to reconstitution (Δ,o), compared to the reactivation kinetics without additional guanidine denaturation (▲,●); (□) reactivation after 5 min denaturation in 6 *M* guanidine·HCl without prior acid incubation

Properties of the Monomeric and Dimeric Intermediates

As previously mentioned, the kinetics of reactivation and reassociation clearly show that the monomeric and dimeric intermediates must be virtually inactive under standard conditions of the assay. Circular dichroism measurements reveal that after guanidine denaturation an appreciable amount of secondary structure returns immediately after diluting out the denaturant.

This proves that the monomeric intermediate (M^*) is structured.

Far more information is available with regard to the properties of the dimeric intermediate, mostly derived from stable "dimers" produced by limited proteolysis with thermolysin during reconstitution.

As shown by SDS polyacrylamide gel electrophoresis, the "dimer" is partially "nicked" to fragments of 12000 and 18000 mol. wt., in accordance with the domain structure of the LDH subunits. The "dimers" can be reconstituted after denaturation by guanidine·HCl, i.e., the individual domains of the enzyme retain the potential of generating the correct tertiary structure, thus, corroborating the idea that protein folding may be understood as "folding by parts", followed by "merging of domains" (Wetlaufer and Ristow 1973).

The "dimers" bind to an NAD-specific affinity column, which indicates that the active center of the enzyme must be almost completed at the dimeric state, although the "dimers" are inactive under standard test conditions. Obviously, the structure is too flexible in the dimeric state to provide catalytic activity. Stabilization of the backbone structure by adding "structure making ions" to the test mixture (e.g., 2 *M* $(NH_4)_2SO_4$) leads to the recovery of an appreciable amount of activity. Considering that the tightening of the structure of native LDH-M_4 by addition of 2 *M* $(NH_4)_2SO_4$ reduces the native activity by 20%, the proteolytic "dimers" possess about 30%-40% of the native activity. That the active "dimers" are in fact dimers in the presence of the stabilizing salt is clearly shown by activity transport experiments using the Cohen method in the analytical ultracentrifuge (Girg 1983).

Catalytic activity of dimeric LDH is not restricted to "dimers" produced by limited proteolysis: If reactivation is monitored in the presence of stabilizing salt (e.g., 1.4 *M* $(NH_4)_2SO_4$ in the test mixture), regain of enzymatic activity precedes tetramer formation due to the intrinsic activity of the dimeric intermediate.

Reactivation vs Aggregation

Complete reactivation turns out to be an exception rather than the rule and "wrong aggregation" as a major side reaction in the in vitro reconstitution seems to represent a general phenomenon characteristic for oligomeric enzymes. Reconstitution of LDH-M_4 after maximum unfolding in 6 *M* guanidine·HCl produces about equal amounts of native enzyme and inactive aggregates. It has been previously proposed that aggregation, with a reaction-order $\geqq 2$, is in kinetic competition with a first-order folding step. Evidence came from the inverse proportion of reactivation and wrong aggregation with increasing enzyme concentration (Zettlmeißl et al. 1979b).

The folding step competing with aggregation is not determined by cis-trans isomerization around X-pro peptide bonds, since aggregation occurs even after very short times of guanidine denaturation, when all X-pro peptide bonds of the denatured chains are expected still to be in their native isomeric conformation (Zettlmeißl et al. 1982b).

Aggregation seems to compete with the slow folding reaction ($M^* \rightarrow M$) on the pathway of reconstitution, indicating that M is resistant towards aggregation. This is in accordance with reconstitution experiments starting from a structure which rapidly forms M, e.g., after acid denaturation in the presence of stabilizing salt. Under these conditions, no aggregates are detectable and deactivation is almost completely reversible. On the other hand, long-term incubation at acid pH in the absence of stabilizing salt leads to a decrease of the rate of formation of M, due to the slow reshuffling of the acid specific structure (see above). As a consequence, aggregation predominates after long-term acid incubation.

Aggregates formed after both guanidine or acid denaturation are expected to be partially structured due to the rapid initial folding of the monomers. In fact, circular dichroism measurements clearly demonstrate that wrong aggregates exhibit quasi-native far UV circular dichroism spectra (Zettlmeißl et al. 1979b). Apparently the encounter of incompletely structured monomers stabilizes incorrect interparticle interactions instead of allowing the formation of the native hydrophobic core of the globular protein.

This mechanism of aggregate formation is corroborated by experiments involving the fully reversible high pressure dissociation of lactate dehydrogenase (M_4 and H_4). In this case, the absence of aggregate formation has been explained by a consecutive dissociation-transconformation mechanism involving the masking of hydrophobic surfaces in the interior of the monomers after subunit separation (Müller et al. 1982).

Lactate Dehydrogenase from Heart Muscle (H_4)

Lactate dehydrogenase from heart muscle has been one of the first enzymes to be studied in detail regarding its reassociation properties (Anderson and Weber 1966; Teipel and Koshland 1971; Jaenicke 1974). The outcome of these early experiments was the strong influence of temperature and enzyme concentration on the rate of reactivation indicating that reconstitution is determined by a second-order reassociation reaction characterized by a high activation energy. Coenzyme and substrates were shown to have a drastic effect on the reactivation reaction, which at first sight appeared to exhibit a higher degree of complexity compared to LDH-M_4. However, more recent results made clear that under appropriate conditions both isoenzymes show rather similar reconstitution properties. Therefore, only a few diverging features of LDH-H_4 have to be added to what has been reported before for LDH-M_4.

Regarding its native quaternary structure, LDH-H_4 is more readily dissociated by acid, guanidine·HCl, or high pressure than LDH-M_4 (Rudolph 1977; Jaenicke et al. 1981c; Müller et al. 1981b). The hybrids of heart and skeletal muscle LDH show a much lower stability indicating that subunit interactions are maximum when all four subunits are identical (Anderson and Weber 1966).

Analysis of the reassociation after acid dissociation in the presence of stabilizing salt by cross-linking follows the same association mechanism previously proposed for LDH-M_4 (Bernhardt et al. 1981). Under this condition, structured monomers are preserved, thus eliminating slow folding steps. Upon reconstitution at neutral pH, association of monomers to dimers is fast. As in the case of LDH-M_4, the dimers are in a dissociation/association equilibrium with monomers, the equilibrium being slightly more on the monomer side compared to LDH-M_4. Slow association of dimers to tetramers is again rate-limiting for the regain of activity; the rate constant for the dimer association is $k'' = 1.4 \cdot 10^4\ M^{-1} \cdot s^{-1}$. After dissociation by guanidine·HCl, urea, or acid in the absence of stabilizing salt, a slow folding step characterized by $k' = 14.5 \cdot 10^{-4}\ s^{-1}$ becomes rate-limiting, in addition to the dimer association (Rudolph et al. 1977c).

The different modes of denaturation differ in the amount of residual structure. Obviously, folding proceeds rapidly to a common intermediate (most probably M* in eqn (4)). Again, the subsequent slow folding step $M^* \rightarrow M$ must be a relatively late event during structure formation.

The attempt to isolate dimeric intermediates by limited proteolysis fails due to the strong tendency of the H_2-"dimer" to aggregate.

Comparison with Other Tetrameric Dehydrogenases

A number of other tetrameric dehydrogenases (alcohol dehydrogenase and glyceraldehyde-3-phosphate dehydrogenase from yeast) have been found to show reconstitution properties similar to those described before, so that the given folding/association mechanism may be considered the general kinetic scheme underlying the formation of the native quaternary structure (Jaenicke and Rudolph 1977; Gerschitz et al. 1978; Rudolph et al. 1977b).

One specific feature of GAPDH which has received much attention in connection with "metabolic control" (Stancel and Deal 1969) is the cold-inactivation of the enzyme in the presence of ATP. As shown by ultracentrifugation and gel filtration, the reaction is accompanied by dissociation to structured monomers, still capable of coenzyme binding. The dissociation kinetics show the typical low rate, characteristic for structural transitions close to the equilibrium, while the kinetics of reconstitution depend on enzyme concentration, as expected for a rate-determining association step. The overall process is again quantitatively described by the above mentioned sequential folding/association scheme (Bartholmes and Jaenicke 1978).

Folding and Association of Multimeric Systems: Pyruvate Dehydrogenase Multienzyme Complex

Successful attempts to reconstitute complex biological structures, such as ribosomes or microtubules suggest that there are no fundamental differences to be expected if we go from oligomeric to multimeric systems. New facets may be a stringent assembly pathway or cooperativity of the constituent polypeptide chains, as well as the requirement for "morphopoietic factors" in order to achieve the correct geometry and stoichiometry of the native quaternary structure.

Among the multienzyme complexes that have been most extensively studied in the past are the 2-oxo acid dehydrogenase complexes from various sources (Koike and Koike 1976). They are of utmost importance because they play a key role at the junction of the metabolic pathways of carbohydrates, lipids, and certain amino acids with the TCA-cycle. Their close relationship is clearly established by the observation that different enzyme complexes are found to share certain components, e.g., the E3 component of the pyruvate and 2-oxo glutarate dehydrogenase complexes of *E. coli*. This may be taken to suggest that assembly in vivo proceeds by nascent cores of the two complexes being complemented from a single pool of E3 component common to both systems. If this hypothesis were true, the individual subunits should fold independently without mutually affecting each other and without requiring external factors for proper folding and association.

Attempts to accomplish in vitro reconstitution after separating the component enzymes in an active form were successful for a number of 2-oxo acid dehydrogenases (Koike and Koike 1976). The result of these experiments proves that even for multimeric enzyme complexes assembly is an intrinsic property of the constituent polypeptide chains. The same holds for the reconstitution and reactivation of the component enzymes after joint dissociation and inactivation at acid pH or in 8 *M* urea. Under these conditions at least partial denaturation occurs so that reconstitution involves folding and association, thus simulating the in vivo process of structure formation.

Considering the pyruvate dehydrogenase complex from *Bac. stearothermophilus* ($M_r \cong 10^7$d), multiple copies of four different polypeptide chains form a lipoate acyl transferase core (E2) surrounded by a coat of pyruvate decarboxylase (E1α, E1β) and lipoamide dehydrogenase (E3) subunits, loosely packed in varying proportions. Reconstitution of the enzyme from the separated components cannot be accomplished because of the instability of E1 (Jaenicke and Perham 1982). Upon "joint dissociation and reconstitution," the presence of all the component enzymes of the complex during folding and self-assembly seems to protect E1 from irreversible denaturation by incorporating the metastable "nascent" form into the growing complex. The effect may be compared to the effect of ligands (like cofactors or substrates) on the yield of reactivation which will be discussed in the subsequent chapter.

As indicated by ultracentrifugation, sucrose gradient centrifugation, SDS polyacrylamide gel electrophoresis, and electron microscopy, the final product of reconstitution that shows PDH activity is an assembled structure akin to the native enzyme. Its quaternary structure resembles the starting material in terms of its hydrodynamic and topological properties, having in mind the above mentioned variation of polypeptide chain stoichiometries observed for different preparations of the native complex.

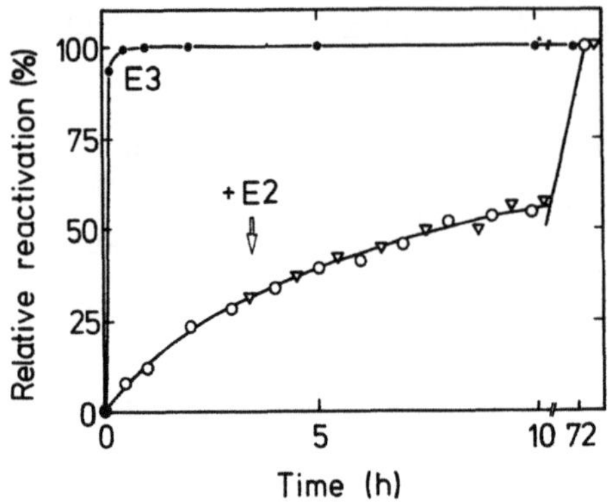

Fig. 7. Kinetics of reactivation of the PDH complex from *B. stearothermophilus* after 2 min incubation in 0.5 *M* glycine/H_3PO_4, pH 2.3 (0°C) and subsequent dilution with 0.2 *M* potassium phosphate, pH 7.0, containing 5 m*M* EDTA and 2 m*M* dithioerythritol, to final concentrations $20 < c_{PDH} < 170$ µg/ml. (O), overall complex activity after reactivation at 53°C, calculated relative to final values determined after 72 h of reactivation. (∇), reactivation after "seeding" with 8 µg of E2 core at the time indicated by the *arrow* ↓. (●), reactivation of E3: aliquots of the reconstituting enzyme (8 µg) were assayed for E3 activity

The kinetics and the yield of reconstitution depend on the mode of denaturation in a similar way to that described for LDH. For example, there is a transition from hyperbolic to sigmoidal reconstitution profiles after long incubation at acid pH. This points to conformational rearrangements within the partially unfolded enzyme in its dissociated state which lead to a decrease in the yield and rate of reactivation. The time course of reactivation is governed by first-order rather than second-order kinetics, thus indicating that rate-limiting folding reactions at the level of the monomeric or multimeric enzyme govern the overall mechanism of reconstitution (Fig. 7). Obviously, bimolecular association reactions are fast; they gain influence only at very low enzyme concentrations where the complex becomes unstable even under native conditions.

With respect to the rate-limiting transconformation step in the overall reconstitution reaction, neither the folding of E3 nor the reshuffling of the E2 core are crucial. Evidence comes from two observations: (1) The folding reactions of both components are fast and independent of each other. The reactivation rate of E3 exceeds the rate of reconstitution of the complex by about two orders of magnitude; similarly, E2 core formation is fast compared to overall reactivation. (2) Adding core particles (E2) to the reactivation mixture during "joint reconstitution" does not alter the overall reactivation pattern (Jaenicke and Perham 1982).

Assuming heterologous interactions between the components of the complex to be involved in mutually induced conformational changes (which finally generate enzymatic activity), the following tentative mechanism of reconstitution may be proposed:

$$n\,(M_i^*) \xrightarrow{\text{folding}} n\,(M_i) \xrightarrow[\text{fast}]{\text{association}} {M_i}_n \xrightarrow{\text{reshuffling}} N \qquad (5)$$

Whether the rate determining step in this reaction is the reactivation of E1 or the reshuffling of the whole complex cannot be decided on the basis of the present data.

The fact that highly complex multienzyme systems can be reconstituted in vitro proves that folding and association occur spontaneously, even in the case of giant molecules like PDH from *Bac. stearothermophilus*. Obviously, the acquisition of the native three-dimensional structure is determined neither by the vectorial cotranslational folding of the nascent polypeptide chain, nor by post-translational processing, or extrinsic factors. The various cofactors participating in the catalytic reactions affect neither the kinetics nor the yield of reconstitution significantly. Whether this finding can be generalized requires further analysis.

Effect of Ligands

Ligands like NAD or substrates are known to stabilize the native structure of dehydrogenases. For certain enzymes, the attainment of their functional state depends on the presence of specific cofactors such as metal ions. Apart from these, additional components, e.g., ions or solvent components, may cause stabilizing or labilizing effects which are expected to promote or inhibit the process of structure formation. In connection with the present discussion, two important questions may be asked: Do cofactors affect the folding and association of enzymes? And if so, at which stage do they enter the game?

Stabilization – Labilization

Dissociation, denaturation, and deactivation parallel each other, the midpoint of the cooperative transition depending on the denaturant as well as on specific ligands. For example, LDH-M_4 has been shown to be stabilized by phosphate ions such that the pH range of the tetramer-monomer transition is shifted by one pH unit (Rudolph and Jaenicke 1976). The effect becomes even more pronounced in the case of the pressure induced dissociation of the enzyme, where dissociation in the presence of phosphate is totally inhibited at pressures up to 2 kbar, while in the presence of Tris/HCl, pressures of $\cong$1.1 kbar are sufficient to cause half-deactivation (Schade et al. 1980). From X-ray crystallographic data, specific binding sites for bivalent ions between the subunits of LDH are well-established, so that there is direct evidence for ligand binding explaining the observed stabilization.

With respect to folding, "structure making ions" have been previously discussed as a means of producing "structured monomers," i.e., of stabilizing the native backbone structure in the process of acid dissociation.

As an example for specific labilizing effects of ligands, the previously mentioned cold-inactivation of yeast GAPDH at low concentrations of ATP may be mentioned (Bartholmes and Jaenicke 1978).

Effects on Folding and Association

Among the effectors that have been suggested as imposing kinetic constraints on the pathway of reconstitution, specific cofactors or ligands have often been suggested. In principle, coenzymes or other effectors may influence the kinetics of reconstitution in two ways: (1) association of the effector with some intermediate state of refolding may lower the activation energy of subsequent rate-limiting steps, or (2) binding of the effector molecule to the native or a quasi-native structure may lower the free energy of this state compared to the local minima of non-native conformations in the free energy hyperspace, in this way shifting the equilibrium towards the native structure. Thus, the effector can influence the reconstitution rate by accelerating rate-limiting steps or by stabilizing either intermediates in the correct folding pathway or the end product of reconstitution itself. In addition to the rate enhancement, both mechanisms may cause the yield of reconstitution to increase if the latter process competes with alternative pathways leading to inactive material.

Thorough kinetic analyses have shown that there is no significant influence of NAD^+ or NADH on the rate of reactivation for a number of dehydrogenases (ADH, LDH-H_4, LDH-M_4, m-MDH, PDH). In the case of GAPDH from yeast, the rate and yield of reactivation are affected by the coenzyme in its reduced and oxidized form. Both enhance the slow transconformation reactions that are rate-limiting at high enzyme concentrations (Krebs et al. 1979). Having in mind the complexity of the two mononucleotide binding units making up the NAD-binding site, it is highly unlikely that the coenzyme serves as a "specific primer" of the initial refolding process thus accelerating a rate-limiting reaction. The effect may be explained either by a decrease in the activation energy of a rate-limiting step or by the stabilization of a correctly folded or associated state.

Effector induced "nucleation" may be promoted by covalently linking the coenzyme to the active center of the enzyme. Applying [3-(3-bromoacetyl pyridinio)-propyl] adenosine pyrophosphate as NAD-analog this has been shown to be the case for yeast GAPDH where the rate of reconstitution of the modified enzyme is found to be increased by a factor of 10, compared to the 2-3-fold acceleration observed with free NAD^+/NADH (Jaenicke et al. 1980). The higher efficiency can be rationalized by facilitated interaction with low affinity sites of folding intermediates, because the bound nucleotide is readily available and does interact without loss of translational entropy.

The previously mentioned external effects on reconstitution are by no means trivial: if a ligand tends to stabilize an incorrect conformation, it may equally well decelerate or even completely inhibit reconstitution, instead of promoting it. This has been clearly shown in the case of alcohol dehydrogenase from horse liver, where the metal ions (though essential for catalysis as well as reconstitution) may *decelerate* reconstitution if applied at high concentrations (Rudolph et al. 1978).

L-ADH has been investigated in detail (Gerschitz et al. 1978; Rudolph et al. 1978). No association is detectable in the absence of Zn^{2+}. Instead, refolding of the Zn^{2+}-free enzyme is characterized by consecutive first-order processes. Reactivation after adding Zn^{2+} depends on enzyme concentration, which indicates that subunit assembly is required for catalytic activity. Folding and association may be analyzed independently by adding Zn^{2+} after completion of the slow first-order processes. Under these conditions, folding of structured monomers ($M^* \rightarrow M$) is no longer rate-limiting for the reactivation reaction which can be easily recognized from the disappearance of sigmoidicity in the kinetic traces. The remaining reaction follows bimolecular kinetics with a pre-equilibrium involving an inactive $M \cdot Zn^{2+}$ complex:

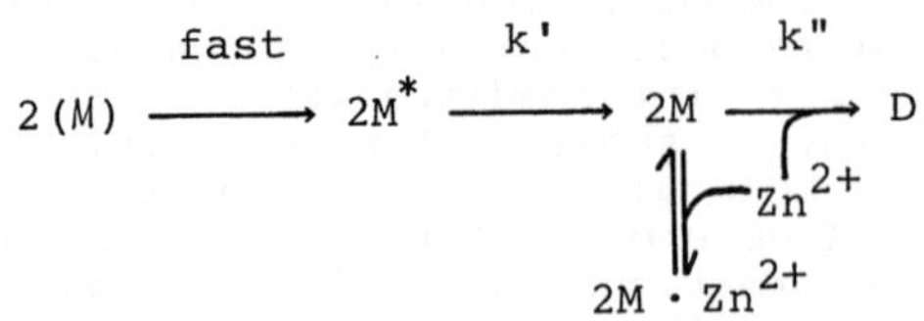

(6)

The rate constants of the separately determined folding and association steps are $k' = 2 \cdot 10^{-3}\ s^{-1}$ and $k'' = 1.6 \cdot 10^3\ M^{-1} \cdot s^{-1}$. Excess of the ligand traps the inactive monomer; at low ligand concentrations Zn^{2+} binds to the folded dimer, in this way shifting the equilibrium to the native product of reconstitution. Regarding the importance of "nucleating ligands," the given example indicates the twofold role of ligands upon reshuffling (besides the well-known participation in catalysis): fixation of inactive intermediates, and stabilization of association products in their proper structure.

Specificity

The most fascinating aspect of ligand effects on protein folding and assembly is the question of whether or not other proteins may function as "ligands" and thus interfere with the formation of the native three-dimensional structure, especially in the case of oligomeric enzymes.

In vivo folding of a given protein takes place in the presence of high concentrations of other proteins, including nascent polypeptide chains coded in the same structural gene. From this fact, the question arises whether chimeric assemblies may exist at least as intermediates of association in the case of structurally related enzymes.

In order to investigate this problem, synchronous reconstitution experiments with two closely related dehydrogenases – LDH-H_4 and m-MDH from pig heart – were performed (Jaenicke et al. 1981b). As discussed in previous chapters, both enzymes possess very similar tertiary structures and their reactivation shows a similar time course. On the other hand, their widely differing fluorescence properties provide a sensitive measure of hybrid formation. Measurements of catalytic activity and fluorescence therefore allow us to determine the formation of active and inactive hybrids as well as the formation of intermediates. Subunit exchange in the native state can be excluded, as monitored by activity measurements and protein fluorescence after 20 h incubation and subsequent separation by gel chromatography. Similarly, there is no hybrid formation after dissociation and unfolding of the two enzymes at acid pH, followed by joint reconstitution at pH 7.6. Thus, under equilibrium conditions neither active nor inactive LDH·MDH hybrids are detectable. Apart from the potential lack of structural complementarity of the subunits, instability of "chimeric intermediates" may be responsible for this finding. With respect to the latter alternative, reactivation kinetics of the two enzymes in the absence and presence of the prospective partner in hybridization have been applied to detect inactive intermediates. The kinetic profiles as well as the final yields of reactivation were found to be unaffected by the presence of the "foreign" subunit in the reconstitution mixture. This shows clearly that no significant formation of metastable hybrids occurs. Obviously, the reactivation rate of a given enzyme depends solely on the subunit concentration of this same enzyme, irrespective of other (even closely related) refolding species present in the mixture. However, this conclusion is confined to reconstitution conditions excluding wrong aggregation. If aggregation takes place, the two dehydrogenases tend to form mixed aggregates, while, e.g., bovine serum albumin (which is often used as protective agent), does not interfere with reconstitution.

A clear-cut case in the present context is the reassembly of PDH from *Bac. stearothermophilus* where independent folding and reactivation has been proved for two out of three activities of the multienzyme complex. No mutual interference of the different enzymes could be detected (Jaenicke and Perham 1982).

In summarizing the present results, correct association of oligomeric enzymes arises from highly specific intersubunit binding interactions. In order to maintain specificity of association, the individual subunits must not collide until they have reached a certain level of backbone structure providing the specific recognition sites required for subunit complementarity in a given quaternary structure. In vivo, this requirement is presumably fulfilled by cotranslational folding of the nascent polypeptide chain. The respective association sites must have been selected over evolutionary time for correct assembly, so that no compartmentation is required to guarantee correct quaternary structure formation (Cook and Koshland 1969).

Conclusions

As in the case of small one-domain proteins, folding of monomeric ODH proceeds via a rapidly formed structured intermediate. Slow reshuffling of this intermediate is rate-limiting for reactivation.

The subunit folding of the oligomeric dehydrogenases exhibits a similar pathway: A fast folding step, which restores a considerable amount of the secondary structure, is followed by the slow formation of the "correctly structured monomers" which can associate further.

The slow folding steps, characterized by halftimes of the order of 10 min are similar in rate for the monomeric, dimeric, or tetrameric dehydrogenases.

In the case of dimeric m-MDH and tetrameric LDH's association steps become rate-determining in addition to slow folding. Despite the close similarity of s-MDH and m-MDH, the rates of monomer association, are markedly different for both enzymes: Association of s-MDH is close to diffusion controlled, whereas reconstitution of m-MDH is slower by about two orders of magnitude. With regard to LDH, dimeric intermediates are formed in a fast reaction. Subsequent association to tetramers is slow , with a rate comparable to the monomer association of m-MDH.

Formation of inactive side products competes with the slow first-order folding to "correctly structured monomers".

Reconstitution starting from correct intermediates (produced by acid dissociation in the presence of a stabilizing salt) eliminates the slow folding step so that reactivation is only determined by association. Under this condition deactivation is fully reversible. If reconstitution starts from "incorrectly structured monomers" (produced by long-term acid dissociation in the absence of stabilizing salt) very slow reshuffling of the wrong structures becomes rate-determining. Consequently, the yield of reactivation decreases.

Since reactivation strictly parallels the formation of the native quaternary structure, it is obvious that all intermediates of association must be catalytically inactive under standard test conditions. In the presence of stabilizing salt, the dimeric intermediates are partially active. Obviously, a reduction in protein flexibility, accomplished upon association to native tetramers, can be mimicked at the level of the dimer by appropriate solvent conditions.

The dimeric association intermediates of LDH, which are in fast equilibrium with monomers, do not possess the complete secondary structure and stability towards proteolysis characterizing the native tetramer.

Stable "dimers" of LDH may be produced by partial proteolysis during reconstitution. These "dimers" contain a low affinity NAD-binding site and show enzymatic activity in the presence of stabilizing salt.

In general, binding of ligands to intermediates influences the kinetics of reconstitution of some of the dehydrogenases, either by stabilizing the intermediate or by accelerating subsequent folding or association steps.

The specificity of recognition of the intermediates must be high, since reconstitution is not affected by the presence of "foreign" subunits. Even in the case of high molecular weight multienzyme complexes (such as PDH) no mutual interference of the component enzymes can be detected.

The present studies provide some insight into the in vitro folding and association of large multidomain proteins. Although high yields of native enzyme can be recovered, it remains open whether folding and association in vitro follow the same pathways as in vivo. Vectorial folding in vivo may eliminate slow reactions of the monomers identified as rate-determining steps in the experiments given. This eventually speeds up structure formation considerably and at the same time prevents side reactions.

Acknowledgments. This study was supported by grants of the Deutsche Forschungsgemeinschaft and the Fonds der Chemischen Industrie. The authors wish to thank Mrs. Ingrid Fuchs and Drs. P Bartholmes, R Girg, R Hermann, FX Schmid, W Teschner, and G Zettlmeißl for fruitful discussions and expert help in performing experiments.

Discussion

Helmreich: In connection with the "unboiling of an egg", you mentioned that incomplete reconstitution is caused mainly by aggregation. Could you comment which step on the renaturation pathway of oligomeric dehydrogenases is involved?

Jaenicke: The competition of aggregation and folding refers to the $M^* \rightarrow M$ transition in the given kinetic schemes. As indicated by the complete reactivation after denaturation in the presence of "structure making ions" or after high pressure denaturation, M itself must be resistant towards aggregation. The fact that wrong aggregates show quasi-native CD characteristics proves that aggregation must be a late event on the folding pathway.

Helmreich: Do aggregates also form with a monomeric dehydrogenase?

Jaenicke: Yes, they do (e.g., octopine dehydrogenase in the transition range at low denaturant concentrations).

Rossmann: Is the proteolytic dimer of LDH-M_4 a Q-axis dimer?

Jaenicke: We cannot answer this question unambiguously. However, indirect evidence may be taken to confirm this assumption.

Rossmann: What is the kinetic difference in comparing the refolding of the proteolytic dimer and the unmodified polypeptide chain of LDH?

Jaenicke: The comparison is complicated by the fact that the "dimer" is partially nicked to smaller fragments. Starting reactivation from *dimeric intermediates* of dissociation leads to the same kinetics as if we would start from monomers, in accordance with tetramer formation rather than dimer formation as the rate-determining step on the pathway of reconstitution.

Schuster: What do you think is the reason for the rather slow assembly of the subunits? One possibility might be that the dimers are not in the correct conformation and have to refold during reconstitution.

Jaenicke: As I pointed out, the assembly may reach the rate of a diffusion controlled reaction. In cases where the assembly is slow, obviously steric effects gain importance, apart from possible slow transconformation reactions as you are proposing.

Schuster: The rate of formation of the Racker band in the case of Y-GAPDH might be used as an indicator for the correct folding of the rapidly formed dimeric species.

Jaenicke: This might be an attractive idea, although nothing is known with regard to the spectral differences in binding NAD^+ to the dimeric intermediate on the one hand and the tetrameric enzyme on the other.

I fully agree that the approach might be used to monitor the formation of the active center. However, in determining the kinetics, one has to have in mind that in the case of Y-GAPDH, the coenzyme causes a significant enhancement of the rate of folding.

Buehner: You mentioned that "arm off"-LDH forms *active* dimers at high salt concentrations. Would you define "high salt"? Could it be that salt mimicks the N-terminal arm by neutralizing some kind of Coulomb repulsion?

Jaenicke: As I mentioned, "high salt" means about 2 *M* $(NH_4)_2SO_4$. If high electrolyte concentrations would "mimick the N-terminal arm", the active species should be the tetramer; however, sedimentation analysis as well as gel filtration prove the product of limited proteolysis to be the "dimer". According to the Debye-Hückel theory shielding of charges would not require the observed high electrolyte concentrations which rather point to changes in the structure of the aqueous solvent as the driving force generating dimer activity. The fact that the activation is not merely caused by ionic strength is stressed by the observation that the formation of active "dimers" is governed by the Hofmeister series.

References

Anderson S, Weber G (1966) Arch Biochem Biophys 116:207-223
Anfinsen CB, Scheraga HA (1975) Adv Protein Chem 29:205-300
Banaszak LJ, Bradshaw RA (1975) Enzymes 3rd ed. 11:369-396
Bartholmes P, Jaenicke R (1978) Eur J Biochem 87:563-567
Bergman LW, Kuehl WM (1979) J Biol Chem 254:8869-8876
Bernhardt G, Rudolph R, Jaenicke R (1981) Z Naturforsch 36c:772-777
Birktoft JJ, Fernley RT, Bradshaw RA, Banaszak LJ (1982) Proc Natl Acad Sci USA 79: 6166-6170
Brandts JF, Halvorson HR, Brennan M (1975) Biochemistry 14:4953-4963
Brückl J (1980) Thesis, University of Regensburg
Burgess AW, Scheraga HA (1975) J Theor Biol 53:403-420
Cassman M, Vetterlein D (1974) Biochemistry 13:684-689
Cook RA, Koshland DE Jr (1969) Proc Natl Acad Sci USA 64:247-254

Dautry-Varsat A, Garel J-R (1981) Biochemistry 20:1396-1401
Eklund H, Nordström B, Zeppezauer E, Söderlund G, Ohlsson I, Boiwe T, Söderberg B-O, Tapia O, Brändén C-I, Åkeson Å (1976) J Mol Biol 102:27-59
Gerschitz J, Rudolph R, Jaenicke R (1978) Eur J Biochem 87:591-599
Girg R (1983) Dr. Thesis, University of Regensburg
Girg R, Rudolph R, Jaenicke R (1981) Eur J Biochem 119:301-305
Glatthaar BE, Barbarash GR, Noyes BE, Banaszak LJ, Bradshaw RA (1974) Anal Biochem 57:432-451
Hamlin J, Zabin I (1972) Proc Natl Acad Sci USA 69:412-416
Hermann R, Jaenicke R, Rudolph R (1981) Biochemistry 20:5195-5201
Hermann R, Rudolph R, Jaenicke R (1979) Nature (London) 277:243-245
Hermann R, Rudolph R, Jaenicke R (1982) Hoppe-Seyler's Z Physiol Chem 363:1259-1265
Holbrock JJ, Liljas A, Steindel SJ, Rossmann MG (1975) Enzymes 3rd ed. 11:191-292
Huber R, Bennet WS Jr (1983) In: Sund H, Veeger C (eds) Mobility and Recognition in Cell Biology. De Gruyter, Berlin, pp 21-48
Jaenicke R (1974) Eur J Biochem 46:149-155
Jaenicke R (ed) (1980) Protein Folding, Elsevier/North-Holland, Amsterdam, pp 587
Jaenicke R (1981) Annu Rev Biophys Bioeng 10:1-67
Jaenicke R, Perham RN (1982) Biochemistry 21:3378-3385
Jaenicke R, Rudolph R (1977) In: Sund H (ed) (1977) pp 351-367
Jaenicke R, Rudolph R, Heider I (1979) Biochemistry 18:1217-1223
Jaenicke R, Krebs H, Rudolph R, Woenckhaus C (1980) Proc Natl Acad Sci USA 77: 1966-1969
Jaenicke R, Müller K, Gäde G (1981a) Naturwissenschaften 68:205-206
Jaenicke R, Rudolph R, Heider I (1981b) Biochem Int 2:23-31
Jaenicke R, Vogel W, Rudolph R (1981c) Eur J Biochem 114:525-531
Kim PS, Baldwin RL (1982) Annu Rev Biochem 51:459-489
Kirschner K, Szadkowski H, Henschen A, Lottspeich F (1980) J Mol Biol 143:395-409
Koike M, Koike K (1976) Adv Biophys 9:187-227
Krebs H, Rudolph R, Jaenicke R (1979) Eur J Biochem 100:359-364
Kuan KN, Jones GL, Vestling CS (1979) Biochemistry 18:4366-4373
Levinthal C (1968) J Chim Phys 65:44-45
Luisi PL, Olomucki A, Baici A, Joppich-Kuhn R, Thomé-Beau F (1977) In: Sund H (ed) (1977) pp 472-484
Monneuse-Doublet M-O, Lefebure F, Olomucki A (1980) Eur J Biochem 108:261-269
Müller K, Jaenicke R (1980) Z Naturforsch 35c:222-228
Müller K, Lüdemann H-D, Jaenicke R (1981a) Naturwissenschaften 68:524-525
Müller K, Lüdemann H-D, Jaenicke R (1981b) Biophys Chem 14:101-110
Müller K, Lüdemann H-D, Jaenicke R (1982) Biophys Chem 16:1-7
Pfeil W (1981) Mol Cell Biochem 40:3-28
Privalov PL (1979) Adv Protein Chem 33:167-241
Privalov PL (1982) Adv Protein Chem 35:1-104
Ptitsyn OB, Finkelstein AV (1980) In: Jaenicke R (ed) (1980) Protein Folding, Elsevier/North-Holland, Amsterdam, pp 101-112
Rao ST, Rossmann MG (1973) J Mol Biol 76:241-256
Richardson JS (1981) Adv Protein Chem 34:168-339
Rossmann MG, Argos P (1981) Annu Rev Biochem 50:497-532
Rossmann MG, Liljas A (1974) J Mol Biol 85:177-181
Rossmann MG, Liljas A, Brändén C-I, Banaszak LJ (1975) Enzymes, 3rd ed. 11:61-102
Rudolph R (1977) Dr. Thesis, University of Regensburg
Rudolph R, Jaenicke R (1976) Eur J Biochem 63:409-417
Rudolph R, Heider I, Jaenicke R (1977a) Biochemistry 16:5527-5531
Rudolph R, Heider I, Jaenicke R (1977b) Eur J Biochem 81:563-570
Rudolph R, Heider I, Westhof E, Jaenicke R (1977c) Biochemistry 16:3384-3390
Rudolph R, Gerschitz J, Jaenicke R (1978) Eur J Biochem 87:601-606
Schade BC, Rudolph R, Lüdemann H-D, Jaenicke R (1980) Biochemistry 19:1121-1126
Stancel GM, Deal WC Jr (1969) Biochemistry 8:4005-4011
Storey KB, Dando PR (1982) Comp Biochem Physiol 73B:521-528
Sund H (ed) (1970) Pyridine Nucleotide Dependent Dehydrogenases. Springer, Berlin, Heidelberg New York, pp 472
Sund H (ed) (1977) 2nd Int Symp Pyridine Nucleotide Dependent Dehydrogenases, FEBS Symp 49. De Gruyter, Berlin, pp 513

Teipel JW, Koshland DE Jr (1971) Biochemistry 10:792-805
Teschner W (1983) Thesis, University of Regensburg
Wetlaufer DB (1981) Adv Protein Chem 34:61-92
Wetlaufer DB, Ristow S (1973) Annu Rev Biochem 42:135-158
Wootton JC (1974) Nature (London) 252:542-546
Zetina CR, Goldberg ME (1980) J Mol Biol 137:401-414
Zetina CR, Goldberg ME (1982) J Mol Biol 157:133-148
Zettlmeißl G (1983) Dr. Thesis, University of Regensburg
Zettlmeißl G, Rudolph R, Jaenicke R (1979a) Eur J Biochem 100:593-598
Zettlmeißl G, Rudolph R, Jaenicke R (1979b) Biochemistry 18:5567-5571
Zettlmeißl G, Rudolph R, Jaenicke R (1981) Eur J Biochem 121:169-175
Zettlmeißl G, Rudolph R, Jaenicke R (1982a) Biochemistry 21:3946-3950
Zettlmeißl G, Rudolph R, Jaenicke R (1982b) Eur J Biochem 125:605-608
Zettlmeißl G, Rudolph R, Jaenicke R (1983) Arch Biochem Biophys 224:161-168

Flavoproteins

Flavoproteins of Known Three-Dimensional Structure

R. H. Schirmer[1] and G. E. Schulz[2]

Flavoenzymes of Known Structure

About 50 years ago FMN (Theorell 1935) and FAD (Warburg and Christian 1938) were discovered as prosthetic groups of the old yellow enzyme and of D-amino acid oxidase, respectively. More than 100 flavoenzymes have been described in the meantime (Dixon and Webb 1979); for four of them, structural data are known. These are glycolate oxidase (Lindquist and Bränden 1980), an FMN-dependent enzyme which probably has a chain fold in common with triose phosphate isomerase and one domain of pyruvate kinase, and three FAD-dependent enzymes: ferredoxin-$NADP^+$ reductase (Sheriff and Herriott 1981; Karplus and Herriott 1982) *p*-hydroxybenzoate hydroxylase (Wierenga et al. 1979; Wierenga et al. 1982; Weijer et al. 1982) and glutathione reductase (Schulz et al. 1978; Thieme et al. 1981; Pai and Schulz 1983). The last mentioned two enzymes have been analyzed in atomic detail. In our presentation, we shall first describe glutathione reductase as a well-understood flavoenzyme and then compare its structure with *p*-hydroxybenzoate hydroxylase. Whenever possible we shall include structural aspects of other flavoproteins in our discussion.

Glutathione Reductase from Human Erythrocytes

General Properties

Human glutathione reductase is a homodimer of 2 × 52 400 mol. wt. (Worthington and Rosemeyer 1974; Schulz et al. 1975; Krauth-Siegel et al. 1982). Its amino acid sequence is known (Untucht-Grau et al. 1981; Krauth-Siegel et al. 1982) and has been fitted to an electron density map of 0.2 nm resolution, which yielded the complete structure of the molecule (Thieme et al. 1981). The enzyme catalyses the reduction of glutathione disulfide by NADPH:

$$NADPH + GSSG + H^+ \rightleftharpoons NADP^+ + 2GSH.$$

This reaction can be shown to proceed in two half-reactions taking place at opposite sides of one subunit (Fig. 1):

$$NADPH + E + H^+ \rightleftharpoons NADP^+ + EH_2 \quad \text{and}$$

$$EH_2 + GSSG \rightleftharpoons E + 2GSH.$$

1 Institut für Biochemie II der Universität Heidelberg, Im Neuenheimer Feld 328, D-6900 Heidelberg, FRG

2 Max-Planck-Institut für Medizinische Forschung, Jahnstraße 29, D-6900 Heidelberg, FRG

34. Colloquium-Mosbach 1983
Biological Oxidations

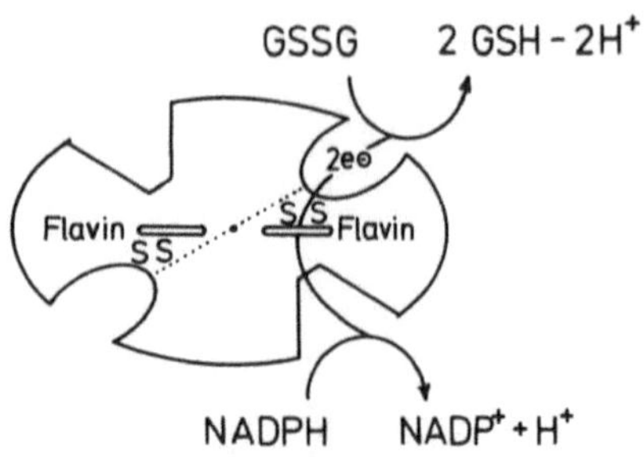

Fig. 1. Topography of glutathione reductase. The diad of the dimeric enzyme (*central dot*) is oriented perpendicular to the paper plane. The *dotted line* represents the interface between the two subunits. A catalytic cycle is shown on the *right-hand side*. The electrons move from the nicotinamide of NADPH via the flavin ring and via the redox-active disulfide Cys-58:Cys-63 to glutathione disulfide. Regarding this isolated catalytic cycle, a difference of 3 H^+ results between upper and lower surface of the enzyme (Fritsch et al. 1979; Mannervik et al. 1980). The difference amounts to 4 H^+ if $NADP^+$ is reduced again at its site of formation, e.g., by a dehydrogenase of the pentose phosphate cycle according to the equation $NADP^+$ + substrate $\cdot$ H_2 $\rightarrow$ H^+ + NADPH + substrate. If glutathione reductase were appropriately oriented in a membrane, the enzyme could establish a transmembrane proton gradient provided that NADPH is available only at one side and GSSG at the other side of the membrane (Untucht-Grau 1983)

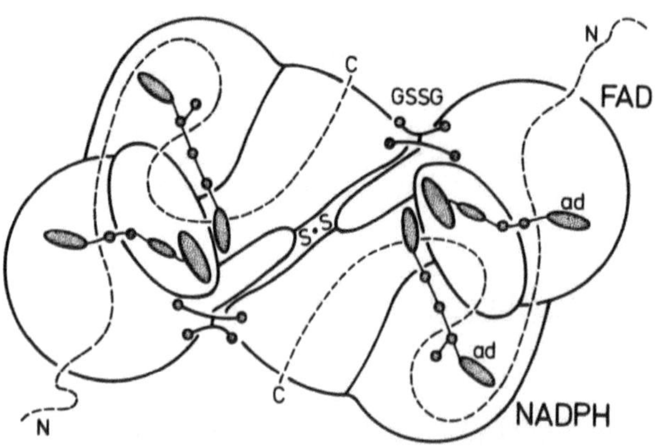

Fig. 2. The dimeric protein glutathione reductase as viewed along the molecular twofold axis. The subunits are connected by a single disulfide bond across this molecular axis. The 18 N-terminal residues have no defined structure in the crystal; the role of this flexible arm is not known. The remaining 460 residues of the polypeptide chain are geometrically organized into four domains. The first two domains along the chain resemble each other structurally (Schulz 1980) and bind FAD and NADPH, respectively. The central and the interface domain follow at the C-terminal side. The ligands FAD, NADPH, and GSSG are shown. The course of the polypeptide chain is sketched as a *dashed line* from N to C. Other details are given in Fig. 7

E represents the oxidized yellow form of the enzyme and EH_2 its two-electron-reduced orange-colored form. Both forms, E and EH_2, are fairly stable so that they can be handled conveniently in the laboratory (Williams 1976; Holmgren 1980).

Glutathione reductase is a functional part of the glutathione redox cycle (Chance et al. 1979) keeping the concentration of GSH at 1-10 mM and the [GSH]/[GSSG] ratio above a value of 100:1 under steady state conditions. Consequently, this enzyme is believed to have contributed to the evolutionary adaptation to molecular oxygen since the glutathione reductase catalysed reaction conserves the reducing milieu of primordial cells. Molecular oxygen by itself would change the [GSH]/[GSSG] ratio to a value of $1:10^{16}$ according to mass action (Fahey 1977).

The arrangement of ligands in glutathione reductase is shown in the structural overview of Fig. 2. The two dinucleotides FAD and NADPH bind in extended conformations to the protein. The nicotinamide moiety of NADPH is in close contact with the flavin moiety of FAD, whereas the adenosines of NADPH and FAD are far apart from each other, the distance between the N1 positions of the two adenines being 2.9 nm.

The Stereochemistry of Catalysis as Revealed by X-Ray Crystallography

In the crystals used for X-ray analysis, glutathione reductase is catalytically active: The yellow crystals of the oxidized enzyme (State E, see page 93) turn orange when NADPH is added and the orange crystals representing EH_2 turn yellow again when NADPH is removed from and GSSG added to the incubation medium.

The binding of both substrates to glutathione reductase has been determined by X-ray crystallography. In addition, two reaction intermediates were prepared in the crystals and their structures were deter-

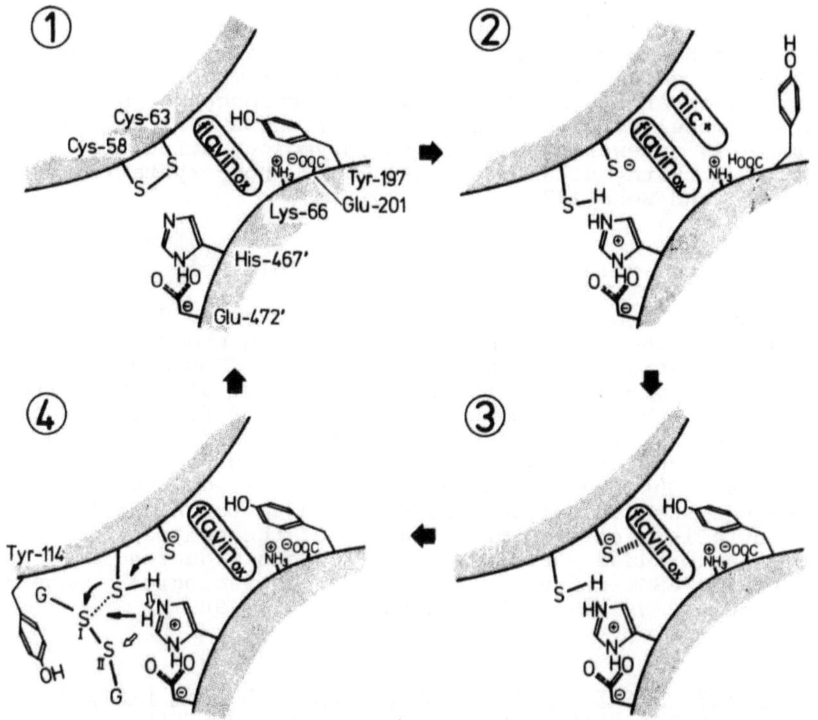

Fig. 3. Four states of the catalytic reaction cycle of glutathione reductase for which crystallographic data are available. State *1* is the oxidized enzyme E. State *2* is the complex between $NADP^+$ and the reduced enzyme EH_2. State *3* is the reduced enzyme EH_2, a stable reaction intermediate. In "State *4*", three catalytic steps are combined: the binding of GSSG to EH_2, the formation of the mixed disulfide between glutathione-I and Cys-58 (*dotted line*) with subsequent release of glutathione-II, and the release of glutathione-I. The *light arrows* refer to assumed hydrogen movements during formation of the mixed disulfide and release of glutathione-II. The *black arrows* refer to splitting of the mixed disulfide and the release of glutathione-I. The hydrogens of all 4 states are postulated, they cannot be located by X-ray diffraction. It should be noted that flavin is oxidized in all four states. The color change, from yellow in State *1* to orange in State *2*, is due to a charge transfer complex between a newly formed thiol(ate) and reoxidized flavin (Kosower 1966; Massey and Ghisla 1974). Further details are discussed by Pai and Schulz (1983)

mined at 0.3 nm resolution. As shown in Fig. 3, these results – obtained with a so-called static method – allow the stereochemical description of the enzyme's mechanism (Pai and Schulz 1983).

NADPH is bound with the nicotinamide ring stacking onto the re-face of the flavin moiety of FAD, which results in the 2-electron reduced form of the enzyme, EH_2. EH_2 has also been analyzed in the absence of bound substrate or product. In EH_2, the flavin ring is reoxidized; the redox-active disulfide/dithiol Cys-58:Cys-63, which is located at the si-face of flavin is open and the sulfur of Cys-58 moves by about 0.1 nm into a new position. From this position it can attack one of the sulfurs of GSSG and a disulfide between Cys-58 and glutathione is formed. This intermediate (State 4 in Fig. 3) could also be stabilized in the crystals and analyzed. The imidazole side chain of His-467' is close to all sulfurs taking part in the thiol-disulfide interchange reactions (Fig. 3) and consequently crucial for catalysis.

In conclusion, the crystallographic results establish that reducing equivalents flow from NADPH via flavin and via a dithiol-/disulfide redox pair to GSSG. It should be emphasized that this mechanism confirms suggestions made earlier on the basis of spectroscopic, biochemical, and chemical model studies (Williams 1976; Massey and Williams 1982). This example is noteworthy because in enzymology structural data are often inconsistent with mechanisms obtained by "non-crystallographic methods" (Dixon and Webb 1979).

Protection of EH_2 Against Molecular Oxygen – A Possible Role of Tyr-197

State EH_2 of glutathione reductase (Fig. 3) can also be produced by GSH. Indeed, because of the high [GSH]/[GSSG] ratio in the cell, EH_2 is probably the prevalent state of the enzyme in vivo (Boggaram et al. 1978). This might explain the role of Tyr-197 (Fig. 3). Like a lid, the phenol ring shields the NADPH-contacting *re*-face of flavin from the solvent and solutes. Thus, it prevents flavin from transferring reduction equivalents to molecular oxygen which happens readily with unprotected isoalloxazines (Gutfreund 1972; Hemmerich 1976). On binding of NADPH, the lid swings back and allows nicotinamide to touch the flavin. The *si*-side of flavin is protected by a segment of the polypeptide chain against O_2; the oxidase activity of glutathione reductase is less than 0.01% of its GSGG-reduction activity at physiological concentrations of GSSG and O_2 (Williams 1976; Krauth-Siegel 1982).

On the Active Site Topography of NAD(P)H-Dependent Membrane Flavoenzymes. Inferences from the Structure of Glutathione Reductase

Compartmentation of the Active Site in Glutathione Reductase

The active center of glutathione reductase is divided into two compartments, one binding NADPH and the other binding GSSG. As shown in Figs. 1 and 3, each compartment corresponds to a deep pocket in the protein which is filled – fully or partly (see above) – with solvent when the respective ligand is absent. The two pockets indent at opposite sides of a subunit and meet each other at the subunit center. If these pockets were not separated from each other by flavin and the disulfide Cys-58:Cys-63, they would form a channel through the subunit. Binding of the substrates to the enzyme in this geometry results in an optimal protection of the reaction center against solvent and solute molecules which otherwise would interfere with the electron transfer process. It might be of interest to consider the possibility that this unique geometry is also present in membrane flavoenzymes, which transfer reducing equivalents from one side of the membrane to the other.

Leukocyte NADPH-Oxidase and Other Membrane Flavoenzymes

Examples of membrane-bound FAD-enzymes are given in Table 1. Only leukocyte NADPH-oxidase shall be described here in some detail. The NADH-dehydrogenase of *Escherichia coli* (Young et al. 1981) will be referred to along with other flavoenzymes of known primary structure on page 107.

Table 1. Some properties of three membrane flavoenzymes

FAD-Containing Enzyme	Location and Function
1. Respiration NADH-Dehydrogenase of *Escherichia coli* The M_r of the FAD-binding unit is 47 304. The sequence of the 434 residues is known (Young et al. 1981).	Inner membrane of *E. coli*; transfers electrons from NADH to the respiratory chain.
2. NADH-Cytochrome b_5 Reductase The M_r of the FAD-binding unit is 44 000 (Hultquist et al. 1981).	Endoplasmic reticulum of liver cells and reticulocytes; transfers electrons from NADH via cytochrome b_5 to fatty acid desaturase and other acceptors (Oshino et al. 1971). A soluble form of the enzyme in red blood cells is probably involved in the reduction of methemoglobin by NADPH or NADH (Hultquist et al. 1981).
3. Leukocyte NADPH-Oxidase The M_r of the flavoenzyme is 150 000 or greater (Light et al. 1981).	Plasma membrane and phagosomal membrane of leukocytes; produces O_2^- by transferring electrons from NADPH to O_2 (Babior 1978).

NADPH-oxidase is the key enzyme of the oxidative processes by which the neutrophilic leukocyte attacks microorganisms and tumor cells or causes tissue injury (Babior 1978; for reviews of the literature see Tauber 1982, and Schirmer et al. 1983). This FAD-protein of $M_r = 150\ 000$ (Light et al. 1981) is located in the plasma membrane or in the phagosomal membrane, which forms by invagination of the plasma membrane. NADPH-oxidase catalyses the reaction

$$NADPH + O_2 + O_2 \rightleftharpoons NADP^+ + H^+ + 2O_2^-,$$

which resembles the reduction of GSSG by NADPH. By inference from the known structure of glutathione reductase (Figs. 1 and 2), we suggest that the topography sketched in Fig. 1 also applies for NADPH-oxidase (Fig. 4). According to this model, NADPH is oxidized at the cytosolic side of the membrane. The reducing equivalents move via the flavin ring to the O_2-binding site oriented to the extracellular compartment or to the phagosomal compartment, respectively. Thus, the cytotoxic oxygen derivatives are formed at the scene of action, namely, at the surface of target cells or foreign particles. The cytosol of the leukocyte is threatened only indirectly, for instance by H_2O_2 diffusing back from the extracellular space.

Extracellular space or phagosomal cavity

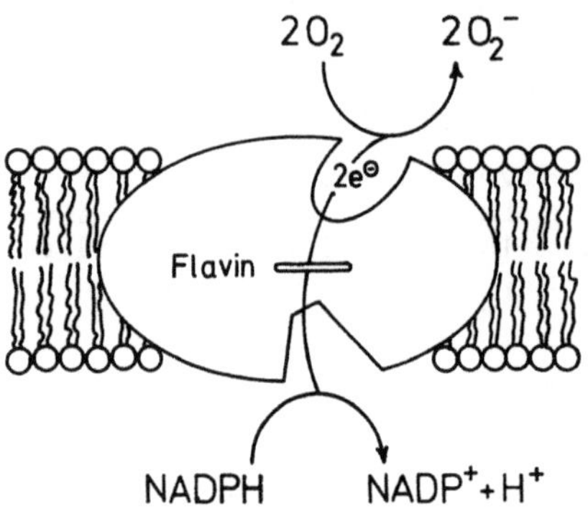

Cytosol

Fig. 4. Proposed topography of leukocyte NADPH-oxidase (Babior et al. 1981, Schirmer et al. 1983). At the cytosolic side of the membrane, the half reaction NADPH → $NADP^+ + H^+$ takes place; two electrons are transferred to the external surface of the membrane where the half-reaction $2\ O_2 + 2\ e^- \rightarrow 2\ O_2^-$ occurs. The possibility that NADPH-oxidase is a dimer with opposite polarities of the subunits-like glutathione reductase (Fig. 1) —would not change our conclusion, since NADPH is available only on one side of the membrane

In this way NADPH-oxidase would transport only reducing equivalents through the membrane and the question of how O_2^- is "secreted" by leukocytes might become irrelevant. A similar mechanism could explain an observation of Eckman and Eaton (1979). They found for malaria-infected red blood cells that GSSG in the parasite compartment can be reduced by NADPH produced in the host compartment. This remarkable process might be brought about by a glutathione reductase-like enzyme located appropriately in a membrane which separates the cytosol of the red blood cell from the parasite compartment.

Flavoenzymes May Establish Transmembrane Proton Gradients

In the reaction catalysed by glutathione reductase probably $2e^-$ and no protons are transferred across the flavin ring (Schulz et al. 1978; Mannervik et al. 1980). As a consequence, a difference of $3H^+$ is generated per catalytic cycle between opposite surfaces of the enzyme (Fig. 1). Since glutathione reductase is a cytosolic enzyme, this gradient breaks down immediately due to proton transfer via the solvent.

Let us now consider the O_2^--production by leukocytes which is metabolically driven by the pentose phosphate cycle (Babior 1978). When $NADP^+$ is reduced in this cycle, one H^+ is formed along with one NADPH in the cytosol (Warburg et al. 1935). According to the model of Figs. 1 and 4, NADPH-oxidase releases a further H^+ (and $NADP^+$) at the cytosolic side of the membrane, while $2e^-$ travel to the external surface. Such a mechanism might, for example, explain the initial drop of $[H^+]$ in neutrophils exposed to chemotactic factors (Molski et al. 1980; Schirmer et al. 1983).

Interactions Between FAD and Apoprotein

In the preceding chapter we have contemplated on flavin as a gross barrier between different binding sites of a protein or even between compartments separated by a membrane. Now we shall turn to the details of the interaction between apoprotein and FAD.

Apo-Glutathione Reductase, a Stable Protein

The FAD-free apoenzyme crystallizes under the same conditions as the holoenzyme (Fritsch et al. 1979; Fritsch 1982). The diffraction patterns of apo- and holoenzyme resemble each other closely at 0.6 nm resolution (Schulz et al. 1982). The major structural difference is that the protein monomer shrinks by 0.5 nm in one dimension when FAD is removed. The observed stability of apo-glutathione reductase supports the validity of the current test for riboflavin deficiency. In this test – which is most important in malnourished women of childbearing age – the amount of apo-glutathione reductase in lysed erythrocytes serves as a quantitative measure of the vitamin deficiency (Glatzle et al. 1968; Prentice and Bates 1981).

In this context we should like to discuss the following fact: The negative charges of the pyrophosphate moiety of FAD are not neutralized by positively charged amino acid residues in holo-glutathione reductase (Table 2). Although this situation is unfavorable in the holoenzyme, it could be advantageous for the apoenzyme: if FAD were fixed by positively charged side chains, these residues would repel each other in the absence of FAD and destabilize the apoenzyme. The stability of the apoenzyme is probably of physiological significance (Fritsch 1982). Under conditions of intermittent or periodic riboflavin deficiency in the diet, it is advantageous that the erythrocytes and other nucleus-free cells are endowed with a stable apo-glutathione reductase. As soon as the vitamin is supplied again, the functional holoenzyme can be assembled which prolongs the life of the cells (Loos et al. 1976).

Table 2. Comparison between the FAD-binding domains of *p*-hydroxybenzoate hydroxylase and of glutathione reductase. (Wierenga et al. 1983)

Feature	*p*-Hydroxybenzoate Hydroxylase	Glutathione Reductase	Comments
Localization of the FAD-domain in the subunit	residues 1 to 163	residues 19-157	Glutathione reductase has an N-terminal extension of 18 residues.
Topology of the FAD-domain	A 4-stranded parallel β-sheet covered by a 3-stranded β-meander on one side and α-helices on the other side $\hat{=}$ a Rossmann fold with an additional sheet strand and an additional β-meander. FAD is bound at the C-terminal edge of the Rossmann fold with flavin at the helix side of the sheet and the adenine at the other side of the sheet.		This topology was also found in the NADPH-binding domain of glutathione reductase, but in no other protein.
Residues chosen for superposition of the two FAD-domains	69 residues contained in the 7 β-strands and in the first α-helix; the r.m.s. C_α-distance of these equivalenced residues is 1.2 Å.		
Total number of superimposed residues	110 residues; the root mean square C_α-distance of equivalent residues being 1.9 Å		

Table 2 (cont.)

Feature	*p*-Hydroxybenzoate Hydroxylase	Glutathione Reductase	Comments
Identical residues at equivalent positions	Ile-8, Gly-9, Gly-11, Gly-14, Leu-15, Gly-25, Glu-32, Pro-137, Ile-155, Gly-160	Ile-26, Gly-27, Gly-29, Gly-32, Leu-33, Gly-43, Glu-50, Pro-138, Ile-152, Gly-157	Ile-8/26 lines the adenine pocket. The φ- and ψ-values for Gly-9/27 are in the region reserved for Gly residues. A longer side chain at position 11/29 would prevent the PP_i-moiety from contacting the main chain, and at position 14/32, it would loosen the contact between β_A and α_A. For Glu-32/50 see below.
Chain excursion between β_B and β_C	62 residues long	48 residues long	The excursions are geometrically not comparable. They may serve as "clips" for keeping the domains of a subunit together.
Localization of flavin relative to the subunits	Buried in one subunit 20 Å away from the other subunit	H-bonded to the other subunit	
Position of FAD in the FAD-binding domain	After superposition of the polypeptide chains the r.m.s. distance of the 53 nonhydrogen atoms of the FAD molecules is 1.9 Å (0.9 Å for the pyrophosphate and the ribose moieties, 1.8 Å for the adenine rings, and 2.6 Å for the flavin moieties.		
Conformation of FAD	Extended in both enzymes; the average difference of equivalent torsion angles amounts to 35° only.		
Environment of flavin	Re-face open to the solvent Si-face is covered by polypeptide	Re-face is the bottom of the NADPH-binding niche; si-face is covered by polypeptide	The binding site for NADPH is not known in *p*-hydroxybenzoate hydroxylase.
	In both enzymes there is a conspicuous α-helix starting close to position N1/O2α of flavin; it comprises segment 298-317; the peptide-NH of residue 299 forms an H-bond to N1	segment 339-354; the peptide-NH of residue 339 forms an H-bond to N1	These helices could stabilize a negative charge close to N1/O2α of the flavin ring. Note that these helices are parts of completely different chain segments in the two enzymes.

Table 2 (cont.)

Feature	*p*-Hydroxybenzoate Hydroxylase	Glutathione Reductase	Comments
Environment of ribitol	The ribityl chain is pleated differently, but its general course is similar in both enzymes; H-bonded to Asp-286 and other side chains	H-bonded to Asp-331 and other side chains	
Environment of pyrophosphate	The charges are compensated by Arg-42 and Arg-44	No charge compensation by side chains; possibly by an Mg^{2+}-ion	Positioned at the loop between β_A and α_A as in other nucleotide-binding proteins.
Environment of ribose	The 2'OH-group and 3'OH-group form strong H-bonds to a negatively charged residue at the C-terminus of β_B, namely, Glu-32	Glu-50	In the NADPH-domain of glutathione reductase the 2' phosphate group of NADPH occupies the position corresponding to Glu-50 in the FAD-domain; this phosphate group is fixed by Arg-218 and His-219, the last residues of β_B.
Environment of adenine	Located in a hydrophobic pocket. N6 points to the solvent. No strong H-bonds between protein and adenine	located in a hydrophobic environment. N6 points to the solvent. H-bonds between N1, N3, and N6 and the protein	

Furthermore, the possibility of reconstituting a holoenzyme from prosthetic group and apoenzyme enables us to replace FAD by FAD-analogs (Ghisla and Krauth-Siegel, unpublished data; Krauth-Siegel 1982). Indeed, the FAD and FMN analogs recently introduced by Massey and his co-workers have greatly enriched the arsenal for the analysis of structure-function relationships in flavoenzymes (Massey and Hemmerich 1980).

When FAD is bound to apo-glutathione reductase, the protein engulfs the FMN moiety completely. Consequently, the protein must open up before or during the binding process. An appropriate entrance gate for FAD is probably formed by a reversible movement of the FAD-binding domain relative to the NADPH-binding domain; these two domains make only few contacts with each other. The solvent accessibility of bound FAD can be summarized as follows. The FMN part is buried in the protein molecule with the ribitol chain and the phosphate group sandwiched between the FAD-binding and the central domain. The AMP moiety makes close contacts only with residues of the FAD-binding domain. One face of adenine, one face of ribose, and the phosphate oxygens O_{A1} and O_{A2} (see Fig. 6) can be reached by the solvent and solutes.

The Conformation of FAD as Bound to Glutathione Reductase

Figure 5 shows a molecular model of FAD as fitted to the corresponding portion of the electron density map. The orientation of the isoalloxazine ring (=flavin) is determined unequivocally: the methyl groups of the benzoid part can be distinguished clearly from the O2α and O4α atoms. Moreover, as indicated in the figure, the hydrogen attached to N3 forms a short, i.e., a strong, hydrogen bond to the carbonyl of His-467'.

The ribitol chain is fully extended. All three hydroxyls can be recognized easily. The phosphate oxygens emerge as distinct humps. The ribose, in particular its hydroxyl groups, are well-shaped. Both rings of adenine as well as its 6α-amino group are outlined which reveals the adenine orientation unambiguously. Altogether, the electron density distribution at O.2 nm resolution clarifies the position of every nonhydrogen atom of FAD as bound to glutathione reductase within an error margin of about O.O3 nm.

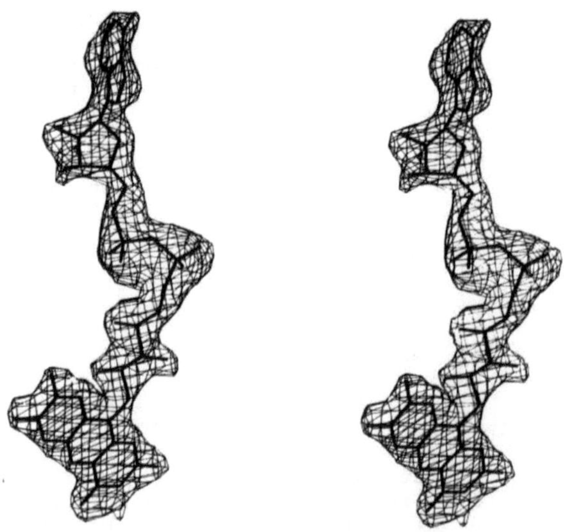

Fig. 5. Stereo view of FAD as bound to glutathione reductase. The electron density is given in bird cage representation. The *bold lines* show the fitted molecular model of FAD. The pyrophosphate moiety gives rise to the highest values of the whole electron density distribution. This cannot be visualized in the figure because the electron density is merely represented by its shape at a rather low cut level (20% of the maximum density.) All structures surrounding FAD have been dissected from the picture; only at N3 of flavin, a stump of the hydrogen-bond to His-467' has remained

The Location of FAD in Glutathione Reductase

Each subunit binds one FAD molecule, the apparent association constant being $1.8 \times 10^6\ M^{-1}$ (Staal et al. 1969). A subunit consists of four domains (Figs. 2 and 7). As shown in Fig. 6, FAD forms most of the polar contacts with the FAD-binding domain; as many as 14 residues of this domain are involved in such polar interactions. FAD also interacts with the central domain, namely with residues Arg-291, Val-329, Asp-331 and Thr-339. The flavin portion of FAD is surrounded by residues from all four domains of one subunit (for example Cys-63, Tyr-197, Thr-339 and Phe-372 in Figs. 6 and 7) as well as from the interface domain of the other subunit (for example, His-467'). This means that as many as five domains contribute to the catalytic site of glutathione reductase assembled around flavin (Figs. 2,3,6, and 7). The remaining part of FAD probably serves as an anchor for the flavin (cf. Jencks 1975).

List of distances [Å]

1	2.9
2	3.0
3	3.7
4	3.6
5	2.6
6	2.4
7	2.7
8	2.9
9	3.0
10	3.3
11	3.7
12	3.6
13	2.7
14	2.4
15	2.8
16	4.2
17	2.7
18	2.3
19	2.9
20	2.8
21	3.8
22	3.1
23	2.7
24	2.7
25	3.3
26	3.6
27	3.4
28	3.1
29	2.4
30	3.3
31	3.0
32	2.5
33	2.4
34	2.2
35	3.0
36	2.7
37	2.8

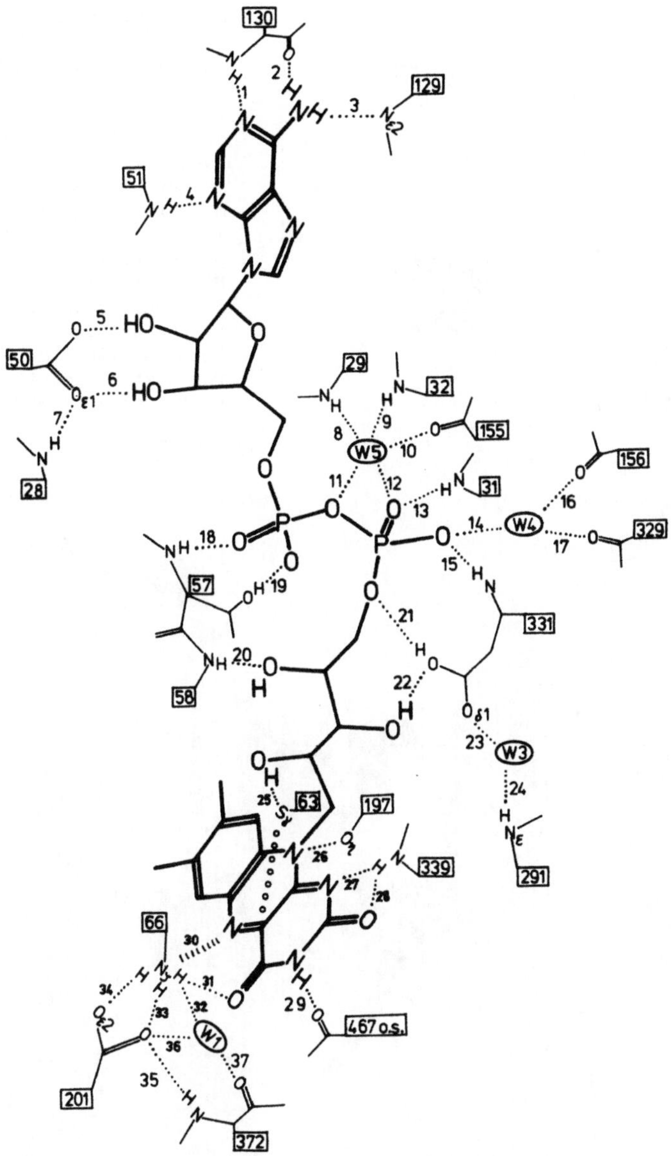

Fig. 6. The polar interactions between FAD and its protein environment (Schulz et al. 1982). FAD is drawn in the same view as in Fig. 5. The *boxed numbers* refer to amino acid residue positions. All hydrogens bound to oxygen and nitrogen atoms are indicated although they do not show up in the electron density map. Interaction distances are given at the *left* side of the figure. The distance between $N_{\varepsilon 1}$ of Arg-291 and C8α of flavin is 0.4 nm (not shown in the sketch). The globular densities W1 and W3 were interpreted as water molecules; W4 and/or W5 probably represent trapped cations

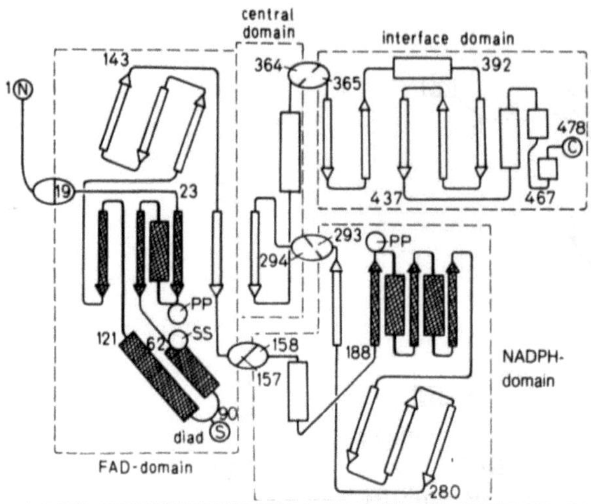

Fig.7a,b. Chainfold of glutathione reductase. (a) The *numbers* denote residue positions and the residues in *ovals* domain boundaries. Strands of β-sheets are given as *arrows*, helices as *rectangles*. In the FAD-domain and in the NADPH-domain, the Rossmann fold (β_A, α_A, β_B, α_B, β_C in Fig. 8a) is hatched. *PP*, pyrophosphate moieties of FAD and NADPH, respectively. *SS*, the redox-active dithiol Cys-58:Cys-63. *S*, the sulfur of Cys-90 which forms a disulfide with Cys-90'. For further details see Krauth-Siegel et al. 1982

◁ Fig. 7a   Fig. 7b ▽

(b) FAD-binding (*left*) and NADPH-binding (*right*) to glutathione reductase. The motifs were taken from Botticelli's *Allegory of the Spring*. From *left to right*: A, adenosine; β, central parallel-stranded sheet of the FAD-binding domain; R, riboflavin; N, nicotinamide nucleoside; β', central sheet of the NADPH-binding domain; A, adenosine. A choreographic arrangement which is present in many dinucleotide-binding proteins is illustrated here twice: The pyrophosphate moiety (pair of hands on top of the wall) binds to the loop between the first strand of the β-sheet and the N-terminal section of the first α-helix. Thus, the negative charge at the pyrophosphate is stabilized by the positive end of the helix-dipole. The dipole moment of the α-helix is indicated by Cupido's arrow in the helix axis. The specific part of the dinucleotide, flavin in the case of FAD and nicotinamide in the case of NAD(P)H, is located on the helix-side of the β-sheet, whereas adenine is on the other side

Although flavin is submerged in the protein, its environment is rather polar (Fig. 6). Three strong dipoles exist: the salt bridge Lys-66: Glu-201 near N5, the trio Arg-291:W3:Asp-331 close to C8α, and the ion pair His-467':Glu-472' (Fig. 3). In all three cases, the positively charged partner is closer to flavin, which may render the isoalloxazine more electrophilic. This assumption is supported by the existence of a semiquinone anion which can be produced in a slow side reaction catalysed by glutathione reductase (Williams 1976). The anionic states of flavin carry a negative charge around N1/O2α, which can be stabilized by the dipole field of an α-helix (Hol et al. 1978). This is indeed the case in *p*-hydroxybenzoate hydroxylase and in glutathione reductase (Table 2). It remains to be seen whether the semiquinone-stabilizing arrangement is of significance in the catalytic mechanism or for any other function of these two enzymes.

Structural Comparisons Between the FAD-Binding Domains of *p*-Hydroxybenzoate Hydroxylase and Glutathione Reductase

p-Hydroxybenzoate hydroxylase from *Pseudomonas fluorescens* (EC 1.14.13.2) is a well-characterized monooxygenase (Müller et al. 1980), which catalyses the reaction

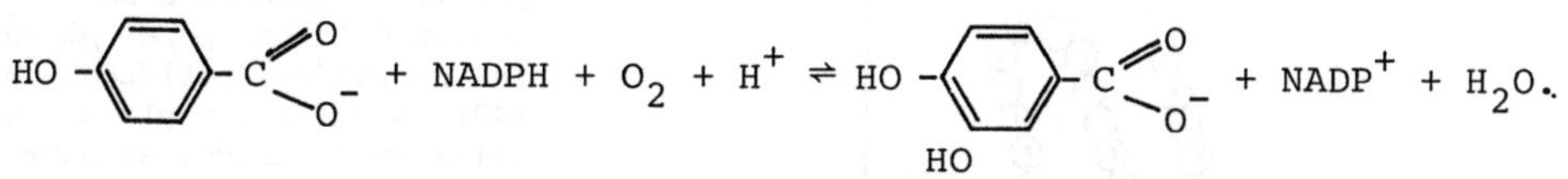

The protein is a homodimer containing 394 amino acid residues and one FAD molecule per subunit. The amino acid sequence is known (Wierenga et al. 1982; Weijer et al. 1982) and has been fitted to an electron density map of 0.25 nm resolution (Wierenga 1979; Hofsteenge 1981).

The FAD-binding domains of *p*-hydroxybenzoate hydroxylase, a bacterial enzyme, and of human glutathione reductase have a chain fold in common which is different from the fold of the structurally known flavodoxins (Watenpaugh et al. 1972; Burnett et al. 1974; Ludwig et al. 1982). As shown in Fig. 8 and in Table 2, this chain fold is characterized by a 4-stranded parallel β-sheet sandwiched between a β-meander and α-helices. Recently, the two structurally known FAD domains have been superimposed and compared in atomic detail (Wierenga et al. 1983). The results of this comparison are summarized in Table 2. The observed distances between equivalent C_α-atoms are remarkably small; just as it has been found in comparisons between the NAD-binding domains of lactate dehydrogenase, malate dehydrogenase, alcohol dehydrogenase, and glyceraldehyde-3-phosphate dehydrogenase (Rao and Rossmann 1973; Ohlsson et al. 1974; Rossmann et al. 1975).

The structural similarities are not restricted to the polypeptide chain, but apply also to the conformation and localization of FAD. Therefore, if one knew the structure of the holoenzyme of *p*-hydroxybenzoate hydroxylase and the structure of the apoenzyme of glutathione reductase (or vice versa), one could predict the position of FAD in the latter enzyme fairly accurately (Table 2 and Fig. 9).

The *general* binding position of the dinucleotide resembles the position of NAD(H) in lactate dehydrogenase and related enzymes (Rossmann et al. 1975). In particular, the pyrophosphate moieties are fixed in each case at the N-terminal side of the first helix, α_A (Figs. 7 and 8). The flavin and the nicotinamide, respectively, i.e., the specific parts

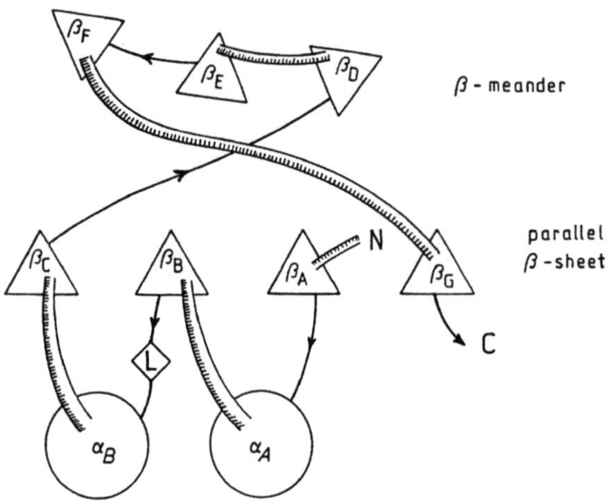

△
Fig. 8a

Fig. 8b
▽

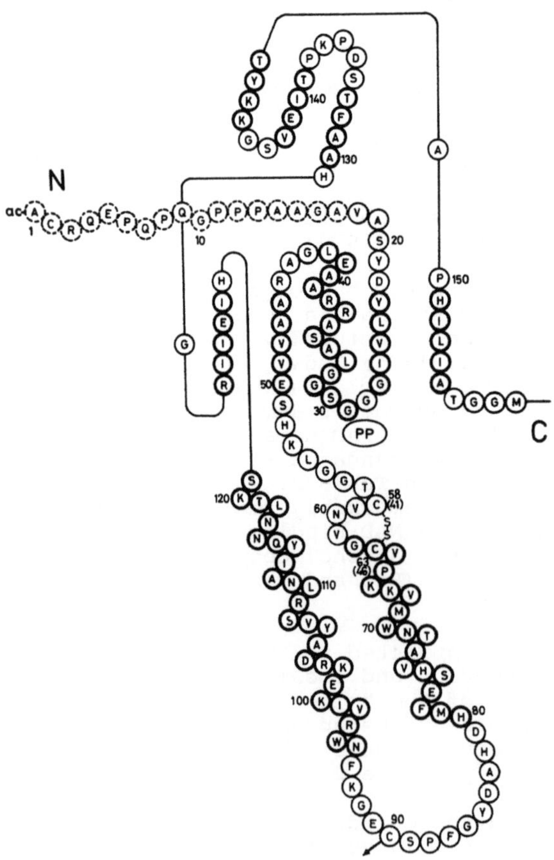

Fig. 8a,b. Structure cartoons of the domain that binds FAD in *p*-hydroxybenzoate hydroxylase and glutathione reductase. (a) The view corresponds to the view from the top in Figs. 7a and 8b. *N* and *C* denote the ends of the domain; the long chain excursion between β_B and α_B is marked by *L*. (b) The chain fold of the FAD-binding domain of glutathione reductase in diagrammatic presentation (Untucht-Grau 1983). The view and the symbols correspond to those of Fig. 7a. The one letter code for amino acid residues is used. Residues which are not visible in the electron density map are drawn with *dotted circles*, residues involved in helical structures and in sheet strands with *heavier circles*. The twisted parallel-stranded β-sheet is depicted in the paper plane; the β-meander (residues 129-148) is below the plane, and helix 29-42 above, as shown in Fig. 8a. Tyr-23, Ala-46 and Ile-123 are the first residues of β_A, β_B, and β_C, respectively

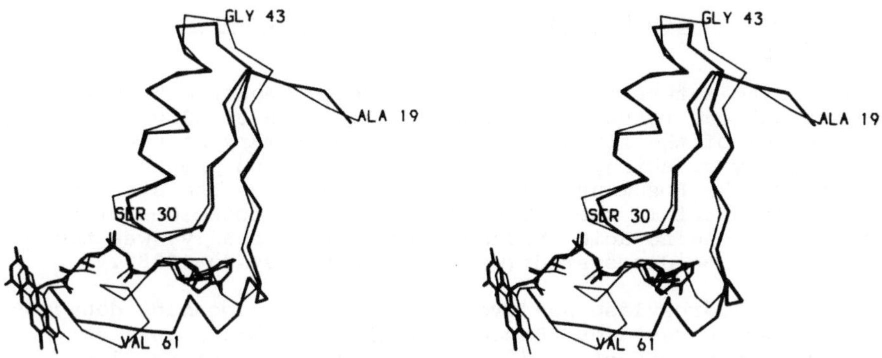

Fig. 9. *p*-Hydroxybenzoate hydroxylase and glutathione reductase: superposition of the β_A α_A β_B- unit together with the FAD molecules (from Wierenga et al. 1983, with permission). Shown are residues 1 to 45 of *p*-hydroxybenzoate hydroxylase (*thick lines*) and residues 19 to 63 of glutathione reductase (*thin lines*; several residues are labelled). The polypeptide chains are reduced to virtual $C_\alpha - C_\alpha$ bonds. The superposition is based on the minimal root mean square (r.m.s.) distance between the C_α-atoms of residues 2 to 34 of *p*-hydroxybenzoate hydroxylase and residues 20 to 52 of glutathione reductase, as well as on the r.m.s. distance between 27 equivalent atoms of FAD (distribution uniformly over the FAD molecules in the two proteins). The resulting r.m.s. distance between the 33 C_α-atoms is 1.2 Å, and the r.m.s. distance between all 53 nonhydrogen atoms of FAD is 1.1 Å

of the dinucleotides FAD and NAD(P)H, are located at the helix side of the parallel β-sheet, whereas the adenines of the dinucleotides, the anchors as it were, are located at the other side of the sheet (Fig. 8a).

Sequence Comparisons Among Flavoenzymes

So far, the FAD-binding chain fold — with its characteristic excursion after the second sheet strand and the β-meander following the third strand of the parallel β-sheet — has been observed only in two proteins (Table 2); possibly it is also present in FAD-dependent ferredoxin-$NADP^+$ oxidoreductase, the structural analysis of which is in progress (Karplus and Herriott 1982). However, on the basis of sequence comparisons among flavoenzymes one may well expect that this chain fold is fairly common. It is probably present in lipoamide dehydrogenase (Williams et al. 1982) and in mercuric reductase (Fox and Walsh 1982; Brown et al. 1983) because these enzymes are homologous to glutathione reductase.

The FAD-containing enzymes D-amino acid oxidase (Ronchi et al. 1982) and respiratory NADH-dehydrogenase (Young et al. 1981) exhibit some sequence similarity with *p*-hydroxybenzoate hydroxylase and glutathione reductase in the region of the $\beta_A\alpha_A\beta_B$-units shown in Fig. 8. - In spite of the close structural relationship between the FAD-binding and the NADPH-binding domains of glutathione reductase (Schulz 1980), only vestigial sequence homologies between these domains could be detected (Untucht-Grau et al. 1981; Untucht-Grau 1983).

Identification of the FAD-Binding Domain in Flavoenzymes of Known Primary Structure

As FAD binds to the N-terminal domain in *p*-hydroxybenzoate hydroxylase and in glutathione reductase it seems reasonable to align the N-terminal sequences of the flavoenzymes of known primary structure (Table 3). Strictly speaking, the last but one residue preceding the first β-strand of the Rossmann fold has been chosen as the starting point for the alignment in the table. This position coincides with the N-terminus of D-amino acid oxidase which, incidentally, was the first FAD-enzyme to be discovered (Warburg and Christian 1938).

p-Hydroxybenzoate hydroxylase has two preceding and lipoamide dehydrogenase, six preceding residues. Because of the long N-terminal excursion in glutathione reductase (Fig. 8b, Table 2), residue 21 of this enzyme coincides with residue 1 of D-amino acid oxidase. Mercuric reductase possesses a domain of approximately 100 residues preceding the N-terminal strand of the Rossmann fold. These 100 residues can be cleaved off using chymotrypsin without any loss of mercuric reductase activity (Fox and Walsh 1983). Consequently, it is the clipped enzyme which can be directly aligned with the enzymes of Table 3 (Brown et al. 1983).

Table 3. Sequence alignment of the N-terminal regions in six FAD-enzymes

	21 25 * *30 35
Glutathione reductase[a,b]	A S Y D [Y L V I G] G [G S G G L A S]
	7
Pig heart lipoamide dehydrogenase[c]	I D A D V T V I G S G P G G Y V A
	3
p-Hydroxybenzoate hydroxylase[d]	M K T [Q V A I I G] A [G P S G L L L]
	1
D-amino acid oxidase[e]	M R V V V I G A G V I G L S T
	6
Respiratory NADH-dehydrogenase[f]	P L K K I V I V G G G A G G L E M
	99
Mercuric reductase[g]	P P V Q V A V I G S G G A A M A A

[a]Untucht-Grau et al. (1981); [b]Thieme et al. (1981); [c]Williams et al. (1982); [d]Hofsteenge et al. (1980); [e]Ronchi et al. (1982); [f]Young et al. (1981); [g]Brown et al. (1983)

In FAD-dependent respiratory NADH-dehydrogenase (Table 3), five residues precede the sequence KKIVIVGGGAGGLEM, but it has not been demonstrated yet that this sequence is involved in the binding of FAD rather than in the binding of NADH. The other chain segment which is a candidate for fixing a dinucleotide begins at residue 171 (Young et al. 1981).

Invariable Residues in Proteins Binding FAD or Other Dinucleotides

In Table 3, the first β-strand (β_A) and the N-terminal portion of the following helix, α_A, of glutathione reductase are marked by broken and continuous lines, respectively. In all proteins, the residues corresponding to the β-strand are mainly Val and Ile, which show a strong propensity of forming β-strands (Schulz and Schirmer 1979). Residues Gly-27 and Gly-29 are conserved in all enzymes given in Table 3. These residues are known to be involved in the binding of FAD in glutathione reductase and *p*-hydroxybenzoate hydroxylase. Moreover, there are a number of other conserved residues in Table 3, so that one may assume that the structures of all given enzymes in this region closely resemble those of glutathione reductase and *p*-hydroxybenzoate hydroxylase.

Hofsteenge et al. (1980) have pointed out that the pattern of Gly-27/Gly-29/Gly-32 (Table 3) has also been found at the NAD-binding sites of alcohol dehydrogenase, lactate dehydrogenase, and glyceraldehyde-3-phosphate dehydrogenase. The glycine patterns at two mononucleotide-binding sites – in FMN-dependent flavodoxin (Watenpaugh et al. 1972; Mayhew and Ludwig 1975) and at the AMP-binding site of adenylate kinase (Pai et al. 1977) – are somewhat different.

The conservation of glycine residues at certain positions (Tables 2 and 3) has been discussed by Schulz et al. (1982). In contrast to the other 19 standard amino acids, glycine has no side chain; therefore, the torsion angles of the main chain (ϕ and ψ) can assume exceptional values, which allows unusual chain conformations (Ramachandran and Sasisekharan 1968; Schulz and Schirmer 1979). Furthermore, a glycine can feed the polypeptide chain through tight places where a side chain would cause steric hindrance (for examples, see Table 2).

Comparisons beyond Ser-35 of Table 2 show that the residue at the C-terminal position of β_B in the Rossmann fold is either a Glu or an Asp: Glu-50 in glutathione reductase (see Fig. 8b), Glu-32 in *p*-hydroxybenzoate hydroxylase, and Glu-36 in lipoamide dehydrogenase. D-amino acid oxidase and the members of the lactate dehydrogenase family have aspartate residues at this position (Ronchi et al. 1982; Rossmann et al. 1975). These acidic residues invariably bind to the 2'-hydroxyl and/or the 3'-hydroxyl of the adenine-linked riboses in the dinucleotides. They are probably responsible for the discrimination against the binding of 2'- or 3'-phosphorylated nucleotides, such as NADPH or CoA.

This hypothesis is corroborated by the fact that in the NADPH-binding domain of glutathione reductase a *basic* residue, namely, Arg-218, occupies the C-terminal position of β_B (Fig. 7). The side chains of Arg-218 and His-219 compensate the negative charges of the 2'-phosphoryl group of NADPH (Thieme et al. 1981; Krauth-Siegel et al. 1982).

Model-Building on the Basis of Sequence Homology? Lipoamide Dehydrogenase as an Example

The comparison of the amino acid sequences of the two mechanistically related flavoenzymes glutathione reductase and lipoamide dehydrogenase had been started with the redox-active dithiol peptide (Williams 1976; Krohne-Ehrich et al. 1977). In spite of the high degree of identity (11 out of 14 residues), some suspicion remained that this similarity might not be caused by divergent evolution, because these peptides are part of the catalytic apparatus in both enzymes. Structural similarities at catalytic sites are often the result of convergent evolution and reflect an optimization according to physical and chemical criteria (Garavito et al. 1977). Recently, almost all tryptic peptides of pig heart lipoamide dehydrogenase were sequenced in William's laboratory and compared with the sequence of glutathione reductase using a computer program based on natural substitution frequencies of amino acids in homologous proteins (Williams, private communication). Three peptides, for example, were fitted to homologous counterparts in the FAD-domain of glutathione reductase where they comprise residues 15-67 (Fig. 8b). The average homology, expressed as degree of identity, is about 40% for the two enzymes. This value is fairly constant along the complete polypeptide chains, which is convincing evidence for a phylogenetic relationship between the two enzyme structures (Williams et al. 1982). In analogy to the classical case of the pair lysozyme/α-lactalbumin (Browne et al. 1969), one expects to gain some insight into the three-dimensional structure of lipoamide-dehydrogenase by fitting the sequence of this enzyme to the three-dimensional model of glutathione reductase. The same applies for mercuric reductase, another flavoenzyme, which is closely related to glutathione reductase (Fox and Walsh 1982, 1983; Brown et al. 1983).

Acknowledgements. We thank Dr. Brown and his colleagues (Bristol, U.K.), Drs. Fox and Walsh (Cambridge, Mass., U.S.A.) and Dr. CH Williams, (Ann Arbor, Mich., U.S.A.) for communicating unpublished results to us. Rosi Förster and Irene König kindly helped with the manuscript. Our work is supported by the Deutsche Forschungsgemeinschaft and by the Fonds der Chemischen Industrie.

Discussion

Rossmann: Looking at the pyrophosphate-binding loops of NAD-dependent dehydrogenase, the strict GXGXXG pattern is not present in all cases.

Schirmer: Mercuric ion reductase (Table 3) is another exception. There is also a pathologically important case: the human protein P21 – whose function is not known – contains Gly-12 (corresponding to Gly-29 of glutathione reductase) in normal cells, but Val-12 in bladder tumor cells. Mutation of Gly-12 has also been described for P21 in other mammalian tumors. The (onco)gene product P21 is dealt with in several research groups (Gay and Walker (1983) Nature 301:262-264; Wierenga and Hol (1983) Nature 302:842-844; Newmark (1983) Nature 303:20).

References

Babior BM (1978) Oxygen-dependent microbial killing by phagocytes. New Engl J Med 298:659-668 and 721-725

Babior GL, Rosin RE, McMurrich BJ, Peters WA, Babior BM (1981) Arrangement of the respiratory burst oxidase in the plasma membrane of the neutrophil. J Clin Invest 67:1724-1728

Boggaram V, Larson K, Mannervik B (1978) Characterization of glutathione reductase from procine erythrocytes. Biochim Biophys Acta 527:337-347

Brown NL, Ford SJ, Pridmore RD, Fritzinger DC (1983) DNA sequence of a gene from the *Pseudomonas* transposon TN501 encoding mercuric reductase. Biochemistry (in press)

Browne WJ, North ACT, Phillips DC, Brew K, Vanaman TC, Hill RL (1969) A possible three-dimensional structure of bovine α-lactalbumin based on that of hen's egg-white lysozyme. J Mol Biol 42:65-86

Burnett RM, Darling GD, Kendall DS, Le Quesne ME, Mayhew SG, Smith WW, Ludwig ML (1974) The structure of the oxidized form of clostridial flavodoxin at 1.9 Å resolution. J Biol Chem 249:4383-4392

Chance B, Sies H, Boveris A (1979) Hydroperoxide metabolism in mammalian organs. Physiol Rev 59:527-605

Dixon M, Webb EC (1979) Enzymes, 3rd edn. Longman, London

Eckman JR, Eaton JW (1979) Dependence of plasmodial glutathione metabolism on the host cell. Nature 278:754-756

Fahey RC (1977) Biologically important thiol-disulfide reactions. Adv Exp Med Biol 86A:1-30
Fox BS, Walsh CT (1982) Mercuric reductase. J Biol Chem 257:2498-2503
Fox BS, Walsh CT (1983) Active site peptide of mercuric reductase: Homology with glutathione reductase and lipoamide dehydrogenase. Biochemistry, in press
Fritsch KG (1982) Zur Charakterisierung und Kristallisation des FAD-freien Apoenzyms der menschlichen Glutathionreduktase. Diplomarbeit, Freie Universität Berlin
Fritsch KG, Pai EF, Schirmer RH, Schulz GE, Untucht-Grau R (1979) Structural and functional roles of FAD in human glutathione reductase. Hoppe-Seyler's Z Physiol Chem 360:261-262
Garavito RM, Rossmann MG, Argos P, Eventoff W (1977) Convergence of active center geometries. Biochemistry 16:5065-5071
Glatzle D, Weber F, Wiss O (1968) Enzymatic test for the detection of riboflavin deficiency. NADPH-dependent glutathione reductase of red blood cells and its activation by FAD in vitro. Experientia 24:1122-1124
Gutfreund H (1972) Enzymes: Physical Principles. New York, Wiley
Hemmerich P (1976) Herz W, Grisebach H, Kirby GW (eds) In: Progress in the chemistry of organic natural products. Vol 33, Springer, Berlin Heidelberg New York, pp 451-527
Hofsteenge J (1981) p-Hydroxybenzoate hydroxylase: Determination of the amino acid sequence and its integration with the crystal structure. PhD thesis, Groningen
Hofsteenge J, Vereijken JM, Weijer WJ, Beintema JJ, Wierenga RK, Drenth J (1980) Primary and tertiary structure studies of p-hydroxybenzoate hydroxylase from Pseudomonas fluorescens. Isolation and alignment of the CNBr peptides. Interactions of the protein with FAD. Eur J Biochem 113:141-150
Hol WGJ, van Duijnen PT, Berendsen HJC (1978) The α-helix dipole and the properties of proteins. Nature 273:443-446
Holmgren A (1980) Pyridine nucleotide-disulfide oxidoreductases. Experientia (suppl) 36:149-180
Hultquist DE, Sannes LJ, Schafer DA (1981) The NADH/NADPH-methemoglobin reduction system of erythrocytes. In: Brewer GJ (ed) The red cell: Fifth Ann Arbor conference. Alan R Liss, New York, pp 291-305
Jencks WP (1975) Binding energy, specificity and enzymic catalysis: The Circe effect. Adv Enzymol 43:220-410
Karplus PA, Herriott JR (1982) The structure of ferredoxin-$NADP^+$ oxidoreductase: A progress report. In: Massey V and Williams CH Jr (eds) Flavins and flavoproteins. Elsevier, New York, pp 28-31
Kosower EM (1966) The role of charge-transfer complexes in flavin chemistry and biochemistry. In: Slater EC (ed) Flavins and Flavoproteins. Elsevier, Amsterdam, pp 1-14
Krauth-Siegel RL (1982) Untersuchungen zur Primärstruktur des Flavoenzyms Glutathionreduktase und zur Wechselwirkung mit seinen Liganden. PhD-thesis, Heidelberg
Krauth-Siegel RL, Blatterspiel R, Saleh M, Schiltz E, Schirmer RH, Untucht-Grau R (1982) Glutathione reductase from human erythrocytes. The sequence of the NADPH domain and of the interface domain. Eur J Biochem 121:259-267
Krohne-Ehrich G, Schirmer RH, Untucht-Grau R (1977) Glutathione reductase from human erythrocytes. Isolation of the enzyme and sequence analysis of the redox-active peptide. Eur J Biochem 80:65-71
Light DR, Walsh C, O'Callaghan, Goetzl EJ, Tauber AI (1981) Characteristics of the cofactor requirements for the superoxide-generating NADPH-oxidase of human polymorphonuclear leukocytes. Biochemistry 20:1468-1476
Lindqvist Y, Brändén CI (1980) Structure of glycolate oxidase from spinach at a resolution of 5.5 Å. J Mol Biol 143:201-211
Loos H, Roos D, Weening R, Houwerzijl J (1976) Familial deficiency of glutathione reductase in human blood cells. Blood 48:53-62
Ludwig ML, Pattridge KA, Smith WW, Jensen LH, Watenpaugh KD (1982) Comparisons of flavodoxin structures. In: Massey V and Williams CH Jr (eds) Flavins and flavoproteins. Elsevier, New York, pp 19-27
Mannervik B, Boggaram V, Carlberg I, Larson K (1980) The catalytic mechanism of glutathione reductase. In: Yagi K and Yamano T (eds) Flavins and Flavoproteins. Japan Scientific Societies, Tokyo, pp 173-187

Massey V, Ghisla S (1974) Role of charge-transfer interactions in flavoprotein catalysis. Ann NY Acad Sci 227:446-465

Massey V, Hemmerich P (1980) Active-site probes of flavoproteins. Biochem Soc Transact 8:246-257

Massey V, Williams CH Jr (eds) (1982) Flavins and flavoproteins. Elsevier, New York

Mayhew SG, Ludwig ML (1975) Flavodoxins and electron-transferring flavoproteins. In: Boyer PD (ed) The enzymes, Vol 12B. Academic, New York, pp 57-118

Molski TFP, Naccache PH, Volpi M, Wolpert LM, Sha'afi RI (1980) Specific modulation of the intracellular pH Of rabbit neutrophils by chemotactic factors. Biochem Biophys Res Comm 94:508-514

Müller F, van Berkel WJH, Drenth J, Wierenga RK, Kalk KH, Hofsteenge J, Vereijken JM, Branno M, Beintema JJ (1980) A multidisciplinary study on p-hydroxybenzoate hydroxylase from Pseudomonas fluorescens: Molecular properties, threedimensional structure and amino acid sequence. In: Yagi K and Yamano T (eds) Flavins and flavoproteins. Japan Scientific Societies, Tokyo, pp 413-422

Ohlsson J, Nordström B, Brändén CI (1974) Structural and functional similarities within the coenzyme-binding domains of dehydrogenases. J Mol Biol 89:339-354

Oshino N, Imai Y, Sato R (1971) A function of cytochrome b_5 in fatty acid desaturation by rat liver microsomes. J Biochem (Tokyo) 69:155-167

Pai EF, Schulz GE (1983) The catalytic mechanism of glutathione reductase as derived from X-ray diffraction analyses of reaction intermediates. J Biol Chem 258:1752-1757

Pai EF, Sachsenheimer W, Schirmer RH, Schulz GE (1977) Substrate positions and induced-fit in crystalline adenylate kinase. J Mol Biol 114:37-45

Prentice AM, Bates CJ (1981) A biochemical evaluation of the erythrocyte glutathione reductase (EC 1.6.4.2) test for riboflavin status. Br J Nutr 45 (N1):37-65

Ramachandran GN, Sasisekharan V (1968) Conformation of polypeptides and proteins. Adv Prot Chem 23:283-437

Rao ST, Rossmann MG (1973) Comparison of supersecondary structures in proteins. J Mol Biol 76:241-256

Ronchi S, Minchiotti L, Galliano M, Curti B, Swenson RP, Williams CH Jr, Massey V (1982) The primary structure of D-amino acid oxidase from pig kidney. J Biol Chem 257:8824-8834

Rossmann MG, Liljas A, Brändén CI, Banaszak LJ (1975) Evolutionary and structural relationships among dehydrogenases. The Enzymes 11:61-102

Schirmer RH, Schulz GE, Untucht-Grau R (1983) On the geometry of leucocyte NADPH-oxidase, a membrane flavoenzyme. Inferences from the structure of glutathione reductase. FEBS-Lett 154:1-4

Schulz GE (1980) Gene duplication in glutathione reductase. J Mol Biol 138:335-347

Schulz GE, Schirmer RH (1979) Principles of Protein Structure. Springer, Berlin Heidelberg New York, 314 pages

Schulz GE, Zappe H, Worthington DJ, Rosemeyer MA (1975) Crystals of human erythrocyte glutathione reductase. FEBS Lett 54:86-88

Schulz GE, Schirmer RH, Sachsenheimer W, Pai EF (1978) The structure of the flavoenzyme glutathione reductase. Nature 273:120-124

Schulz GE, Schirmer RH, Pai EF (1982) FAD-binding site of glutathione reductase. J Mol Biol 160:287-308

Sheriff S, Herriott JR (1981) Structure of ferredoxin-NADP oxidoreductase and the location of the NADP-binding site. Results at 3.7 Å resolution. J Mol Biol 145:441-451

Staal GEJ, Visser J, Veeger C (1969) Purification and properties of glutathione reductase of human erythrocytes. Biochim Biophys Acta 185:39-48

Tauber AI (1982) The human neutrophil oxygen armory. Trends Biochem Sci 7:411-414

Theorell H (1935) Über die Wirkungsgruppe des gelben Ferments. Biochem Z 275:37

Thieme R, Pai EF, Schirmer RH, Schulz GE (1981) Three-dimensional structure of glutathione reductase at 2 Å resolution. J Mol Biol 152:763-782

Untucht-Grau R (1983) Zur Primärstruktur der Glutathionreduktase: Die FAD-bindende Domäne und das C-terminale Bromcyanfragment: PhD thesis, Heidelberg

Untucht-Grau R, Schirmer RH, Schirmer I, Krauth-Siegel RL (1981) Glutathione reductase from human erythrocytes. Amino-acid sequence of the structurally known FAD-binding domain. Eur J Biochem 120:407-419

Warburg O, Christian W (1938) Isolierung der prosthetischen Gruppe der d-Aminosäureoxydase. Biochem Z 298:150-168

Warburg O, Christian W, Griese A (1935) Wasserstoffübertragendes Co-Ferment, seine Zusammensetzung und Wirkungsweise. Biochem Z 282:157-205

Watenpaugh KD, Jensen LH, Legall J, Dubourdieu M (1972) Structure of the oxidized form of a flavodoxin at 2,5 Å resolution: Resolution of the phase ambiguity by anomalous scattering. Proc Natl Acad Sci USA 69:3185-3188

Weijer WJ, Hofsteenge J, Vereijken JM, Jekel PA, Beintema JJ (1982) Primary structure of *p*-hydroxybenzoate hydroxylase from Pseudomonas fluorescens. Biochim Biophys Acta 704:385-388

Wierenga RK, de Jong RJ, Kalk KH, Hol WGJ, Drenth J (1979) Crystal structure of *p*-hydroxybenzoate hydroxylase. J Mol Biol 131:55-73

Wierenga RK, Kalk KH, van der Laan JM, Drenth J, Hofsteenge J, Weijer WJ, Jekel PA, Beintema JJ, Müller F, van Berkel WJH (1982) The structure of *p*-hydroxybenzoate hydroxylase. In: Massey V and Williams CH Jr (eds) Flavins and flavoproteins. Elsevier, New York, pp 11-18

Wierenga RK, Drenth J, Schulz GE (1983) Comparison of the three-dimensional protein and nucleotide structure of the FAD-binding domain of *p*-hydroxybenzoate hydroxylase with the FAD as well as the NADPH-binding domains of glutathione reductase. J Mol Biol 167:725-739

Williams CH Jr (1976) Flavin-containing dehydrogenases. In: Boyer PD (ed) The Enzymes, Vol 13, 3rd ed. Academic, New York, pp 89-173

Williams CH Jr, Arscott LD, Schulz GE (1982) Amino acid sequence homology between pig heart lipoamide dehydrogenase and human erythrocyte glutathione reductase. Proc Natl Acad Sci (USA) 79:2199-2202

Worthington DJ, Rosemeyer MA (1974) Human glutathione reductase: Purification of the crystalline enzyme from erythrocytes. Eur J Biochem 48:167-177

Young IG, Rogers BL, Campbell HD, Jaworowski A, Shaw DC (1981) Nucleotide sequence coding for the respiratory NADH dehydrogenase of Escherichia coli. UUG initiation codon. Eur J Biochem 116:165-170

The Mechanism of Action of Flavoprotein – Catalyzed Reactions

V. Massey[1] and S. Ghisla[2]

Introduction

1983 is not only the 100th anniversary of the birth of Otto Warburg, but is also the 50th anniversary of the determination of the structure of the flavin chromophore by Kuhn and Wagner-Jauregg (1930) and the 50th anniversary of the isolation of the first flavoprotein, the Old Yellow Enzyme of brewers bottom yeast (Warburg and Christian 1933). In the intervening 50 years, some 200 different flavoproteins have been recognized, making this one of the largest single groups of related enzymes. These enzymes function in the catalysis of key steps in virtually every metabolic pathway in all life forms. They catalyze a variety of different types of chemical reaction in which the flavin is intimately involved. With a few exceptions, where the role of the flavin is not clear, e.g., glyoxylate carboligase (Cromartie and Walsh 1976) or oxynitrilase (Jorns 1980), flavoproteins carry out oxidation-reduction reactions, where one substrate is oxidized and a second is reduced. For all these enzymes, each catalytic cycle consists of two distinct processes, the acceptance of redox equivalents from a reducing substrate and the transfer of these equivalents to an oxidized acceptor. Accordingly, catalysis is comprised of two separate half-reactions: (1) the reductive half-reaction where the flavin is reduced and (2) the oxidative half-reaction, where the reduced flavin is reoxidized. This feature is very convenient experimentally, since it is possible to study each half-reaction separately. In this way, it is generally possible to identify individual steps in each half-reaction, often with the determination of the absorption spectra of intermediates and the rate constants of their formation and decay. This information can then be combined and fitted to catalytic turnover data, sometimes even employing the spectral properties of the flavin itself to monitor enzyme catalytic turnover (Gibson et al. 1964).

The nature of the substrates involved in the two separate half-reactions has been used as the basis for a classification scheme for flavoenzymes. Thus, Hemmerich et al. (1977) have defined five broad classes of flavoenzymes:

1) *Transhydrogenase*, where two-electron equivalents are transferred along with the appropriate hydrogen ions, from one organic substrate to another.

2) *Dehydrogenase-oxidase*, where two-electron equivalents are transferred to the flavin from an organic substrate and molecular oxygen is the oxidizing substrate, being reduced to H_2O_2.

1 Department of Biological Chemistry, University of Michigan, Ann Arbor, Michigan 48109, U.S.A.

2 Fakultät für Biologie, Universität Konstanz, D-7750 Konstanz, FRG

34. Colloquium - Mosbach 1983
Biological Oxidations

3) *Dehydrogenase-monooxygenase*, where the flavin is reduced, generally by a reduced pyridine nucleotide, and where on oxidation with O_2 in the presence of a cosubstrate one atom of oxygen is inserted into the cosubstrate, while the other is reduced to H_2O.

4) *Dehydrogenase-electron transferase*, where the flavin is reduced by 2-e^- transfer from a reduced substrate and then reoxidized in sequential single electron transfers to acceptors, such as cytochromes and iron-sulfur proteins.

5) *Electron transferase*, where the flavin is reduced and reoxidized in 1-e^- steps.

Model studies have shown that flavins are versatile catalysts, being able to be reduced and reoxidized in single- or 2-electron steps, being capable of forming adducts at various positions in the flavin ring system, and undergoing very facile photochemical reactions. A dramatic example of how this versatility is modulated and controlled by binding to specific proteins is the reactivity of various flavoproteins toward molecular oxygen. The free coenzyme in its reduced state reacts with O_2 fairly rapidly and by a complex series of reactions involving flavin-oxygen adducts, flavin radicals, and the superoxide anion, O_2^-, with the ultimate products being oxidized flavin and hydrogen peroxide (Gibson and Hastings 1962, Massey et al. 1973, Kemal et al. 1977). Flavoproteins, on the other hand, show very different responses to O_2, depending on the particular class of enzyme. Some enzymes, such as those of the electron transferase class, react with O_2 to generate almost quantitatively the blue flavin neutral radical, and O_2^-. This reaction may be quite fast as in the case of flavodoxin (Massey et al. 1969); it is generally the second 1-electron oxidation of the semiquinone by O_2 which is slow and which accounts for the overall slow catalytic reaction of such enzymes with O_2. In another group of enzymes, the oxidases, the overall reaction with O_2 is fast. In all cases examined of simple, nonmetal containing oxidases, there is no evidence for formation of either O_2^- or flavin semiquinone; the reaction seems to go smoothly and monophasically to the products, H_2O_2 and oxidized flavin (Massey et al. 1969). In a third group of enzymes, the monooxygenases, the ability of the reduced flavin to react rapidly with O_2 is retained, but now the oxygen molecule is utilized so that one atom is reduced to H_2O and the other is incorporated into a primary substrate of the enzyme. All the enzymes of this class, which have been examined, share a common property; the first observed product in the oxidative half-reaction is a covalent adduct between flavin and O_2, which has been identified as the flavin C(4a)-hydroperoxide (for a recent review, see Ballou 1982).

Over the years, it has become clear that members of each of these classes of flavoenzymes share other common properties in addition to their type of behavior towards O_2. Thus, practically all members of the oxidase class form red flavin anion radicals on reduction by artificial reducing agents, but never by their normal substrates. They also readily form adducts with sulfite at the flavin N(5)-position and they all stabilize the benzoquinoid form of 8-mercaptoflavin, when the latter is incorporated into apoenzyme (Massey et al. 1979; Massey and Hemmerich 1980). Electron-transferase enzymes, on the other hand, generally stabilize the blue neutral flavin radical, do not form sulfite adducts, and exhibit the spectrum of the thiolate form of 8-mercaptoflavin. By contrast, flavoprotein monoxygenases, in general, do not stabilize any flavin radical form or any particular form of 8-mercaptoflavin, they fail to react with sulfite, and as already mentioned, all form observable flavin C(4a)-hydroperoxides with O_2. It is clear that such common characteristics within a particular functional class of flavoproteins must reflect common features of protein structure in the flavin-binding pocket, which direct the versatile chemistry of the flavin coenzyme along particular paths. Current concepts of these structure-function relationships are based on studies of modified flavins and modified proteins and on the initial three-dimensional structures that have been determined for flavodoxins (Ludwig et al. 1982), for p-hydroxybenzoate hydroxylase (Wierenga et al. 1979; Wierenga et al. 1982), and for glutathione reductase (a C-S transhydrogenase) (Thieme et al. 1981; Schulz et al. 1982). In the case of *Clostridium MP* flavodoxin, the stabilization of the neutral semiquinone can be accounted for by hydrogen-bonding of the flavin N(5)H to a backbone carbonyl group of the protein (Ludwig et al. 1976); such H-bonding could account for the characteristics of electron transferases in general (Massey and Hemmerich 1980). In glutathione reductase, where the radical flavin is not stabilized, it is Lys 66 which makes the closest approach to N(5) (Schulz et al. 1982). Detailed structures are not yet available for any flavoprotein oxidases, but studies of chemically modified enzymes provide strong evidence that the common characteristics of the oxidases are imposed by a positively charged residue in the vicinity of the flavin N(1)-C(2α) locus (Müller and Massey 1969; Massey et al. 1979; Fitzpatrick and Massey 1983). Similarly, the observed stabilization of the flavin C(4a)-hydroperoxides in the case of flavin monooxygenases implies a common regiospecific interaction with the protein which is different from those in either the oxidases or the electron transferases. In the structure determined for the oxidized complex of p-hydroxybenzoate hydroxylase with substrate (Wierenga et al. 1979), the substrate orientation is reasonable for attack by a C(4a) hydroperoxide, but the structure does not offer definitive evidence for the way in which the protein controls the oxygen reactivity of the flavin. The most informative structure would be that of the reduced enzyme-substrate complex, which is the form that reacts with oxygen to form the flavin hydroperoxide (Entsch et al. 1976a); unfortunately, one cannot extrapolate too much from the oxidized enzyme structure, since it is clear that a substantial conformational change occurs on substrate binding and probably also on reduction (Wierenga et al. 1979; Claiborne et al. 1982).

General Considerations of Flavin-Protein Interactions

Most of the binding energy in the interaction of the flavin coenzyme with its specific protein is associated with the side chain at position N(10). Thus, flavoproteins in general are specific for binding either FMN or FAD and will generally accommodate artificial flavins with a number of structural modifications in the isoalloxazine ring system just as readily as the native flavin. But while it is the N(10) side chain which provides the main anchor to the protein, the type of reaction catalyzed by the particular protein must be determined by specific interactions of the protein with the isoalloxazine ring system. This is illustrated in Scheme 1, where possible hydrogen bonding positions between the flavin and protein or possible charge interactions, are indicated.

The pyrimidine ring of oxidized flavins contains the structural elements of barbituric acid or alloxan and is strongly electron deficient. The amide functions, N(1)-C(2)O, N(3)H-C(4)O (and also to a much smaller extent N(5) of the central pyrazine ring) provide sites for H-bonding interactions with appropriate groups in the protein. From 2-electron reduction, the resulting 1,5-dihydroflavin has in addition, an ionizable function, N(1)H, with a pK in the region of 7 (Dudley et al. 1964). In the course of dehydrogenation of certain carbonyl-containing

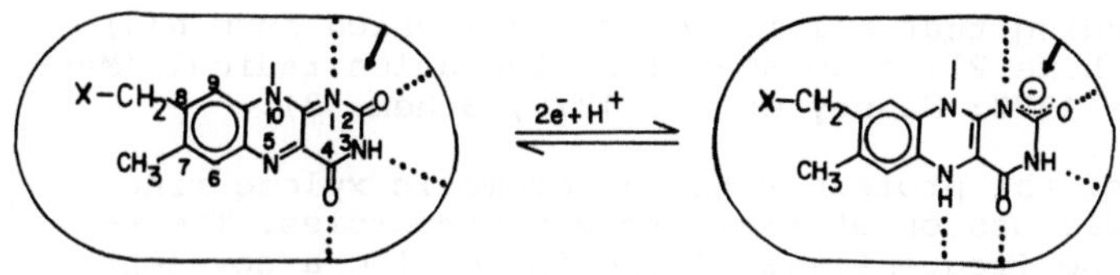

<u>Scheme 1.</u> Positions of possible flavin-protein hydrogen bonding (.....) or of charge interactions (*arrows*)

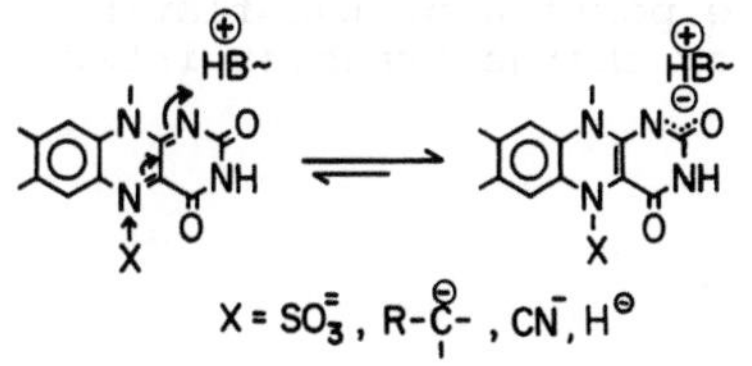

<u>Scheme 2.</u> Charge stabilization of negatively charged substrates by flavin adduct formation

substrates, the abstraction of the substrate α-proton by a protein base generates a negatively charged (carbanion) transition state. This negatively charged species, with a very high pK, can interact with oxidized flavin to form a negatively charged reduced molecule (Ghisla and Massey 1980). The flavin thus plays the role of a charge sink in the stabilization of this negatively charged transient (Scheme 2).

The middle pyrazine ring with its free electron pair at N(5) has scarcely any basic function ($pK < 0$) and in the oxidized state therefore cannot have any strong interaction with the protein. This position is on the other hand electrophilic. The presence of a positive charge in the neighborhood of N(1)C(2)O enhances the electrophilic character of position N(5) and makes it much more susceptible to attack by a nucleophile X (Massey et al. 1969; Müller and Massey 1969). In model systems, X can be for example, sulfite (Müller and Massey 1969), phosphines (Müller 1972), cyanide (Jorns 1980), borohydride (Müller et al. 1969); in biochemical systems, it can be a hydride equivalent, sulfite or a carbanion (Massey et al. 1979; Ghisla and Massey 1980; Scheme 2).

The one-electron reduction of flavins yields the radical HFl$^{\bullet}$. The position N(5) of this species is some 10 orders of magnitude more basic ($pK \sim 8$) than it is in Fl_{ox} (Ehrenberg et al. 1967; Land and Swallow 1969). With this function, the radical (and the fully reduced flavin) has an added possibility of being stabilized by hydrogen bonding to the protein. In the case of flavodoxin, such an effect is the apparent cause of thermodynamic stabilization of the blue neutral radical (Ludwig et al. 1976; see Scheme 3). On the other hand, the existence of a protein positively charged group in the vicinity of the N(1)C(2)O region would be expected to stabilize the red anion radical,

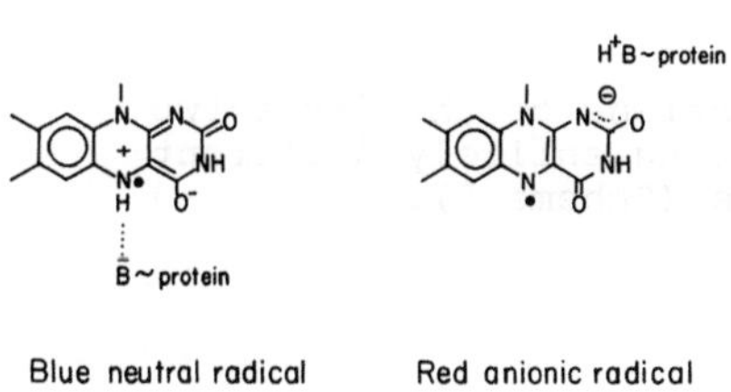

<u>Scheme 3.</u> Modes of protein stabilization of neutral and anionic flavin radicals

thus explaining the finding that all flavoproteins, which form N(5) sulfite adducts (see Scheme 2), also stabilize the anion radical (Massey et al. 1969; Müller 1972; Massey et al. 1979; Scheme 3).

Specific interactions of the protein with the aromatic xylene ring are rarely invoked in discussion of flavoprotein properties. The 8-methyl group is of course frequently employed in forming a covalent linkage of the flavin to the protein, where histidyl, tyrosyl, and cysteinyl derivatives have been found (Edmondson and Singer 1976). These groups can directly affect the chemical properties of the isoalloxazine nucleus and also offer the possibility of secondary charge and hydrogen bonding interactions with the protein. But in terms of noncovalent interactions, the dimethyl benzene portion of the flavin would be restricted primarily to hydrophobic or charge transfer interactions.

Mechanisms of Flavoprotein Catalysis

Dehydrogenation Reactions (The Reductive Half-Reaction)

In a biochemical sense, the term "dehydrogenation" means the breaking of two bonds involving hydrogen in the system XH-YH and the transfer of two electrons (and protons) to an acceptor, A.

$$XH\text{-}YH + A \rightleftharpoons X = Y + AH_2$$

The acceptor A in our case if Fl_{ox}. According to the chemical nature of the substrate, the mechanism of the dehydrogenation reaction will be distinct. We can categorize the following types of substrate dehydrogenations (Scheme 4).

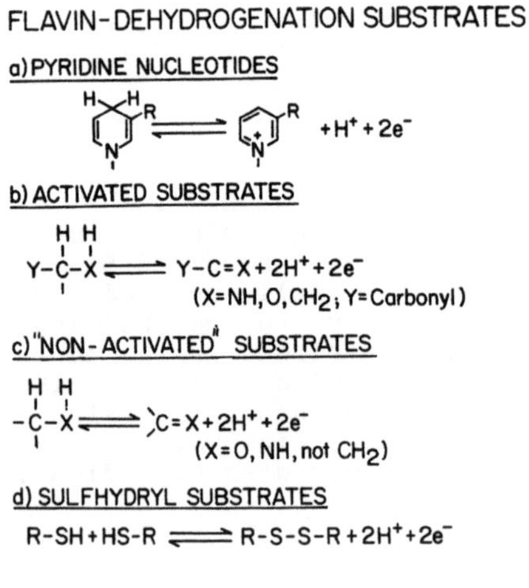

Scheme 4. Types of substrates oxidized by flavin enzymes, classified according to their chemical properties. "Nonactivated" substrates denotes substrates which cannot stabilize α-carbanions by delocalization

The mechanisms for the dehydrogenation reactions of the first two classes are of special interest. The following entirely different mechanisms are possible for these reactions (Scheme 5).

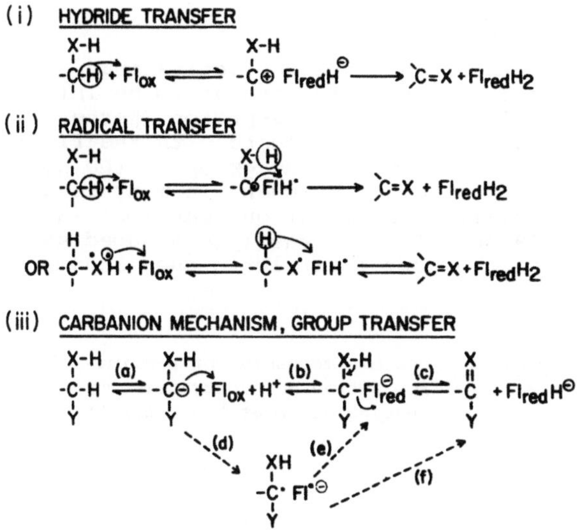

Scheme 5. Chemical mechanisms for the oxidation of substrates containing kinetically stable C-H bonds. Note that mechanisms (i) and (iii) can be combined to yield a carbanion initiated hydride transfer, cf. also Scheme 11

Pyridine Nucleotides

In the case of reversible dehydrogenation of reduced pyridine nucleotide, the present evidence is strongly in favor of direct transfer of a hydride equivalent between the C(4)H of the pyridine nucleotide and the flavin N(5)-position, according to mechanism (i) (Brustlein and Bruice 1972; Walsh 1979). The earlier mechanism studies are strongly supported by the crystallographic data on glutathione reductase, which show that the positions N(5) of the oxidized flavin and C(4)H of NADPH are precisely in the position for such a transfer (Pai and Schulz 1982; Scheme 6). Such a juxtaposition is essential for a hydride transfer mechanism, in contrast to a radical mechanism, where orbital overlap between flavin and pyridine nucleotide would need to be much less restrictive. The same crystallographic data with glutathione reductase also show that the flavin and pyridine nucleotide ring systems are juxtaposed in an approximately parallel fashion, as was predicted from the earlier observations of charge transfer complexes in pyridine nucleotide-flavoprotein interactions (Massey and Ghisla 1974).

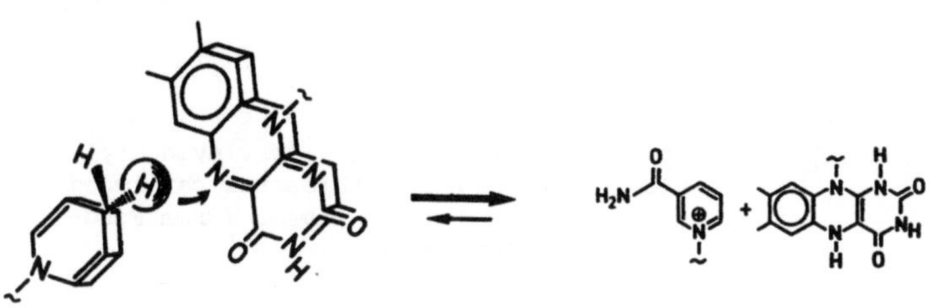

Scheme 6. Hydride transfer between pyridine nucleotides and flavins. Note that the hydrogen to be transferred as hydride must be located *above or below* the flavin plane, juxtaposed to the flavin N(5)

Substrates Activated for Carbanion or Group Transfer

The oxidation of substrates with electron withdrawing activating groups adjacent to the position of dehydrogenation appear to be initiated by the abstraction of the relatively acidic α-proton (Porter et al. 1973; Ghisla and Massey 1980). This is in accord with the finding that substrate analogs which have a good leaving group in the position β- to the oxidizing C-H function, e.g., halogen, in many cases are capable of undergoing enzyme catalyzed elimination reactions (Walsh et al. 1971; Walsh et al. 1973a, 1973b). This is easily envisaged as a side reaction running parallel to the normal catalytic one (Scheme 7).

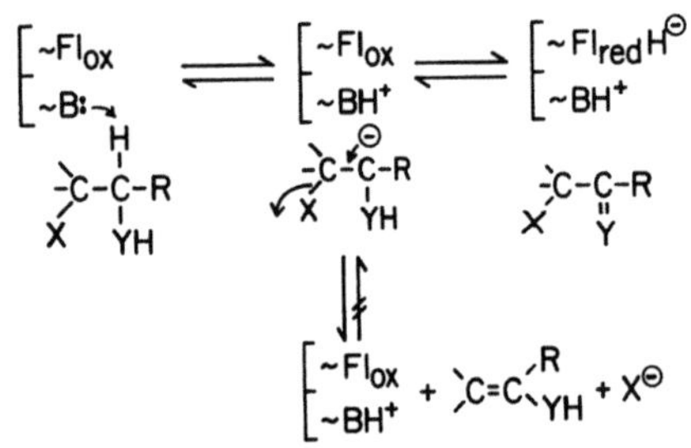

Scheme 7. Possible mechanism of β-elimination reactions from activated substrates containing a good leaving group at the β-position

Following the initial event of proton abstraction, several routes are possible for the formation of the final products, as illustrated in Scheme 5, mechanism (iii). The carbanion may make a direct nucleophilic attack at the flavin N(5)-position to yield a covalent adduct on the route to reduced flavin and oxidized substrate (reactions iii b and c). Alternatively, the reaction could proceed via the radical mechanism (iii d) to yield a radical pair, which could react further, either to collapse into the covalent intermediate or by 1e⁻ transfer yield directly the overall products of the reaction.

In the special case of α-β-oxidation of acyl substrates by acyl CoA dehydrogenases and possibly also of succinate by succinate dehydrogenase, the reaction sequence appears to involve a hydride transfer initiated by deprotonation of the substrate at the α-position (combination of sequences iii a) and (i), Scheme 5). This hypothesis is supported by the following experimental evidence with acyl CoA dehydrogenases:

a) Bacterial butyryl CoA dehydrogenase was found (Fendrich and Abeles 1982) to catalyze the 2→4 tautomerization of vinylacetyl CoA to crotonyl CoA (Scheme 8)

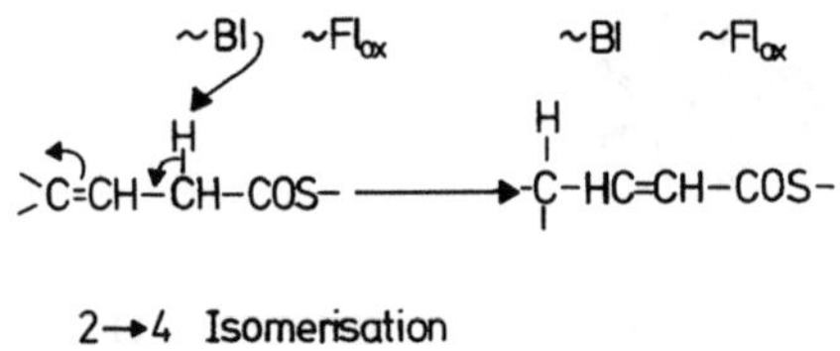

Scheme 8. Tautomerization of vinylacetyl thioester to crotonyl thioester catalyzed by butyryl CoA dehydrogenase. (From Fendrich and Abeles 1982)

b) β-Fluoro-substituted acyl CoAs eliminate fluoride in the presence of butyryl CoA dehydrogenase without the enzyme apparently undergoing reduction (cf. Scheme 7) (Fendrich and Abeles 1982).

c) 3-Pentynoyl pantetheine is first isomerized by the bacterial enzyme to 2,3 pentadienoyl pantetheine, which subsequently leads to irreversible inactivation (Fendrich and Abeles 1982; Frerman et al. 1980; Scheme 9).

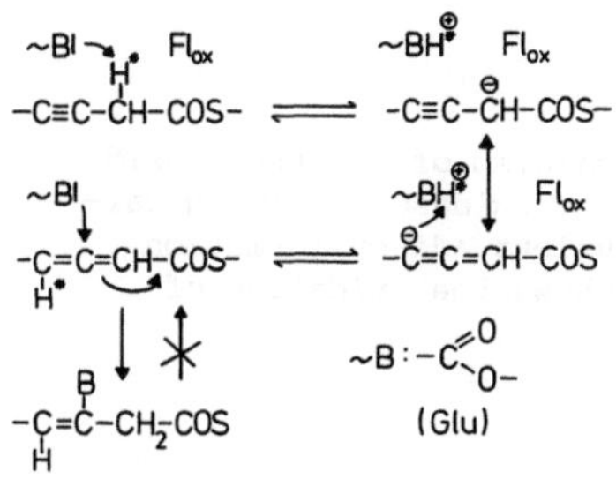

Scheme 9. Inactivation of butyryl CoA dehydrogenase by pentynoyl pantetheine. (From Fendrich and Abeles 1982)

d) 3,4 Pentadienoyl CoA is converted to the 2,4-tautomer by pig kidney general acyl CoA dehydrogenase (Wenz et al. 1982; Scheme 10).

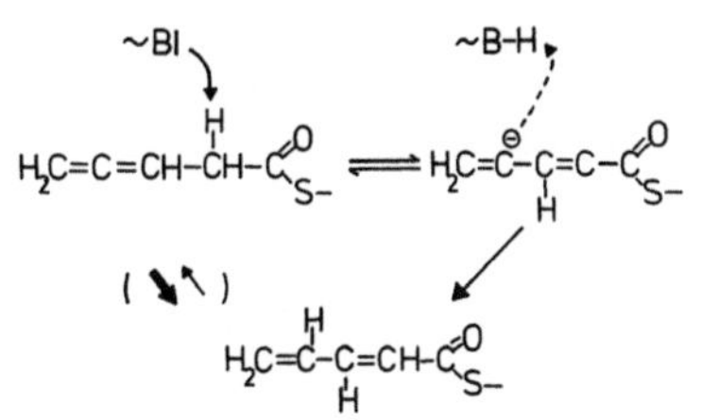

Scheme 10. Tautomerization of 3,4-pentadienoyl CoA to the 2,4-isomer by general acylCoA dehydrogenase

The sum of this evidence leads to the firm conclusion that catalysis is initiated by proton abstraction from the α-carbon of the substrate. For elucidating the rest of the mechanism, the normal coenzyme, FAD, was replaced by reduced 5-deaza-FAD, tritium-labelled at position C(5). This replacement of coenzyme was necessary in order to prevent exchange of hydrogen from the flavin 5-position with solvent protons. Incorporation of one ^{3}H-equivalent into crotonyl CoA to form 3(^{3}H)-butyryl CoA was found (Ghisla, unpublished). The stereochemistry of the oxidation of butyryl CoA by normal enzyme was found to be 2(R), 3(R) (antiperiplanar) (Bielemann and Hirt 1970; LaRoche et al. 1971; Kawaguchi et al. 1980). Thus, the mechanism of Scheme 11 can be proposed, which is also in agreement with the large isotope effects observed by Reinsch et al. (1980).

In the case of dehydrogenation of substituted carboxylic acids such as α-amino or α-hydroxy acids, the catalytic sequence is also initiated by the abstraction of the α-hydrogen as a proton, according to step (iii a) of Scheme 5. The substrate carbanion-Fl_{ox} pair created during the transition state can collapse into a covalent N(5) adduct, which then fragments into the products. This reaction sequence has been shown directly during the oxidation of glycollate by lactate oxidase (Ghisla and Massey 1980) and is illustrated in Scheme 12.

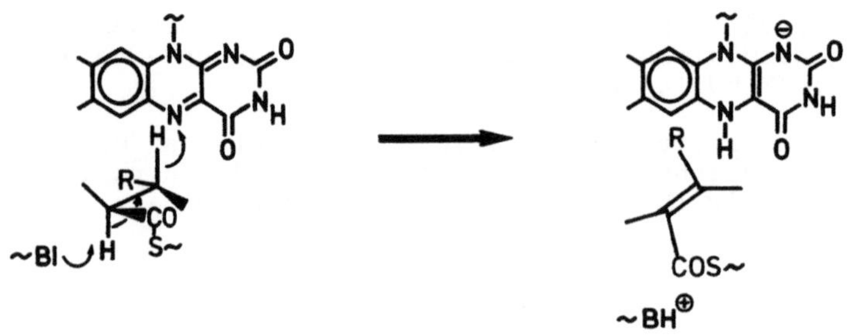

Scheme 11. Mechanism of α-β oxidation catalyzed by acyl CoA dehydrogenase

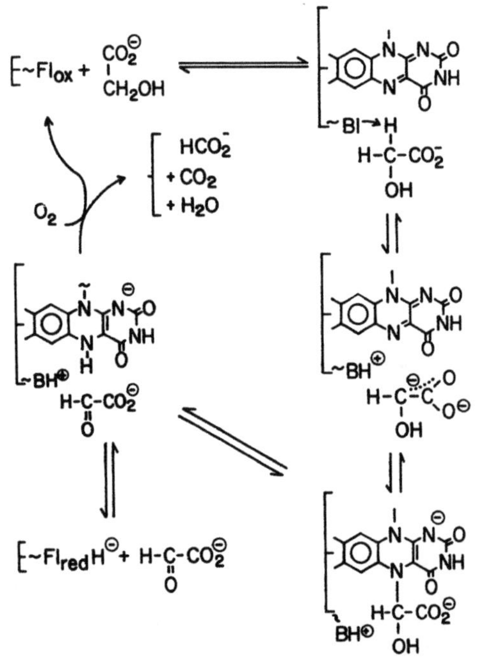

Scheme 12. Mechanism of substrate oxidation by lactate oxidase involving carbanion and covalent adduct formation. The sequence shows the oxidation of glycollate

A similar mechanism is favored strongly for D-amino acid oxidase. While the direct formation of a flavin N(5)-substrate adduct has not been demonstrated for any natural substrate, it has been shown in the case of the enzyme-catalysed oxidation of nitroalkane carbanions (Porter et al. 1973). This enzyme has been the subject of intensive mechanistic study by several groups over the years (Massey and Gibson 1963; Walsh et al. 1971; Porter et al. 1973); its amino acid sequence has recently been determined (Ronchi et al. 1982) and specific active site residues identified (Swenson et al. 1982, 1983). These are illustrated diagrammatically in Scheme 13. Its mechanism and the role played by specific protein residues may be envisaged as shown in Scheme 14.

Of particular interest is the role played by an arginine residue shown in the above scheme. An active site arginine residue was discovered by Ferti et al. (1981) by reaction of the enzyme with cyclohexanedione. We have found that when enzyme is modified in this way and the native flavin, FAD, is replaced by 8-mercapto-FAD, the spectrum of the resulting 8-mercapto-FAD enzyme is dramatically different from that of unmodified enzyme containing 8-mercapto-FAD (Fitzpatrick and Massey 1983).

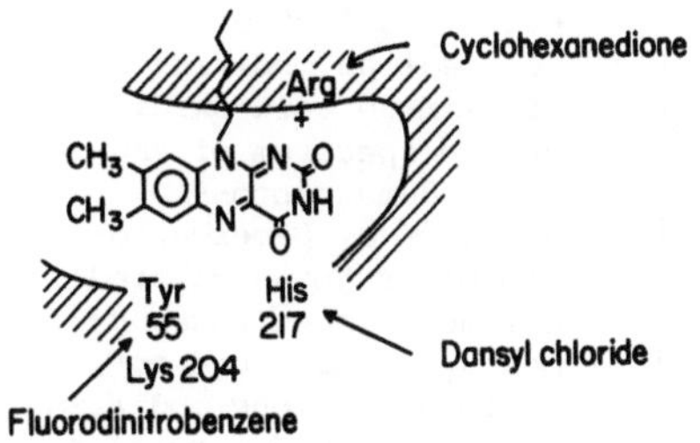

Scheme 13. Active site of D-amino acid oxidase

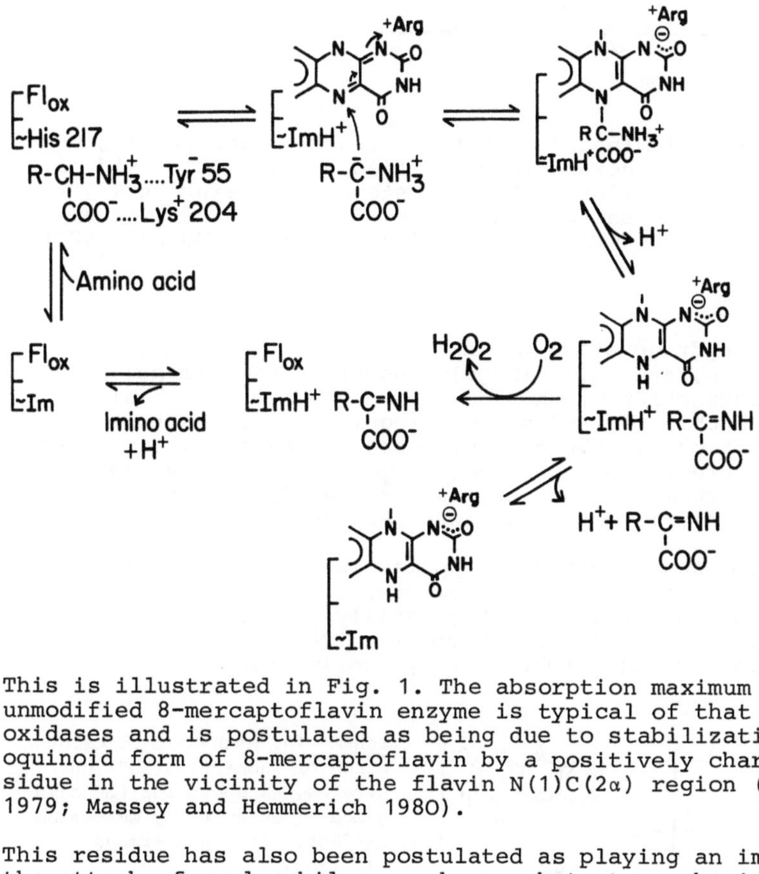

Scheme 14. Proposed mechanism of action of D-amino acid oxidase

This is illustrated in Fig. 1. The absorption maximum at 595 nm of the unmodified 8-mercaptoflavin enzyme is typical of that found with other oxidases and is postulated as being due to stabilization of the benzoquinoid form of 8-mercaptoflavin by a positively charged protein residue in the vicinity of the flavin N(1)C(2α) region (Massey et al. 1979; Massey and Hemmerich 1980).

This residue has also been postulated as playing an important role in the attack of nucleophiles, such as substrate carbanions and sulfite, at the flavin N(5) position and for stabilization of the anionic flavin semiquinone (cf. Schemes 2 and 3). These hypotheses are strongly supported by the results with cyclohexanedione-modified D-amino acid oxidase. With the active site arginine modified, the spectrum of the 8-mercapto-FAD enzyme now shows a maximum at 535 nm, typical of the 8-thiolate form of the flavin (Fitzpatrick and Massey 1983). Furthermore, it has been found that flavoprotein oxidases containing 8-mercapto-FAD all form N(5)-sulfite adducts just like the native flavin enzymes. On the other hand, nonoxidase flavoenzymes, which lack the inductive effect of the positive charge around the flavin N(1)C(2α) region, do not form sulfite adducts. However, when the native flavin of these enzymes is replaced by 8-mercaptoflavins, sulfite is found

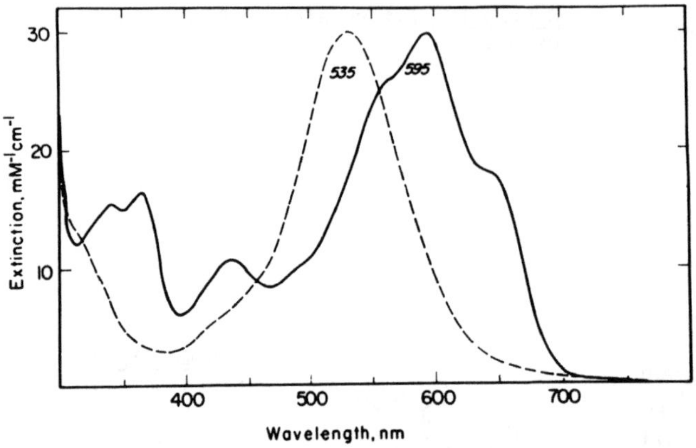

Fig. 1. Effect of cyclohexanedione modification of D-amino acid oxidase on the spectrum of enzyme-bound 8-mercapto-FAD. *Solid line*, 8-mercapto-FAD bound to unmodified enzyme. *Dashed line*, 8-mercapto-FAD bound to cyclohexanedione modified enzyme (From Fitzpatrick and Massey, 1983)

to react irreversibly, eliminating the thiol function and forming 8-sulfonylflavins. With cyclohexanedione-modified D-amino acid oxidase, sulfite now reacts in a similar fashion; instead of forming an N(5)-sulfite adduct as with the unmodified enzyme, it now reacts irreversibly to form the 8-sulfonyl-FAD enzyme (Fitzpatrick and Massey 1983).

"Nonactivated" Substrates

The mechanisms of dehydrogenation of "nonactivated" substrates, such as alcohols and amines, are still largely unsettled experimentally. In these cases, the abstraction of the α-hydrogen to form a carbanion is unlikely, since such a carbanion would require a large activation energy for its formation. In the case of monoamine oxidase, Silverman et al. (1982) therefore suggested a radical mechanism similar to that of equation (ii) of Scheme 5. The reaction for dehydrogenation of amines as formulated by Silverman is shown (Scheme 15).

An analogous reaction scheme probably applies in the case of oxidation of alcohols by flavoproteins. It is noteworthy that most alcohol oxidase preparations show a variable, but generally high content of stable anionic flavin radical (Geissler and Hemmerich 1981; Mincey et al. 1980). However, this radical form is catalytically inactive and it

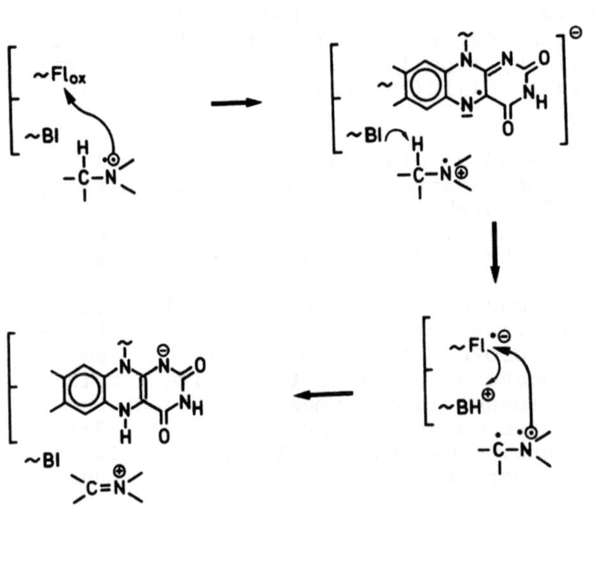

Scheme 15. Mechanism of oxidation of "nonactivated" substrates, as shown for the oxidation of an amine by monoamine oxidase. (From Silvermann et al. 1982)

would seem likely that it originates as a side reaction of catalysis, possibly due to a finite rate of dissociation of substrate radical from the substrate radical-flavin radical pair as a competing reaction to the collapse of the radical pair to yield aldehyde and reduced flavin.

It should be emphasized, however, that up to now no compelling arguments against a hydride mechanism have been advanced. Therefore, while we think that a radical mechanism is more likely, the possibility of oxidation of amines and alcohols proceeding via a hydride mechanism cannot be dismissed.

Thiol-Disulfide Substrates

The reversible oxidation-reduction reactions of thiol-disulfide pairs, such as lipoic acid, glutathione, and thioredoxin have been examined extensively in several laboratories, notably that of Williams (1976). An important factor for the understanding of this group of flavoproteins was the discovery of an internal redox active disulfide of the protein, which is in contact with the flavin, and mediates the transfer of redox equivalents between the thiol-disulfide substrate and the enzyme bound flavin (Searls 1960; Massey and Veeger 1960; Massey and Williams 1965; Zanetti and Williams 1967; Fox et al. 1982). The reaction catalyzed by lipoyl dehydrogenase as postulated by Arscott et al. (1981) is shown in part in Scheme 16. Dihydrolipoic acid reduces the active site disulfide via formation of a mixed protein-substrate disulfide in a reaction involving proton abstraction from the dihydrolipoic acid by an active site base, believed to be a histidyl residue. Under the influence of bound NAD^+, the proximal active site thiolate makes a nucleophilic attack on the flavin to form the covalent C(4a)-adduct of structure 5. The distal thiolate anion then attacks the sulfur of the flavin adduct with reformation of the active center disulfide and leaving behind the reduced flavin (structure 7). The latter now participates in a hydride transfer to the bound pyridine nucleotide, as discussed earlier (Scheme 6). This mechanistic proposal is nicely supported by the X-ray crystallographic studies on the closely related enzyme glutathione reductase, where all of the structural elements required for this mechanism are present (Pai and Schulz 1982). Similarly, Loechler and Hollocher (1975) concluded from kinetic studies in model systems that C(4a) adducts also occur during the oxidation of thiols by free flavins. In addition, with lipoyl dehydrogenase, a flavin C(4a)-S-R adduct was identified spectrally when NAD^+ was

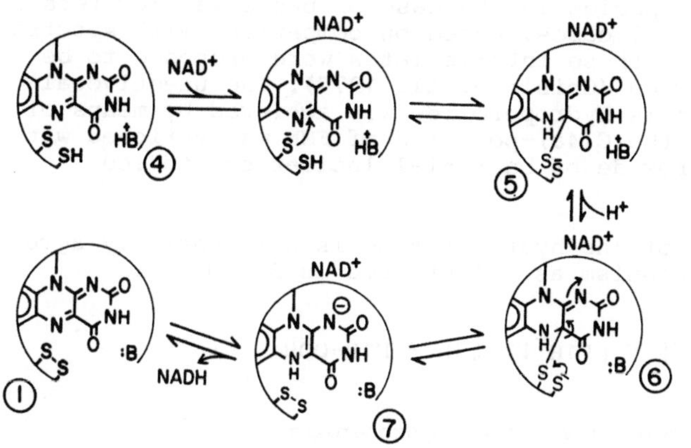

Scheme 16. Proposed mechanism of lipoyl dehydrogenase, showing the intermediate formation of a flavin-thiolate C(4a)-adduct

added to enzyme which was alkylated at one of its thiol functions (Thorpe and Williams 1981). This alkylation prevented frank electron transfer and so trapped this putative intermediate.

The Oxidative Half-Reaction

As indicated in the introduction, reduced flavoenzymes produced by the pathways already discussed, can complete their catalytic cycle in the oxidative half-reaction, by reaction with a variety of one-electron or two-electron acceptors. The nature of the oxidizing substrate is specific to the particular enzyme, as indicated in the classification scheme shown in the introduction. The crossover between two-electron and one-electron transfer, which is a biochemical phenomenon restricted almost entirely to flavin enzymes, is a subject which has not yet received the same depth of experimental investigation as the simple two-electron transfers, which have been discussed here. In this review, we are restricting our coverage of the oxidative half-reaction to that of reduced flavins and flavoproteins with molecular oxygen. For reviews of the role of flavin in $2e^-/1e^-$ crossover, see Kamin and Lambeth 1982; Hemmerich et al. 1982.

The Activation of Molecular Oxygen

Flavins are one of the few catalysts used by living cells which have the capacity to activate molecular oxygen. In its reaction with reduced flavin, O_2 can be reduced to the $1e^-$ reduced state, superoxide anion (O_2^-), to the $2e^-$ reduced state, peroxide (H_2O_2), or to the $4e^-$ reduced state, H_2O. In the case of flavoproteins, the inherent reactivity of the singlet reduced flavin molecule with triplet O_2 is modulated in various ways, as described in the introduction, and has long served as a basis for useful differentiation of flavoenzymes into broad classes.

Impressive progress has been made in the last few years in the elucidation of the primary steps in the reaction of the reduced flavin with O_2. The results of early rapid reaction kinetic studies with free flavins had indicated the existence of oxygenated flavin species in the overall reaction (Gibson and Hastings 1962; Massey et al. 1973; Kemal et al. 1977). The first decisive evidence for such intermediates came from studies with flavoprotein monooxygenases, where species with absorbance maxima near 400 nm were observed as transients in the reaction with O_2 of reduced p-hydroxybenzoate hydroxylase (Spector and Massey 1972) and melilotate hydroxylase (Strickland and Massey 1973) and as a stable isolatable species in the case of bacterial luciferase (Hastings et al. 1973). In all cases, based on comparison with spectra of known flavin derivatives, these intermediates were proposed to be the flavin C(4a)-hydroperoxide (Ghisla et al. 1977). The unequivocal proof of the structure of this hydroperoxide was achieved by means of ^{13}C-NMR spectroscopy where the C(4a)-position of FMN was enriched with ^{13}C and the stable hydroperoxide of bacterial luciferase studied (Ghisla et al. 1978).

The mechanism of formation of the hydroperoxide is now generally agreed to result via a radical mechanism as illustrated in Scheme 17.

$$Fl_{red}H^- + \uparrow O_2 \uparrow \rightleftharpoons [FlH^{\cdot}\uparrow\uparrow O_2^{\cdot -}] \rightleftharpoons [FlH^{-}\downarrow\uparrow O_2^{\cdot -}] \rightleftharpoons FlH\text{-}OOH$$

Scheme 17. Mechanism of formation of flavin C(4a)-hydroperoxide

First a one-electron transfer occurs to form a flavin radical-superoxide pair having parallel spins. After spin inversion, which is supposed to be the rate-limiting step, the biradical pair collapses to form the flavin C(4a)-hydroperoxide, as expected from the spin density of the flavin radical (Platenkamp et al. 1980). In model studies, the pair of radicals, F1H$^{\bullet}$ and O_2^{-}, generated by pulse radiolysis, has been shown to collapse into the hydroperoxide (Anderson 1982). Similar conclusions have been reached by Nanni et al. (1981). While the collapse of the radical pair is very rapid, it clearly can form the C(4a)-hydroperoxide only when this position of the flavin is unimpeded by the surrounding protein. Recent studies with chemically reactive flavins (5-deazaflavin and 4-thioflavin) substituted for the native flavin suggest strongly that with electron transferases, the flavin region N(5)-C(4a)-C(4) is rather inaccessible to solvent-borne reagents, as illustrated in Scheme 18 (Schopfer, Claiborne, Detmer, and Massey, unpublished). Therefore, the possibility is very real that in this group of enzymes, the flavin hydroperoxide does not form and that the radical pair F1H$^{\bullet}$$O_2H^{\bullet}$ instead largely dissociates into the free flavin radical and O_2^{-}, the experimentally observed products of this group of enzymes.

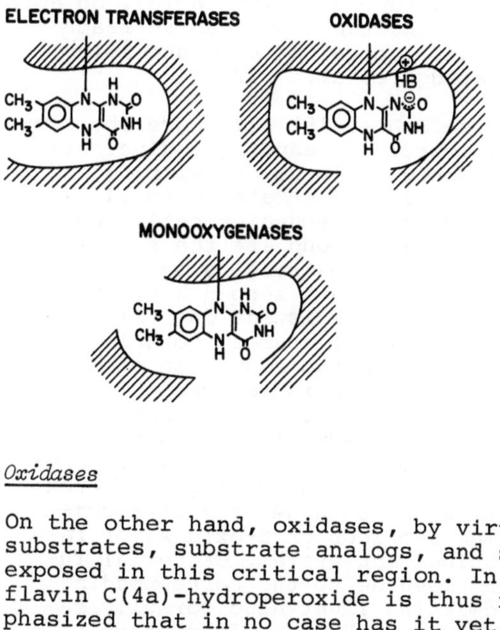

Scheme 18. Active site topography of different classes of flavoproteins, as determined by replacement of the native coenzymes with flavins containing chemically reactive substituents

Oxidases

On the other hand, oxidases, by virtue of their known reactivity with substrates, substrate analogs, and sulfite, clearly have their flavin exposed in this critical region. In this class, the formation of the flavin C(4a)-hydroperoxide is thus feasible, although it must be emphasized that in no case has it yet been observed experimentally. However, this may be simply for kinetic reasons. Bruice and colleagues have found that even with the chemically stabilized N(5)-substituted C(4a)-flavin hydroperoxides, the stability is limited to apolar solvents; in aqueous solution, the hydroperoxides break down rapidly at neutral pH to yield H_2O_2 and oxidized flavin (Kemal and Bruice 1976). On the basis of the available evidence, we consider it likely that a similar sequence of reactions occurs with this group of enzymes, as illustrated in Schemes 18 and 19.

Monooxygenases

The third main group of flavoenzymes is the monooxygenases where the O_2 molecule is cleaved, with one atom being incorporated into a substrate and the other converted to H_2O. Four main subclasses of this group can be differentiated. The first of these is sometimes known as

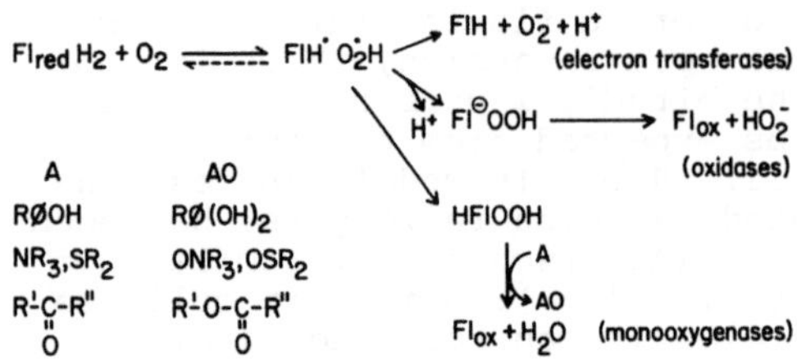

Scheme 19. Proposed explanation for the different types of reaction with O_2 of flavoprotein electron transferases, oxidases, and monooxygenases

internal monooxygenases, because the source of reducing equivalents for reduction of the flavin is the substrate which is monooxygenated. This class is exemplified by L-lactate monooxygenase and L-lysine monooxygenase, which carry out oxidative decarboxylations of L-lactate to yield acetate, CO_2 and H_2O, and of L-lysine to yield δ-aminovaleramide, CO_2 and H_2O. However, convincing evidence has been produced that these enzymes belong better to the oxidase class of flavoproteins, being in fact "adventitious" monooxygenases, because the normal oxidase products, H_2O_2 and the corresponding keto and imino acids, do not diffuse rapidly from the enzyme surface, and there proceed to react to produce the oxidative decarboxylations typical of keto and imino acids with H_2O_2 free in solution (Lockridge et al. 1972).

The remaining "true" monooxygenases (see Scheme 19) all share the common property of requiring an external source of reducing equivalents to produce the reduced flavin, which subsequently reacts with O_2 and the substrate which is oxygenated. This third substrate is in all cases either NADH or NADPH. There appear to be mechanistically two different types. The largest and most studied group are the so-called aromatic or phenolic hydroxylases, where the substrate is an unsubstituted or substituted phenolic compound and where a second hydroxyl group is introduced, generally ortho to the preexistent one. All of the enzymes of this class share certain common features, notable among which is the effector role of the substrate on the reduction of the enzyme flavin by pyridine nucleotide; the flavin is reduced 10^3-10^5 times more rapidly by NAD(P)H when the substrate is bound than it is when the substrate is absent (see Massey and Hemmerich, 1975 for a review). The other types of flavoprotein monooxygenase carry out oxygen transfer reactions typical of those carried out by organic peroxides in free solution, converting either nitrogenous or sulfur containing substrates to the corresponding N-oxides or S-oxides, or inserting an oxygen atom into a preexisting ring system. Bacterial luciferase, which converts a long chain aldehyde to the corresponding carboxylic acid at the same time as emitting light also formally belongs in this group. This group of enzymes lacks the effector role of the substrate displayed by the phenolic hydroxylases; the rate of flavin reduction by NAD(P)H is uninfluenced by prior substrate binding (Beaty and Ballou 1981a; Ryerson et al. 1982).

With all groups of "true" monooxygenases, the first observed product of reaction of O_2 with the reduced flavin is the C(4a)-hydroperoxide. In the case of the phenolic hydroxylases, this is a transient species, requiring rapid reaction studies for its detection and generally requiring the presence of an effector for its (partial) stabilization. With the other monooxygenases, however, the flavin C(4a)-hydroperoxide is generally remarkably stable and usually requires the presence of oxygenatable substrate to be discharged rapidly.

Considerations on the Molecular Mechanism of Substrate "Hydroxylation"

As might be expected, both from the different properties exhibited by each group of oxygen activating flavoenzymes, as well as from the different types of reactions catalyzed, these enzymes appear to operate by distinctly different mechanisms. The most studied, and still not completely understood, are the reactions catalyzed by the phenolic hydroxylases. Soon after the initial discovery of the flavin C(4a)-hydroperoxide with p-hydroxybenzoate hydroxylase (Spector and Massey 1972), it was recognized that two other distinct intermediates were involved in the oxygenation reaction and their spectra determined (Entsch et al. 1974, 1976a). The third intermediate has a spectrum similar to that of the hydroperoxide and as rapid quench studies showed that oxygen atom transfer to substrate had already occurred at the stage of intermediate II (Entsch et al. 1974), the structure of intermediate III could logically be ascribed to that of the C(4a)-hydroxyflavin, which by elimination of H_2O, would yield oxidized flavin ready for the next cycle of catalysis. The structure of intermediate II has remained without satisfactory explanation and so the details of the catalytic mechanism have remained incomplete. Entsch et al. (1976a) proposed the mechanism shown in Scheme 20, which involves nucleophilic attack by the delocalized electrons of the phenolate ion on the terminal oxygen of the hydroperoxide. Cleavage of the O-O-bond is synchronous with formation of a ring-opened form of the flavin and the cyclohexadienone tautomer of the product. The ring-opened flavin next recloses to form the pseudobase, C(4)-hydroxyflavin (intermediate III), while the oxygenated substrate rearomatizes. Finally H_2O is split out from the hydroxyflavin and oxidized enzyme is regenerated. This step, in which the product is also released from the enzyme, is generally the rate-limiting step in catalysis. In the above formulation, intermediate II, which in the case of p-hydroxybenzoate hydroxylase has

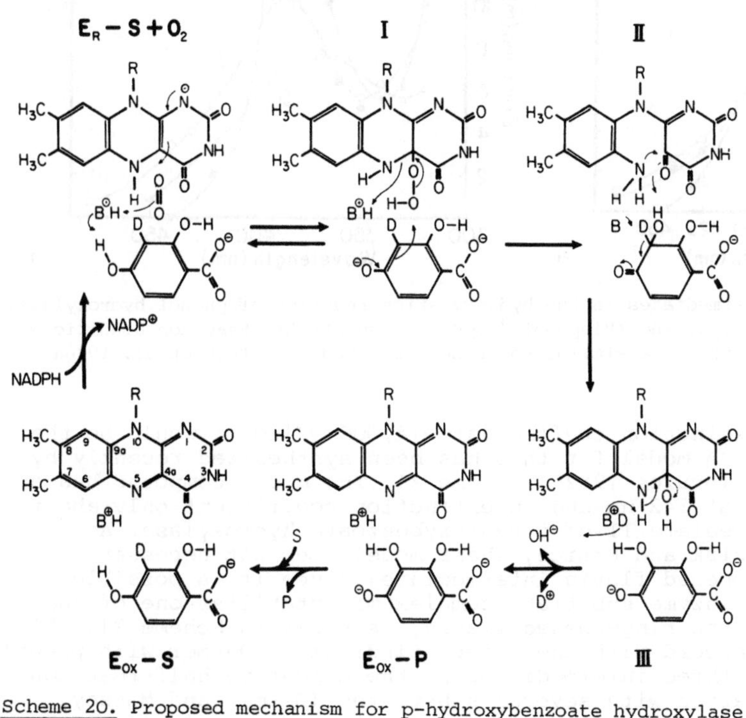

Scheme 20. Proposed mechanism for p-hydroxybenzoate hydroxylase. (From Entsch et al. 1976a)

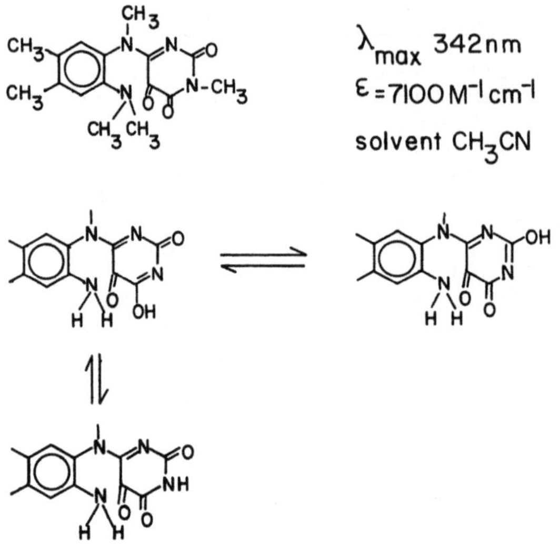

Scheme 21. Model for the ring-opened flavin proposed in Scheme 20 (*top*). The structures at the *bottom* show tautomeric forms possible for an enzyme-bound species. (After Bruice 1982)

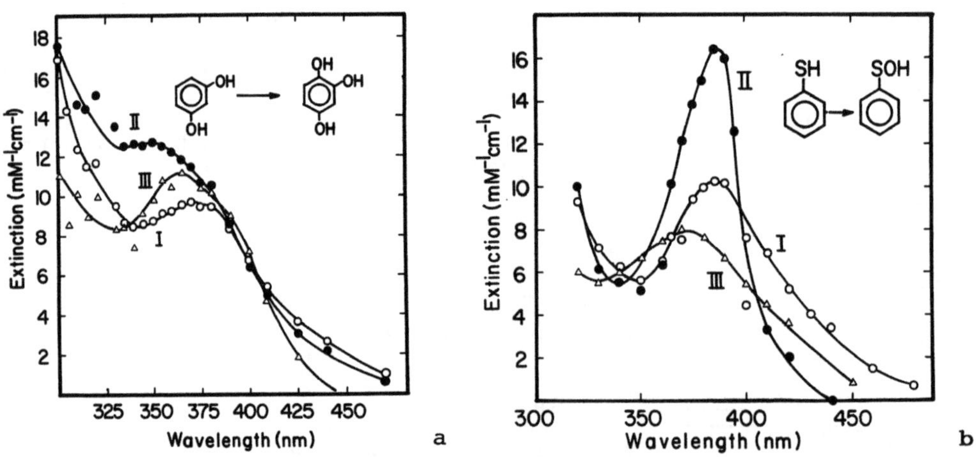

Fig. 2. Spectra of intermediates in the hydroxylation reaction of phenol hydroxylase, employing resorcinol (*left*) and thiopenol (*right*) as substrate. Reaction conditions and calculation of spectra were similar to those described in Entsch et al. 1976a (Detmer, unpublished)

λ_{max} ~400 nm and extinction coefficient ~14,000 $M^{-1}cm^{-1}$, would be the ring-opened flavin. A model for this has been synthesized recently by Wessiak and Bruice (Bruice 1982), Scheme 21. The model compound has a wavelength maximum at 342 nm and an extinction coefficient only about half that of intermediate II of p-hydroxybenzoate hydroxylase. As pointed out by Wessiak and Bruice, their model does not necessarily rule out the ring-opened flavin intermediate, since it is possible that the enzyme or enzyme substrate complex may stabilize one of the other tautomers of the ring-opened flavin, as shown in Scheme 21. This possibility is in accord with the recent findings of Detmer with phenol hydroxylase, where three intermediates in the oxidative half-reaction have also been observed with several substrates (Detmer and Massey, unpublished). In all cases, the spectra of intermediates I and III are similar to those found with p-hydroxybenzoate hydroxylase and melilolate hydroxylase (Entsch et al. 1976a; Schopfer and Massey 1980). Now, however, the spectrum of intermediate II differs markedly, depending on the substrate. This is illustrated in Fig. 2 for the substrates resorcinol and thiophenol. The finding of a high extinction intermediate II with thiophenol as substrate is similar to that with p-hydroxybenzoate hydroxylase and p-mercaptobenzoate, where hydroxylation was shown to take place at the sulfur rather than in the aromatic ring, to yield the unstable sulfenate (Entsch et al. 1976b). These findings effectively rule out the possibility that the high extinction of intermediate II is due to a complex between hydroxyflavin and a dienone tautomer of the product, since such compounds would not be formed with the sulfur-containing substrates.

Recent model studies by Bruice and colleagues (Ball and Bruice 1980; Ball and Bruice 1981) have shown that flavin hydroperoxides are surprisingly high in their reactivity towards various nucleophiles and are more comparable to organic peracids than to alkyl peroxides or H_2O_2. The finding that the p*K* of the hydroperoxide is unexpectedly low, in the region of pH 9 (Bruice 1983), would indicate a strong -I influence of the C(4a)-substituents. A partial explanation why 1-deazaflavin substituted p-hydroxybenzoate hydroxylase (Entsch et al. 1980) is able to form a C(4a)-hydroperoxide, but that this is incompetent to carry out the oxygen transfer step, might be related to its lower redox potential (Spencer et al. 1977), i.e., its smaller activation of the peroxide group. The inability of 1-deazaflavin to carry out hydroxylation, even though it is very competent in catalyzing substrate-dependent NAD(P)H oxidase activity, appears to be common with all the phenolic hydroxylases investigated (orcinol hydroxylase (Walsh 1980), p-hydroxybenzoate hydroxylase (Entsch et al. 1980), and melilotate and phenol hydroxylases (Schopfer and Detmer, unpublished). However, the fact that 2-thio FAD, which has a higher redox potential (Light and Walsh 1980), when bound to the apoenzyme of p-hydroxybenzoate hydroxylase, also forms a C(4a)-hydroperoxide, which is incompetent in normal catalysis, points toward a more specific function of the N(1)C(2)O moiety in the hydroxylation reaction (Claiborne and Massey 1983). The spectral properties of 8-mercapto-FAD p-hydroxybenzoate hydroxylase suggest that the benzoquinoid form is partially stabilized in the presence of substrate (Massey et al. 1979). The spectral shifts observed were not so marked as in the case of flavoprotein oxidases, indicating the lack of a full positive charge in the N(1)C(2) region as appears to be present in the oxidases. This is in keeping with the 3-dimensional structure of the p-hydroxybenzoate hydroxylase-substrate complex; Wierenga et al. (1982) in fact suggest that the dipole of helix 5, which starts near the flavin N(1)-position, could exert a partial positive charge towards this region of the flavin. A similar helix-induced electric dipole is present in glutathione reductase (Thieme et al. 1981), where the same type of p-quinoid 8-mercaptoflavin stabilization is found (Krauth-Siegel 1982).

Such a dipole would enhance the oxidizing power of the flavin C(4a) hydroperoxide. At the same time, it might be related to the spectral properties of intermediate II as indicated in Scheme 22. Here several species are depicted, in which, by different modes of interaction with the protein, different chromophoric species could be stabilized, possibly depending on different protein conformations being induced by different substrates. It must be emphasized that this is speculation, as spectral properties of the species proposed as contributing to intermediate II are difficult to correlate with those of known compounds. An interaction of position N(5) with the protein, such as shown in Scheme 22, structure A, to produce the cationic species, would further enhance the electrophilicity of the flavin C(4a) hydroperoxide, and,

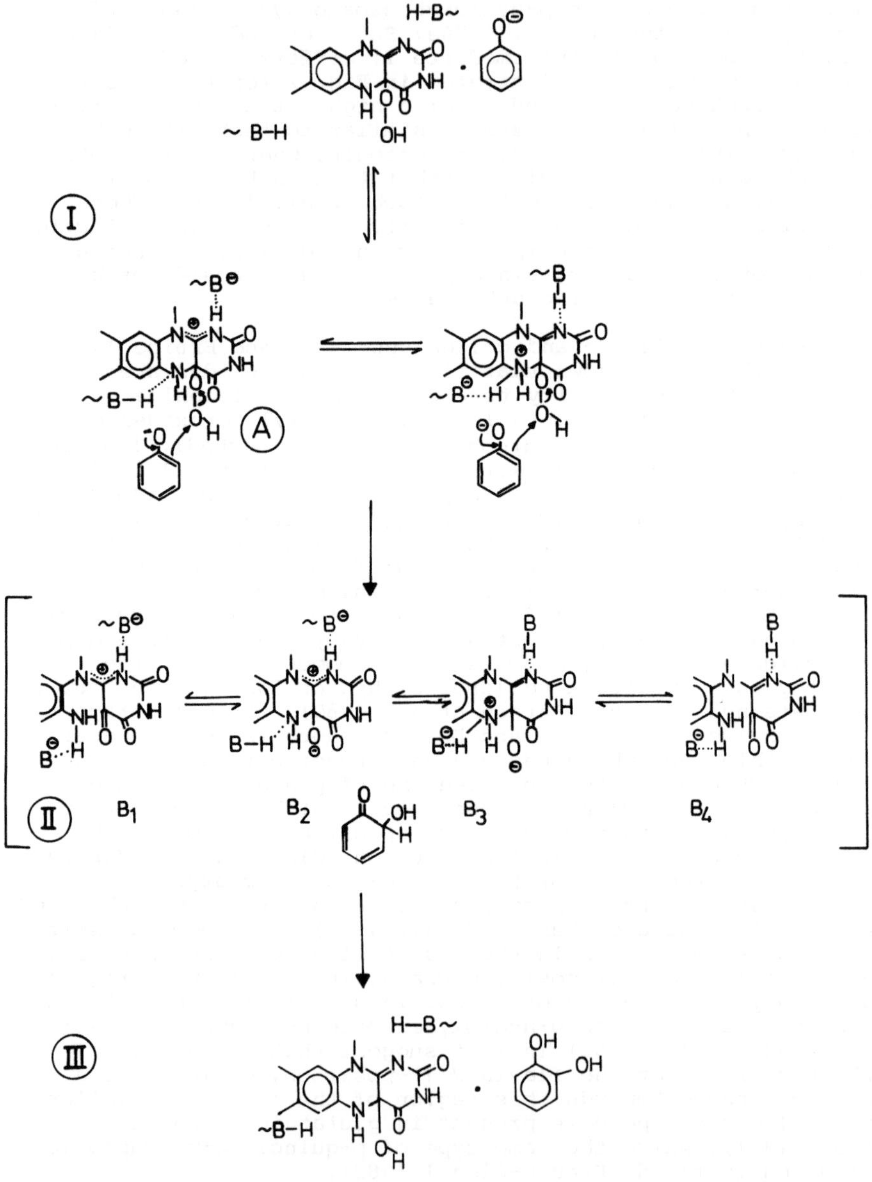

Scheme 22. Possible mechanisms of hydroxylation carried out by flavoprotein aromatic hydroxylases. The transition from intermediate I to II is shown to proceed via the protonated flavin of structures A. Such forms would be expected to enhance the electrophilicity of the hydroperoxide and could be induced by specific protein interactions, such as those shown. Intermediate II could be composed of any of a number of different chromophoric structures, as indicated

after oxygen transfer to substrate has occurred, lead to species, such as B_1 and B_3. The pK of protonation of 4a,5 dihydroflavins is known to be in the region 2-4 (Ghisla, unpublished) making the speculation of Scheme 22 not unreasonable. Alternatively, a hydrogen bridge to N(5) could constitute a driving force for the ring-opened mechanism already discussed (Scheme 20). Protonated 4a,5-dihydroflavins have spectra with maxima in the range 310-400 nm and with extinction coefficients from 3000-10000 $M^{-1}cm^{-1}$ (Ghisla et al. 1973).

Protein interactions with the flavin, such as those shown in Scheme 22, could also account for the fact that most enzymes of the aromatic hydroxylase class do not consistently stabilize either the neutral of the anionic semiquinone and that semiquinone forms, when observed, are generally dependent on the nature of the bound substrate. In the case of the remaining monooxygenases (cf. Scheme 19), the flavin hydroperoxide is remarkably stable, often with t 1/2 values in the range of minutes or even hours in the absence of the substrate to be monooxygenated. This stabilization is correlated with a corresponding stabilization of the blue neutral radical. In the case of p-hydroxybenzoate hydroxylase, the decay rate of the pseudobase intermediate III is of the order of 0.1-2 s^{-1} at 4° (Entsch et al. 1976a). This compares with a value of ~100 s^{-1} for decay of the flavin hydroperoxide in free solution (Anderson 1982) and ≤ 1 s^{-1} for the N(5)-substituted pseudobases studies by Ball and Bruice (1980, 1981). Hence it appears that no pronounced stabilization of the oxygenated flavin species is required to account for the catalytic sequence of the aromatic hydroxylases.

Mechanism of N,S-Monooxygenation

In the case of the microsomal flavoprotein N,S-monooxygenase, the enzyme carries out the insertion of oxygen in a wide variety of nitrogen and sulfur containing compounds, making it a valuable adjunct to the cytochrome P-450 system in drug detoxification. Its mechanism has been investigated recently by the research groups of Ziegler (Poulsen and Ziegler 1979) and Ballou (Beaty and Ballou 1981a,b). In the presence of NADP, the reduced flavoprotein forms a remarkably stable C(4a)-hydroperoxide, which is discharged rapidly only when substrate is added (Beaty and Ballou 1981b). An intermediate with spectral properties similar to those of intermediate III of the phenolic hydroxylases has also been reported (Beaty and Ballou 1981b). It would appear likely that this reaction involves a simple nucleophilic attack of the N or S electron pair on the terminal oxygen of the flavin hydroperoxide to yield the oxygenated product and the C(4a)-hydroxyflavin. The model studies of Ball and Bruice (1980, 1981) have shown that N(5)-substituted flavin C(4a)-hydroperoxides behave in similar fashion to organic peracids in carrying out such reactions even in the absence of enzyme. Significantly, in these model studies, no aromatic ring hydroxylation by peracids or alkyl peroxides has been reported, indicating, as expected, that some extra inductive effect, such as that discussed above in Scheme 22 or in the ring-opened mechanism of Entsch et al. (1976a) is required in this type of reaction.

Mechanism of "Nucleophilic" Monooxygenation

With the remaining types of flavoprotein monooxygenases, exemplified by cyclohexanone monooxygenase (Ryerson et al. 1982) and bacterial luciferase (Hastings et al. 1973), the occurrence of a flavin C(4a)-hydroperoxide in the reaction pathway has also been demonstrated. Now, however, the hydroperoxide (anion) appears to function as a nucleophile, rather than as an electrophile as in the case of the phenolic hydroxylases. Ryerson et al. (1982) propose the following mechanism for cyclohexanone monoxygenase based on a steady state and rapid reaction kinetics study (Scheme 23).

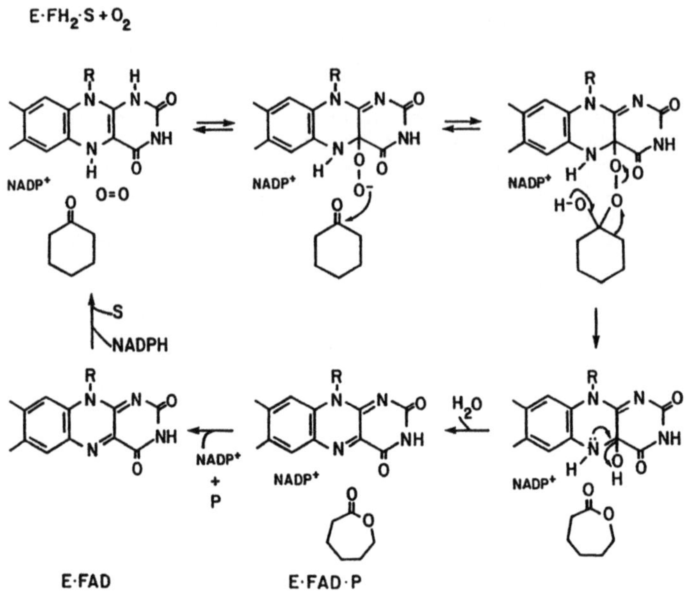

Scheme 23. Proposed reaction mechanism of cyclohexanone monooxygenase, involving the flavin hydroperoxide as a nucleophile. (From Ryerson et al. 1982)

In this mechanism, the terminal oxygen of the hydroperoxide makes a nucleophilic attack on the electrophilic carbonyl carbon of the substrate. The tetrahedral adduct formed then decomposes to C(4a)-hydroxyflavin with a Baeyer-Villiger C → O bond migration to incorporate the terminal oxygen atom into the substrate ring. With this enzyme, it is interesting to note that enzyme with 1-deazaflavin incorporated instead of the native FAD is fully competent in carrying out the complete catalytic reaction, distinguishing it in yet another way from the phenolic hydroxylases.

Similar mechanisms have been proposed for the bacterial luciferase reaction, but details of this intriguing biological phenomenon are still uncertain. It has been well established that the flavin C(4a)-hydroperoxide is the primary product of reaction of the reduced luciferase flavin with O_2. This hydroperoxide is remarkable in its stability in the absence of long chain aldehyde substrate (Hastings et al. 1973) and can be further stabilized by addition of long chain alcohols and/or high electrolyte concentration (Tu 1979; Becvar et al. 1978). The stabilization of the hydroperoxide is essentially kinetic in its nature and was found to parallel a similar stabilization of the blue flavin semiquinone (Kurfürst et al. 1982b); the mechanisms of the two processes were proposed to involve a strong H-bridge from the protein to the flavin position N(5)-H as shown on Scheme 24.

For the mechanism of substrate oxidation and concomitant generation of an excited state, a series of imaginative proposals have been put

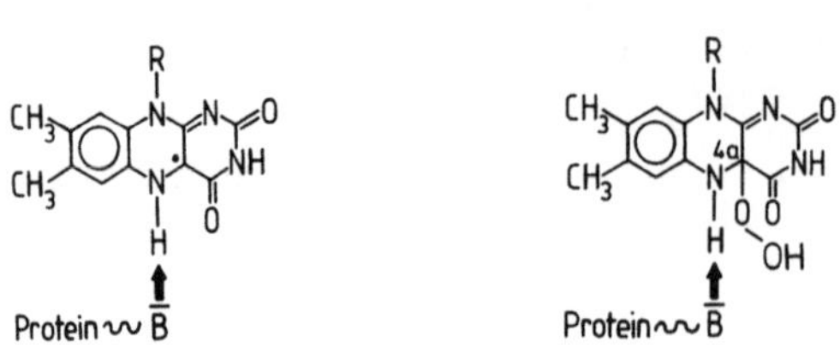

Scheme 24. Proposed stabilization of luciferase flavin C(4a)-hydroperoxide and neutral flavin radical by hydrogen bonding of the flavin N(5)H with the protein

forward in the past (Kosower 1980; Mager and Addink 1979; Wessiak et al. 1980). They have mechanistic features, such as migration of the peroxide residue, ring openings, and closure, and the generation of an N(1) protonated flavin as the emitter. Many of these proposals have been addressed by a specific experiment involving the use of 1-deaza-FMN, a flavin analogue which cannot undergo the type of reactions proposed which involve position 1 or protonation at this locus. This analoge was found to be competent in light emission and the spectrum of the emitted light clearly excludes a protonated species as the emitter (Kurfürst et al. 1982a).

Recent work has provided evidence that the oxidation of aldehyde indeed follows the pattern already demonstrated for the monooxygenases discussed above. Kinetic analysis of the steps following addition of aldehyde to the peroxide, shows that a species is present after most of the light has been emitted, which shares the spectral properties common to 4a,5-dihydroflavins (Kurfürst and Ghisla, unpublished). Its structure should logically be that of the 4a-pseudobase. The mechanism of bacterial luciferase can be envisaged as shown in Scheme 25. It involves a nucleophilic attack of the peroxide anion at the aldehyde function followed by a Baeyer-Villiger type reaction leading to the corresponding carboxylic acid and the flavin C(4a) hydroxy derivative (pseudobase). The question as to what is the primary excited chromophore is still unsettled. The flavin pseudobase clearly can exert this function, as shown also by experiments involving the use of specifically modified flavin chromophores (Kurfürst et al. 1982a; Tu 1982). In these experiments, it was demonstrated that the emission profile of bioluminescene was closely matched by the fluorescence emission spectrum of the corresponding flavin C(4a)-hydroperoxide. In some cases, (Leisman and Nealson 1982; Koka and Lee 1979), a second protein chromophore can accept the (primary) excitation and serve as the emitter. Whether the flavin pseudobase is an obligatory intermediate in such energy transfer will have to await further experimental analysis.

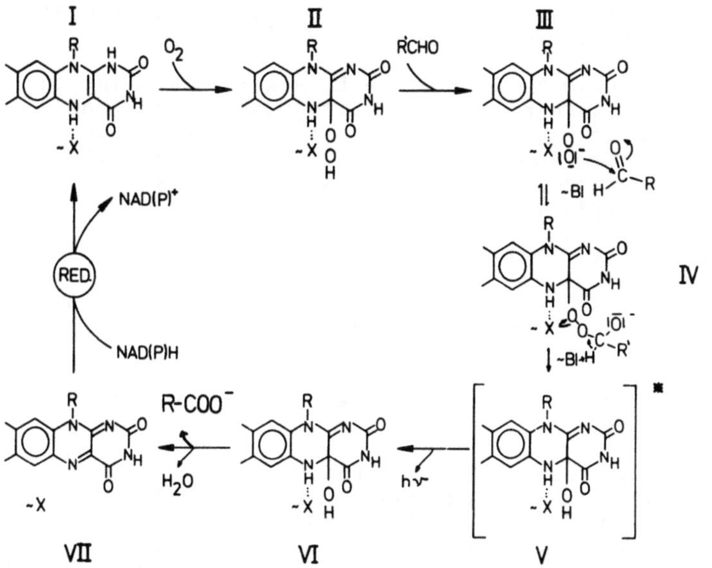

Scheme 25. Proposed reaction mechanism of bacterial luciferase. Enzyme-bound $FMNH_2$ (I) reacts with O_2 to yield the stable hydroperoxide (II). On addition of long chain aldehydes a Baeyer-Villiger type reaction is envisaged (III and IV), which leads to the excited emitter V. Upon emission of light, the pseudobase VI is formed, which generates oxidized flavin VII by the elimination of H_2O. Formation of reduced flavin I from VII requires an external NAD(P)H-FMN reductase. (From Kurfürst and Ghisla, unpublished)

Acknowledgements. The authors wish to acknowledge support from the U.S. Public Health Service, Grant GM 11106 and from the DFG, Grant Gh 2/4-4. We wish to thank our colleagues, Drs. A Claiborne, K Detmer, M Kurfürst, LM Schopfer, A Wenz, and A Wessiak for permission to quote unpublished work and for many valuable discussions.

Discussion

Biellmann: Glucose oxidase is inactivated after 10^7 turnovers, methionine being oxidized to sulfoxide. Since intermediates in flavoprotein catalyzed reactions should be very reactive, are flavoproteins inactivated after a certain number of turnovers?

Massey: Sometimes inactivation does occur, but this is generally the result of reaction with active products rather than intermediates. For example, flavoenzymes, such as glucose oxidase, produce H_2O_2 as a result of turnover with O_2 as acceptor and H_2O_2 can often oxidize exposed sensitive amino acid residues. Perhaps the inactivation you refer to is of this sort.

Mason: Your elucidation of the biological oxygen reactions catalyzed by flavoproteins is very beautiful. It is very interesting that the same range of reactions is catalyzed by the other O_2-activating prosthetic groups, Cu, Fe, and heme, except that two classes of enzyme reactions are missing: reversible O_2 transport and dioxygenation. Are these possible for flavoproteins too?

Massey: Dioxygenation is certainly possible; a flavoprotein dioxygenase has been studied by Snell and co-workers over a number of years. Their enzyme catalyzes ring-opening of the pyridine ring in the bacterial metabolism of pyridoxal. The second property, of reversible O_2 transport, seems to be one which flavoproteins are not designed for.

Ruf: The leukocyte pyridine nucleotide oxidase generates O_2^- during phagocytosis at high rates. Schirmer showed this enzyme as a protein with one flavin. Can you conceive a mechanism for rapid one-electron reduction of O_2 by such a protein without the need for an additional one-electron acceptor, such as a cytochrome?

Massey: In principle this should be possible with a single flavin-containing protein. Some flavoproteins, such as flavodoxin, react rapidly in their reduced form with O_2 to produce the flavin semiquinone and O_2^-. In this case, the reaction of the semiquinoid protein with O_2 is slow, so that the overall reaction $Fl_{red}H_2 + 2O_2 \rightarrow Fl_{ox} + 2O_2^- + 2H^+$ is slow. But in principle there seems to be no real reason why the reaction of flavoprotein semiquinone with O_2 need be slow. We have shown that the neutral radical form of glucose oxidase reacts rapidly with O_2 to generate O_2^-. So, while nature may choose often to use a second redox acceptor as an electron sink, I would not be surprised that some day a simple flavoprotein will be found to do this conversion efficiently in a catalytic fashion.

Schirmer: Could you please comment on the turnover numbers of the different types of flavoenzymes?

Massey: I cannot think of any real generalization about the various classes. Some oxidases, like glucose oxidase, have very high turnover numbers, of the order of 50000 min^{-1} at 25°, while others, such as D-amino acid oxidase have turnover numbers around 1000 min^{-1}. The monooxygenases have medium catalytic rates, of the order of 1000-10000 min^{-1}. Some transhydrogenases, such as lipoyl dehydrogenase, have very high turnover numbers, again in the region of 50000 min^{-1}. But in fairness to the catalytic efficiency of the flavin molecule, it should be mentioned that in some cases, such as that of D-amino acid oxidase, the rate-limiting step is the dissociation of product from the enzyme!

Ullrich: With regard to the structure of intermediate II, do you think that it could correspond to either the peroxide anion or an endoperoxide-type intermediate with the substrate on one side?

Massey: It is unlikely that intermediate II could be the flavin hydroperoxide anion, since our rapid chemical quench data indicate that oxygen transfer to substrate occurs coincident with the formation of intermediate II. With an endoperoxide the rapid quench data could be explained by acid breakdown of the endoperoxide so that dihydroxy-product would be found. However, I doubt that this would have a spectrum significantly different from that of the hydroperoxide (intermediate I).

References

Anderson RF (1982) In: Massey V, Williams CH (eds) Flavins and flavoproteins. Elsevier North Holland, Amsterdam, pp 278-283
Arscott LD, Thorpe C, Williams CH Jr (1981) Biochemistry 20:1513-1519
Ball S, Bruice TC (1980) J Amer Chem Soc 102:6498-6503
Ball S, Bruice TC (1981) J Amer Chem Soc 103:5494-5503
Ballou DP (1982) In: Massey V, Williams CH (eds) Flavins and flavoproteins. Elsevier North Holland, Amsterdam, pp 301-310
Beaty NB, Ballou DP (1981a) J Biol Chem 256:4611-4618
Beaty NB, Ballou DP (1981b) J Biol Chem 256:4619-4625
Becvar JE, Tu S, Hastings JW (1978) Biochemistry 17:1807-1812
Biellmann JF, Hirth CG (1970) FEBS Letters 9:335-336
Bruice TC (1982) In: Massey V, Williams CH (eds) Flavins and flavoproteins. Elsevier North Holland, Amsterdam, pp 265-277
Bruice TC (1983) J Chem Soc Chem Commun 1983:14-15
Brüstlein M, Bruice TC (1972) J Amer Chem Soc 94:6548
Claiborne A, Massey V, Fitzpatrick PF, Schopfer LM (1982) J Biol Chem 257:174-182
Claiborne A, Massey V (1983) J Biol Chem 258:4919-4925
Cromartie TH, Walsh CT (1976) J Biol Chem 251:329-333
Dudley KH, Ehrenberg A, Hemmerich P, Müller F (1964) Helv Chim Acta 47:1354-1383
Edmondson DE, Singer TP (1976) FEBS Letters 64:255-265
Ehrenberg A, Müller F, Hemmerich P (1967) Eur J Biochem 2:286-293
Entsch B, Massey V, Ballou DP (1974) Biochem Biophys Res Comm 57:1018-1025
Entsch B, Ballou DP, Massey V (1976a) J Biol Chem 251:2550-2563
Entsch B, Ballou DP, Husain M, Massey V (1976b) J Biol Chem 251:5367-7379
Entsch B, Husain M, Ballou DP, Massey V, Walsh C (1980) J Biol Chem 255:1420-1429
Fendrich G, Abeles RH (1982) Biochemistry 21:6685-6695
Ferti C, Curti B, Simonetta MP, Ronchi S, Galliano M, Minchiotti L (1981) Eur J Biochem 119:553-557
Fitzpatrick PF, Massey V (1983) J Biol Chem 258:9700-9705
Frerman FE, Miziorko HM, Beckmann JD (1980) J Biol Chem 255:11192-11198
Fox B, Walsh CT (1982) J Biol Chem 257:2498-2503
Geissler J, Hemmerich P (1981) FEBS Letters 126:152-156
Ghisla S, Massey V (1980) J Biol Chem 255:5688-5696
Ghisla S, Hartmann U, Hemmerich P, Müller F (1973) Liebigs Ann Chem 1973:1388-1415
Ghisla S, Entsch B, Massey V, Husain M (1977) Eur J Biochem 76:139-148
Ghisla S, Hastings JW, Favaudon V, Lhoste JM (1978) Proc Nat Acad Sci USA 75: 5860-5863
Gibson QH, Hastings JW (1962) Biochem J 68:368-377
Gibson QH, Swoboda BEP, Massey V (1964) J Biol Chem 239:3927-3934

Hastings JW, Balny C, Le Peuch C, Douzou P (1973) Proc Natl Acad Sci USA 70:3468-3472
Hemmerich P, Massey V, Fenner H (1977) FEBS Letters 84:5-21
Hemmerich P, Massey V, Michel H, Schug C (1982) Structure and bonding 48:93-123
Jorns MS (1980) In: Yagi K, Yamano T (eds) Flavins and flavoproteins. Univ Park Press, Baltimore, pp 161-172
Kamin H, Lambeth JD (1982) In: Massey V, Williams CH (eds) Flavins and flavoproteins. Elsevier North Holland, Amsterdam, pp 655-666
Kawaguchi A, Tsubotani S, Seyama Y, Yamakawa T, Osumi T, Hashimoto T, Kikuchi T, Ando M, Okuda S (1980) J Biochem 88:1481-1484
Kemal C, Bruice TC (1976) Proc Natl Acad Sci USA 73:995-999
Kemal C, Chan TW, Bruice TC (1977) J Amer Chem Soc 99:7272-7286
Koka P, Lee J (1979) Proc Natl Acad Sci USA 76:3068-3072
Kosower EM (1980) Biochem Biophys Res Commun 92:356-364
Kuhn R, Wagner-Jauregg T (1933) Ber 66:1577-1582
Krauth-Siegel RL (1982) Ph D Thesis, Heidelberg p 44
Kurfürst M, Ghisla S, Hastings JW (1982a) In: Massey V, Williams CH (eds) Flavins and flavoproteins. Elsevier North Holland, Amsterdam, pp 353-358
Kurfürst M, Ghisla S, Presswood R, Hastings JW (1982b) Eur J Biochem 123:355-361
Land EJ, Swallow AJ (1969) Biochemistry 8:2117-2125
La Roche HJ, Kellner M, Günther H, Simon H (1971) Hoppe-Seyler's Z Phys Chem 352: 399-402
Leisman G, Nealson KH (9182) In: Massey V, Williams CH (eds) Flavins and flavoproteins. Elsevier North Holland, Amsterdam, pp 383-386
Light DR, Walsh C (1980) J Biol Chem 255:4264-4277
Lockridge O, Massey V, Sullivan PA (1972) J Biol Chem 247:8097-8106
Loechler EL, Hollocher TC (1975) J Amer Chem Soc 97:3225-3237
Ludwig M, Burnett RM, Darling GD, Jordan SR, Kendall DS, Smith WW (1976) In: Singer TP (ed) Flavins and flavoproteins. Elsevier North Holland, Amsterdam, pp 393-404
Ludwig M, Pattridge KA, Smith WW, Jensen LH, Watenpaugh KD (1982) In: Massey V, Williams CH (eds) Flavins and flavoproteins. Elsevier North Holland, Amsterdam, pp 19-27
Mager HJ, Addink R (1979) Tetrahedron Letters 37:3545-3548
Massey V, Veeger C (1960) Biochim Biophys Acta 40:184-185
Massey V, Gibson QH (1963) Federation Proc 23:18-29
Massey V, Williams CH Jr (1965) J Biol Chem 240:4470-4480
Massey V, Ghisla S (1974) Ann NY Acad Sci 227:446-465
Massey V, Hemmerich P (1975) In: Boyer PD (ed) The enzymes. Academic, New York, Vol 12, pp 191-252
Massey V, Hemmerich P (1980) Biochem Soc Transactions 8:246-257
Massey V, Müller F, Feldberg R, Schuman M, Sullivan PA, Howell LG, Mayhew SG, Matthews RG, Foust GP (1969) J Biol Chem 244:3999-4006
Massey V, Palmer G, Ballou DP (1973) In: King TE, Mason HS, Morrison M (eds) Oxidases and related redox systems. University Park Press, Baltimore, pp 25-49
Massey V, Ghisla S, Moore EG (1979) J Biol Chem 254:9640-9650
Mincey T, Tayrien G, Mildvan AS, Abeles RH (1980) Proc Natl Acad Sci USA 77: 7099-7101
Müller F (1972) Zeitschrift Naturforschung 27b:1023-1026
Müller F, Massey V (1969) J Biol Chem 244:4007-4016
Müller F, Massey V, Heizmann C, Hemmerich P, Lhoste JM, Gould DC (1969) Eur J Biochem 9:392-401
Nanni EJ, Sayer DT, Ball SS, Bruice TC (1981) J Amer Chem Soc 103:2797-2802
Pai EF, Schulz GE (1982) In: Massey V, Williams CH (eds) Flavins and flavoproteins. Elsevier North Holland, Amsterdam, pp 3-10
Platenkamp RJ, Palmer MH, Visser AJ (1980) J Mol Structure 67:45-64
Porter DJT, Voet JB, Bright HJ (1973) J Biol Chem 248:4400-4416
Poulsen LL, Ziegler DM (1979) J Biol Chem 254:6449-6455
Reinsch J, Katz A, Wean J, Aprahamian G, McFarland J (1980) J Biol Chem 255:9093-9097
Ronchi S, Minchiotti L, Galliano M, Curti B, Swenson RP, Williams CH Jr, Massey V (1982) J Biol Chem 257:8824-8834
Ryerson CC, Ballou DP, Walsh C (1982) Biochemistry 21:2644-2655
Searls RL (1960) Federation Proc 19:36
Schopfer LM, Massey V (1980) J Biol Chem 255:5355-5363

Schulz GE, Schirmer RH, Pai EF (1982) J Mol Biol 160:287-308
Silverman RB, Hoffman SJ, Catus WB (1982) In: Massey V, Williams CH (eds) Flavins and flavoproteins. Elsevier North Holland, Amsterdam, pp 213-216
Swenson RP, Williams CH Jr, Massey V (1982) J Biol Chem 257:1937-1944
Swenson RP, Williams CH Jr, Massey V (1983) J Biol Chem 258:497-502
Spector T, Massey V (1972) J Biol Chem 247:5632-5636
Spencer R, Fisher J, Walsh C (1977) Biochemistry 16:3586-3594
Strickland S, Massey V (1973) J Biol Chem 248:2953-2962
Thieme R, Pai EF, Schirmer RH, Schulz GE (1981) J Mol Biol 152:763-782
Thorpe C, Williams CH Jr (1981) Biochemistry 20:1507-1513
Tu S (1979) Biochemistry 18:5940-5945
Tu S (1982) J Biol Chem 257:3719-3725
Walsh C (1979) Enzymatic reaction mechanisms. WH Freeman, San Francisco, Chapters 11 and 12
Walsh C (1980) Acc Chem Research 13:148-155
Walsh C, Schonbrunn A, Abeles RH (1971) J Biol Chem 246:6855-6866
Walsh C, Krodel E, Massey V, Abeles RH (1973a) J Biol Chem 248:1946-1955
Walsh C, Lockridge O, Massey V, Abeles RH (1973b) J Biol Chem 248:7049-7054
Warburg O, Christian W (1933) Biochem Z 266:377-414
Wessiak A, Trout GE, Hemmerich P (1980) Tetrahedron Letters 21:739-742
Wenz A, Ghisla S, Thorpe C (1982) In: Massey V, Williams CH (eds) Flavins and flavoproteins. Elsevier North Holland, Amsterdam, pp 605-608
Wierenga RK, de Jong RJ, Kalk KH, Hol WGJ, Drenth J (1979) J Mol Biol 131:55-73
Wierenga RK, Kalk KH, van der Laan JM, Drenth J, Hofsteenge J, Weisser WJ, Jeker PA, Beintema JJ, Müller F, van Berkel WJH (1982) In: Massey V, Williams CH (eds) Flavins and flavoproteins. Elsevier North Holland, Amsterdam, pp 11-18
Williams CH Jr (1976) In: Boyer PH (ed) The enzymes, 3rd ed Vol 13. Academic, New York, pp 89-172
Zanetti G, Williams CH Jr (1967) J Biol Chem 240:5232-5236

Mechanistic Studies on Mercuric Ion Reductase and Cyclohexanone Oxygenase: Pharmacologic and Toxicologic Aspects

C. Walsh, B. Branchaud, B. Fox, and J. Latham[1]

Introduction

Since the initial detection of a riboflavin-based coenzyme, FMN, as the prosthetic group in Old Yellow Enzyme by Warburg and his colleagues some 50 years ago, a vast amount of biochemical and chemical information has accumulated on the redox roles of flavins in biological systems. Flavoenzymes are unmatched in their redox versatility and in the diversity of substrate structural types oxidized and reduced, reflecting the finely balanced chemical pathways for electron flow in and out of the flavins' tricyclic isoalloxazine ring system.

This laboratory has been involved during the last decade in examination of flavoenzyme-catalyzed reactions where the substrate processing has pharmacological or toxicological significance. In particular, several flavoenzymes which activate molecular oxygen for its reductive metabolism are of interest both in analysis of how specific substrates become oxygenated during such catalyses and because of the potential toxic consequences of reduced oxygen metabolites. These studies have included oxygenation of aromatic compounds in bacterial catabolic pathways (Dagley 1978), drug sulfoxidation pathways as in metabolism of the antiarthritic drug sulindac (CLINORIL) (Light, Waxman, Walsh 1982), and the enzyme system responsible for the killing burst of superoxide ions released by human white blood cells when they meet up with invading bacteria (Light et al. 1981). One flavoprotein monooxygenase of recent special study has been a bacterial cyclohexanone monooxygenase (Ryerson, Ballou, Walsh 1982), the first purified enzyme known to carry out a Baeyer-Villiger type oxygen insertion, here shown in conversion of cyclohexanone to ring expanded, oxygen-inserted lactone ε-caprolactone (Eq. 1). This is one of the enzymes discussed in this paper.

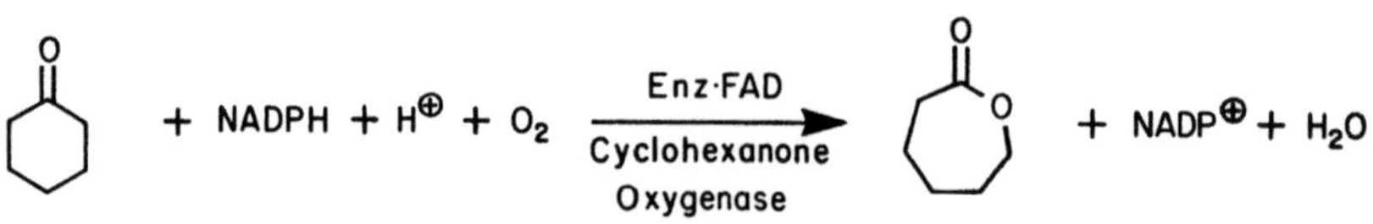

(1)

A second flavoenzyme under current study in this laboratory is one with obvious toxicological significance, specifically the pseudomonal mercuric ion reductase, a key component of the mercurial detoxification system found in mercury-resistant bacteria (Summers and Silver 1978). We have studied the enzyme encoded on transposon Tn501, found on plasmid pVS1, carried by *Pseudomonas aeruginosa* PAO9501. This enzyme catalyzes the NADPH-dependent reduction of Hg^{II} to elemental mercury,

1 Departments of Chemistry and Biology, Massachusetts Institute of Technology, Cambridge, Massachusetts 02139, USA

34. Colloquium - Mosbach 1983
Biological Oxidations

Hg^{O}, which, with high vapor pressure, volatilizes efficiently out of the bacterial cell (Eq. 2). This is the second topic of this paper.

$$H^{+} + NADPH + RS\text{-}Hg^{II}\text{-}SR \xrightarrow{E\text{-}FAD} NADP + 2RSH + Hg^{O} \quad . \qquad (2)$$

Bacterial Cyclohexanone Oxygenase

Cyclohexanone oxygenases were purified and characterized from Acinetobacter and Nocardia species by Trudgill and colleagues (Donoghue et al. 1976) and attracted our interest because the ketone to lactone conversion bears some formal analogy to the bacterial luciferase conversion of long chain aldehyde to acid (and blue-green photon), but appeared different from the characteristic bacterial flavoenzyme monooxygenases which typically take phenols to catechols. In particular, we have commented on the fact that if the ketone to lactone conversion is indeed a Baeyer-Villiger sequence, as confirmed by studies where D_4 cyclohexanone yields D_4-lactone (Ryerson et al. 1982) as well as retention of configuration at the migrating carbon center in chiral 2-D_1- and 2-CH_3-cyclohexanones (Schwab 1981), then one can portray the oxygen transfer sequence as follows. The distal oxygen transferred from a putative flavin-4a-OOH intermediate would behave as nucleophile towards substrate's electrophilic carbonyl carbon. The resultant adduct would decompose toward product with C-C bond migration and expulsion of 4a-hydroxyflavin (Eqs. 3 and 4).

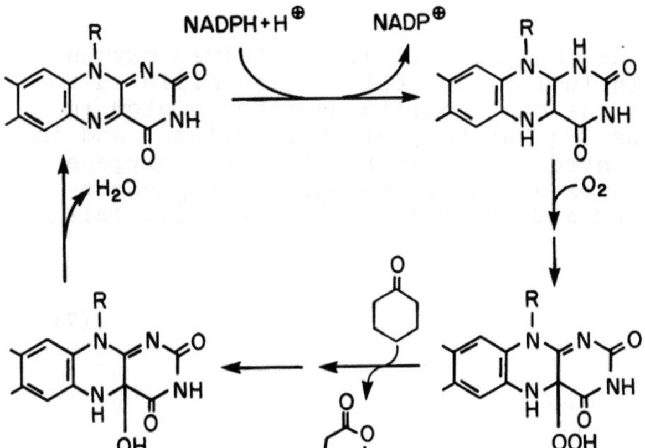

(3)

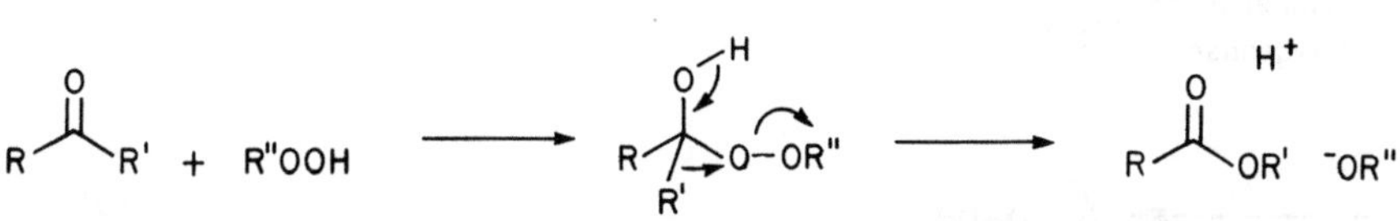

(4)

In contrast, the well-studied phenol to catechol flavoenzyme hydroxylases are generally interpreted as delivering an electrophilic oxygen equivalent (from flavin-4a-OOH) to a nucleophilic substrate site, a clear reversal of polarity from the above enzyme.

(5)

In both flavoenzyme monooxygenase types, fast kinetic studies using single turnover analyses have revealed formation of enzyme-bound FAD-4a-OOH species as kinetically competent intermediates (Ryerson et al. 1982; Entsch et al. 1976).

We have been pursuing the question of ambident reactivity of flavin 4a-OOH as biological oxygen donor further with Acinetobacter cyclohexanone oxygenase. In particular, we have observed this enzyme has a remarkably broad capacity to carry out oxygen transfer with saturation kinetics on a wide range of substrate functional group types. We have shown for example that the six membered ring cyclic sulfide, thiane, is converted to the S-oxide (Eq. 6) at 85% the Vmax of cyclohexanone, a reaction likely to be initiated by nucleophilic attack of sulfur on the flavin-4a-OOH as electrophile, demonstrating ambident reactivity for the first time in a single flavoenzyme monooxygenase (Ryerson et al. 1982). Further with p-tolylethylsulfide, S-oxidation is stereoselective with a 4:1 ratio of S(-)-sulfoxide to R-(+) sulfoxide product (Light et al. 1982).

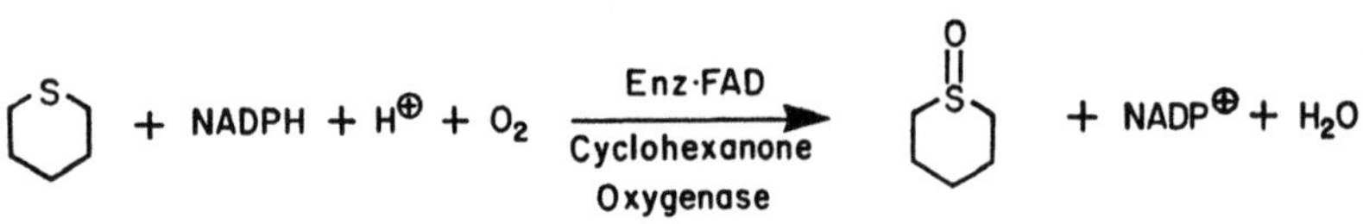

(6)

We have now observed that the pure enzyme will also catalyze oxygen transfer to other heteroatoms including the still nucleophilic sulfur of thiane sulfoxide (Eq. 7), to the nucleophilic phosphorus atom in triethyl phosphite, to selenides (Eq. 8) to yield selenoxides, and in a novel and previously unprecedented enzymatic reaction the oxygenative rearrangement of arylboronates and alkyl boronates to phenols and alcohols, respectively (Branchaud and Walsh 1983 (Eq. 9)). This

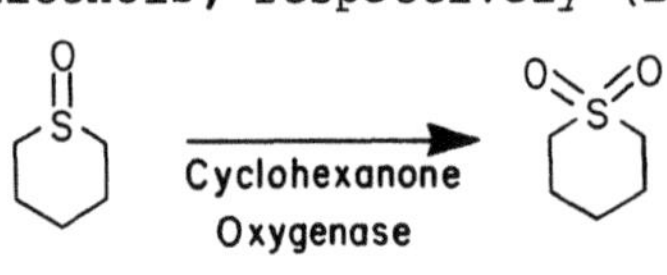

(7)

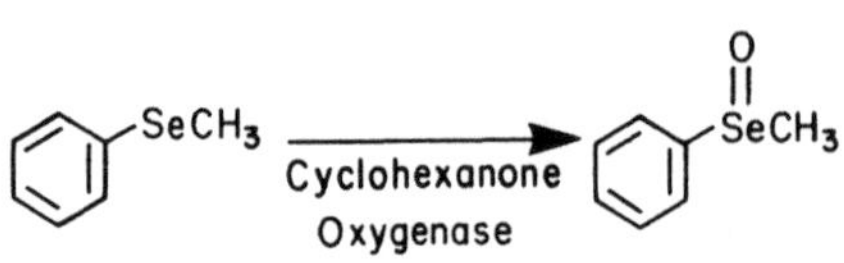

(8)

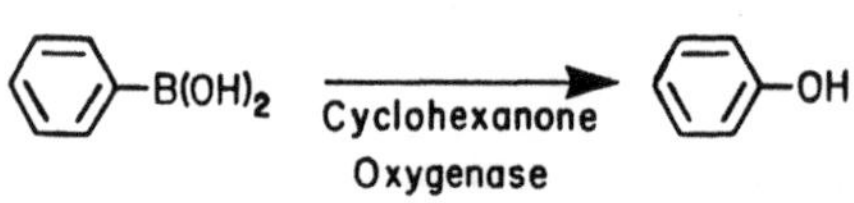

(9)

last category may be formulated as initial nucleophilic attack of flavin-4a-OOH on the boron atom followed by migration of the carbon substituent to oxygen with O-O bond scission to yield the net result of oxygen insertion into the carbon-boron bond (Eq. 10).

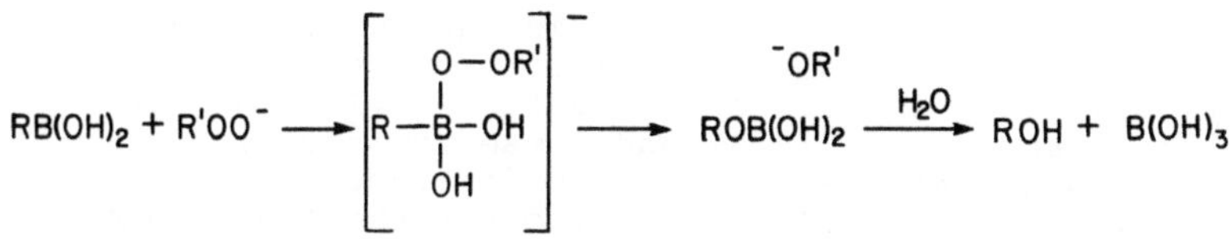

(10)

The borate ester products are hydrolytically unstable. All of these substrates show Michaelis-Menten kinetics. This panoply of oxygenations reactions, some requiring formal "electrophilic" oxygen delivery and others requiring formal "nucleophilic" oxygen delivery is mirrored by the chemistry of various organic hydroperoxides for which such a mechanistic dichotomy is unexceptional (Ruggero and Edwards 1970; Hiatt 1971; Pfesnicar 1978). Coupling these enzymic data with Bruice's model chemistry, demonstrating that N^5-alkyl-4a-OOH flavins are indeed 10^5-fold more reactive than t-butyl hydroperoxide in S- and N-oxygenations (Bruice 1980), leads us to conclude that FAD-4-a-OOH at the enzyme's active site is a sufficiently reactive ambident oxygenation catalyst that no more active biological "oxygen gun" structure need be invoked to account for flavoenzymes as oxygenation catalysts (Branchaud and Walsh 1983 (Eq. 11)).

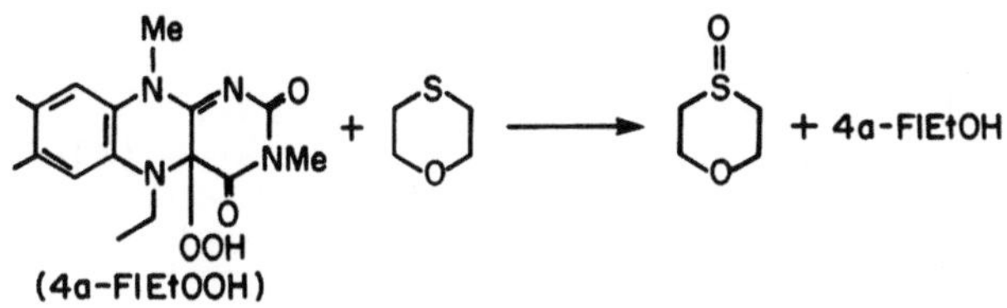

(11)

Hydroperoxide	Krel
+ OOH	1
H_2O_2	24
4a-FlEtOOH	180,000

Returning to the ability of cyclohexanone oxygenase to carry out either sulfur oxygenation or oxygen insertion at ketone sites, we have tested thiolactones, such as the 5-ring and 6-ring thiolactones, both as substrates and mechanism-based inactivators (Eq. 12). Both hopes were realized with each substrate.

(12)

$\frac{k_{cat}}{k_{inact}} \approx \frac{20}{1}$ $\frac{k_{cat}}{k_{inact}} \approx \frac{50}{1}$

Oxygen consumption in competition with enzyme inactivation indicates partition ratios of 50/1 and 20/1 as noted, reflecting a 2% and 5% probability, respectively, that in any catalytic cycle enzymic processing will result in autoinactivation (Latham, Branchaud, and Walsh, unpublished observations). Preliminary evaluation shows essentially stoichiometric labeling of enzyme with S^{35}-γ-thiolbutyrolactone and that it is the apoprotein, not the flavin cofactor, that has been modified. While studies continue, it is reasonable that enzymic S-oxygenation may occur, converting thiolactones into product acylsulfoxides (Eq. 13). Acylsulfoxides appear to be reactive acylating agents (Mirato et al. 1977; Barton et al. 1975) and these may covalently modify some amino acid residue at the active site in kinetic competition with release (Eq. 14). It is possible that analogous chemical sequences account for the toxicity of thiocarbamate herbicides on S-oxygenation by liver microsomal cytochrome P_{450} monoxygenases (Schuphan et al. 1979; Menard et al. 1979) and the known autodestructive behavior of adrenal and testicular P_{450}s in processing the diuretic spirolactone. Thus, the strategy of using thiolactones or thiolester groups with oxygenases may have general applicability in mechanism-based inactivator design.

Possible Mechanism for Inactivation
Direct Sulfur Oxidation Generates Acylsulfoxide

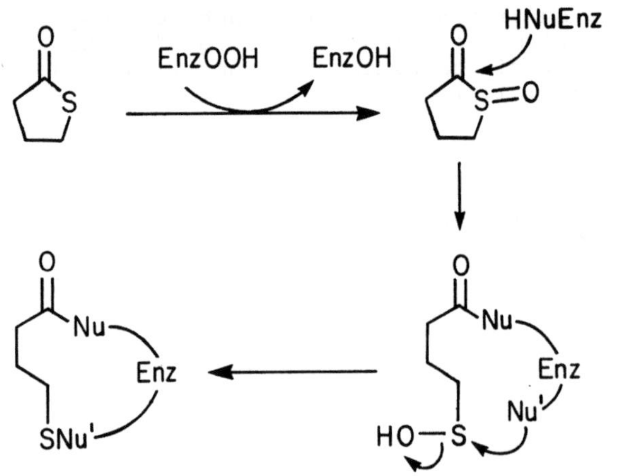

(13)

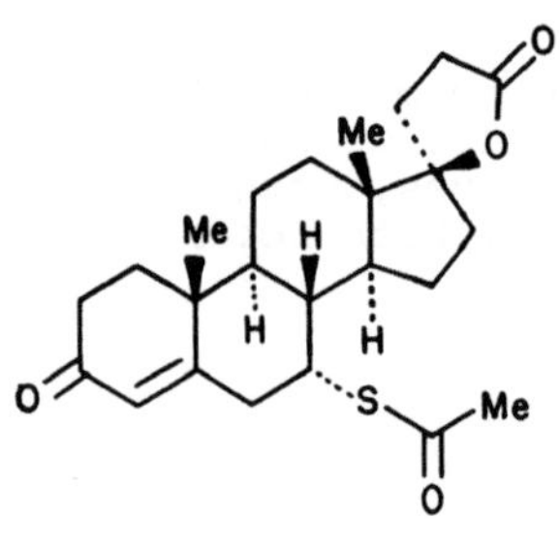

Spironolactone

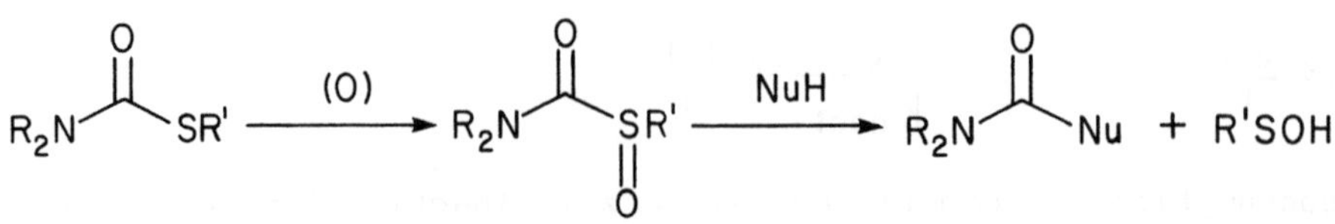

(14)

Mercuric Ion Reductase

We initiated study of the bacterial enzyme mercuric ion reductase for several reasons: because it is the molecular basis for bacterial resistance to inorganic mercurial salts and so of toxicological significance; because it is the only enzyme known to be able to reduce Hg^{II} catalytically (at a rate of ~1000 molecules per minute per molecule of enzyme), in contrast to most other proteins which bind mercuric ions in sulfhydryl chelation and become inhibited; because earlier work had identified mercuric ion reductase as an FAD-enzyme, suggesting Hg^{II} as a specific dihydroflavin reoxidant in this enzyme (Schottel 1978).

From the Tn501-encoded mer A gene on pVS1 in *P. aeruginosa* induction with merbromin causes greater than 1000-fold selective production of mercuric ion reductase. We have developed a two step (65° heat step, Orange A matrex affinity column) high yield (80%) purification (Fox and Walsh 1982). We estimate 1 g of enzyme would be available from 160 g of grams of cells. Elution of enzyme from the Orange A matrex column with NADPH yielded a green enzyme solution, reflective of a nicotinamide-enzyme charge transfer complex. This led us to examine the UV-visible spectrum of the enzyme-FAD complex upon reduction.

As we have recently detailed (Fox and Walsh 1982), the enzyme-FAD complex showed the ability to take up four electrons before generation of an $FADH_2$-type spectrum, suggesting an additional two electron redox component. Furthermore the two electron-reduced enzymes had the optical spectrum of a thiolate-oxidized FAD species, reminiscent of lipoamide dehydrogenase and glutathione reductase (Williams 1976). These latter two flavoproteins are prototypic of a small class of flavoenzymes with a redox active cystine disulfide at the active site (Eq. 15). Indeed, subsequent thiol titration of oxidized and two electron-reduced mercuric reductase with DTNB revealed an additional two sulfhydryls in the EH_2 form of the enzyme. At this point it seemed mercuric reductase was structurally a new member of the disulfide-reducing class of flavoenzymes, despite the fact that it displays no catalytic activity towards lipoamide, oxidized glutathione, or thioredoxin and they in turn no mercuric ion reduction capacity.

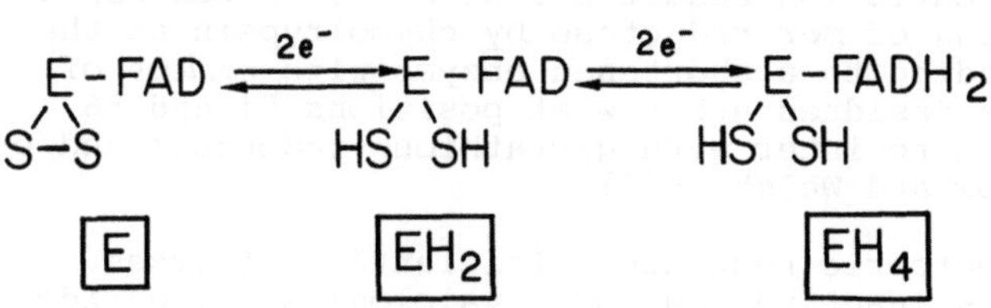

(15)

At this point, we began to examine the extent of the structural similarities between the enzymes. We turned to iodoacetamide as a potentially specific active site alkylating agent based on the work of Williams and his colleagues on this reagent with the disulfide reductases. They determined that oxidized enzyme was insensitive to alkylating agents at low molar ratios to enzyme, while EH_2 was dramatically sensitive. One label per mole of enzyme was incorporated and led to complete inactivation (Thorpe and Williams 1976; Arscott et al. 1981). With [^{14}C]-iodoacetamide, the radiolabeled tryptic peptide from either glutathione reductase or lipoamide dehydrogenase on isolation was found to contain the active site pair of cysteines and the amino proximal cysteine of the pair was selectively labeled on inactivation in EH_2. The other cysteine, five residues along in the sequence, was protected either by its charge transfer interaction with the FAD or by inaccessibility. The recent solution of the human enthrocyte glutathione reductase X-ray structure at 2.0 Å resolution has clearly revealed this relative orientation of the active site cystine disulfide and FAD in the oxidized form of the enzyme (Thieme et al. 1981).

Our studies with [^{14}C]-iodoacetamide on mercuric ion reductase prove the structural similarity hypothesis. Only EH_2 is inactivated with molar stoichiometry and the major tryptic peptide has been isolated and sequenced. This active site peptide has two Cys residues with 12 out of 13 residues identical to glutathione reductase, including the invariant positioning and spacing of the Cys-Val-Asn-Val-Gly-Cys sequence. Furthermore, it is again the proximal cysteine that is predominantly alkylated, here with an 18/1 ratio of proximal/distal cysteine modification. The monoalkylated enzyme also retains a thiolate-FAD charge transfer complex indicating that the distal cysteine is the one in close contact with the FAD at the active site (Fox and Walsh 1983).

							T7					T12					[^{14}C]-Iodoacetamide Labeling (T7/T12)
Mercuric Reductase	Gly	Thr	Ile	Gly	Gly	Thr	Cys	Val	Asn	Val	Gly	Cys	Val	Pro	Ser	Lys	18/1
Glutathione Reductase	His	Lys	Leu	Gly	Gly	Thr	Cys	Val	Asn	Val	Gly	Cys	Val	Pro	Lys	Lys	8/1
Lipoamide Dehydrogenase	Asn	Thr	Leu	Gly	Gly	Val	Cys	Leu	Asn	Val	Gly	Cys	Ile	Pro	Ser	Lys	13/1

In parallel with the active site peptide sequence work, we have analyzed the amino terminal sequence (somewhat complicated by some in vivo proteolytic clipping which does not affect catalytic activity) with the intent that this would facilitate identification of the start of the mer A gene on Tn501. This sequence was provided to Nigel Brown and his colleagues at Bristol to assist in aligning the mer A gene sequence with the protein sequence. They have just completed the nucleotide sequence (Brown et al. 1983); the gene specifies an encoded protein of 561 amino acids with a molecular weight of 59 K. The active site cystine bridge is at residues 136 and 141 in comparison with residues 58 and 63 in human enthrocyte glutathione reductase (Untucht-Grau et al. 1981) and 45 and 50 in pig heart lipoamide dehydrogenase (C Williams, personal communication), revealing a long N-terminal extension in mercuric ion reductase. However, we can effect specific proteolytic clipping of mer reductase by chymotrypsin at the amide bond after Trp-85 leading to a shortened enzyme with *no loss* of catalytic activity. The Cys residues are now at positions 51 and 56, respectively, more nearly in register with glutathione reductase and/or lipoamide dehydrogenase (Fox and Walsh 1983).

As yet we do not know how mercuric reductase efficiently catalyzes Hg^{II} reduction nor why it is inactive against physiological disulfide substrates. Perhaps relevant to the latter point, the mer A gene sequence (Brown et al. 1983) reveals the absence of an active site histidine equivalent to His467 in glutathione reductase, which is thought to act as general base catalyst in shuttling protons around purposefully in that active site. The mercuric ion reductase, available now in quantity and of known primary sequence seems a good candidate for X-ray crystallography for comparison.

We suspect the mercuric reductase operates catalytically between oxidized (E) and EH_2 states and that EH_4 is a kinetically irrelevant and catalytically incompetent oxidation state in analogy to the disulfide oxidoreductases (Williams 1976). It is likely that NADPH on binding reduces FAD at the active site by hydride transfer to N^5. Then the dihydroflavin rapidly passes electrons to the active site cystine disulfide (likely via a flavin 4a-thiol adduct) producing as predominant form of EH_2, the oxidized FAD, active site dithiolate with Cys-141 in charge transfer to FAD and Cys-136 available for initial chelation with RSHgSR at it diffuses in. Exchange of one thiol ligand to mercuric ion would yield initially a mixed E-Cys-S-Hg-SR species. What occurs next is not yet clear (Eq. 16). If the Cys-41 thiolate then forms the bidentate chelation to Hg^{II} is that catalytically active or is it a dead-end complex as it would be in every other enzyme that binds Hg^{II} by dithiol ligation? If

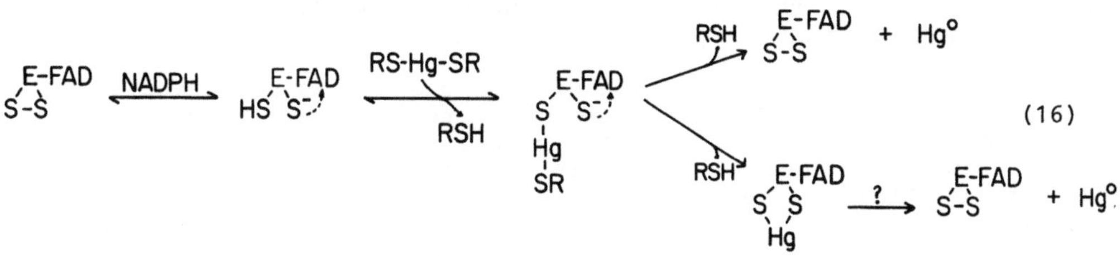

(16)

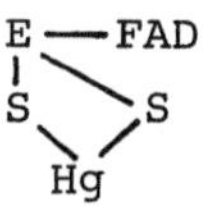

E—FAD forms as the active EH_2 complex, how is the net two electron transfer to mercury effected, by one electron path or by some two electron process? If the mercuric ion stays as monodentate to Cys_{136}, how is electron transfer effected from that complex? We anticipate studies with FAD analogs, such as 5-deazaFAD and fast reaction kinetic studies on native FAD-enzyme, may address some of these questions and begin to unravel the mechanism of action of this unusual flavoprotein catalyst.

Discussion

Massey: The early mechanistic studies of the other flavoprotein disulfide reductases made very successful use of rapid reaction kinetics in the reductive and oxidative half reactions. As you can also prepare a stable half reduced form (EH_2) of mercuric reductase, it would be very instructive to see what happens when you mix it with Hg^{2+} or other mercurials. Have you done such experiments?

Walsh: Such experiments are in progress and we hope to have answers soon as to whether EH_2 is the catalytically competent intermediate.

Ghisla: The enzyme which is eluted from the affinity column with $NADP^+$ is green, has the absorption of the oxidized flavin, contains one equivalent $NADP^+$, and you suggest the green color is due to a charge transfer transition. Now, what should be the donor? Could it be the -S- ($-S^-$) on the other side of the flavin?

Walsh: While there is a stoichiometric amount of NADP bound in the green enzyme form, we do not yet know whether one of the active site thiols is the donor. On dialysis against high salt one obtains free oxidized enzyme not the EH_2 species.

Ullrich: You suggested that epoxidation may be the privilege of iron-oxo-intermediates and not within the capacity of flavin-hydroperoxides.

This is conceivable in view of the radical attack versus polar attack at the FeO$^{\cdot}$ and ROOH, respectively. However, if one would take a more reactive olefin, especially with polarizing groups, I would think that you also would succeed in an epoxidation by flavin-dependent monooxygenases.

Walsh: I agree that we have not pushed the case to the limit with more reactive olefins as potential substrates for this enzyme and that is a good suggestion.

Schirmer: Is Hg^{2+} or the bidentate GS^{-}-Hg^{2}-^{-}SG the "true" substrate of mercuric ion reductase?

Walsh: We believe the mercuric ion dithiol chelate is presented from solution. Whether the Hg^{2+} is handed over to an intervening set of thiols, e.g., Cys-555, Cys-556, on the way to the active site Cys-136, Cys-141 remains to be seen. There is no reduction of the oxidized glutathione. Whether there are binding sites for reduced glutathione remains to be seen.

References

Barton D, Manly D, Widdowson D (1975) J Chem Soc Perkin I:1568-1594
Brauchaud B, Walsh C (1983) J Am Chem Soc (submitted)
Brown NL, Ford SJ, Pridmore RD, Fritzinger DC (1983) Biochemistry (submitted)
Bruice TC (1980) In: Dolphin D, McKenna C, Murakami Y, Tabushi I (eds) "Biomimetic Chemistry" American Chemical Society in Chemistry Series 191:89-118
Dagley S (1978) The Bacteria 6:305-388
Donaghue NA, Norris DB, Trudgill PW (1976) Eur J Biochem 63:175-192
Entsch B, Ballou DP, Massey V (1976) J Biol Chem 251:2550-2563
Fox B, Walsh CT (1982) J Biol Chem 257:2498-2503
Fox B, Walsh CT (1983) Biochemistry (submitted)
Light D, Waxman D, Walsh C (1982) Biochemistry 21:2490-2498
Light D, Walsh C, O'Callaghan A, Goetel E, Tauber A (1981) Biochemistry 20:1468-1476
Menard R, Guenthner T, Kon H, Gillette J (1979) J Biol Chem 254:1726-1731
Mirato H, Kodoma H, Msura T, Kobayashi M (1977) Chem Lett 413-416
Plesnicar B (1978) In: Trahanovsky W (ed) "Oxidation in Organic Chemistry, Part C". Academic, New York, pp 211-296
Ruggero C, Edwards JO (1970) In: Swern D (ed) "Organic Peroxides, Vol I". Wiley, New York, pp 199-264
Ryerson C, Ballou D, Walsh C (1982) Biochemistry 21:2644-2655
Schottel JL (1978) J Biol Chem 253:1341-????
Schuphan I, Rosen J, Casidu J (1979) Science 205:1013-1015
Schwab J (1981) J Am Chem Soc 103:1976-1878
Summers AO, Silver S (1978) Ann Rev Microbial 32:637-672
Thieme R, Pai EF, Schirmer RH, Schulz GE (1981) J Mol Biol 152:763-782
Thorpe C, Williams CH Jr (1976) J Biol Chem 251:3553-3557
Arscott LD, Thorpe C, Williams CH Jr (1981) Biochemistry 20:1513-1520
Untucht-Grau R, Schirmer RH, Schirmer I, Krauth-Siegel RL (1981) Eur J Biochem 120:407-419
Williams CH Jr (1976) In: Boyer PD (ed) The enzymes, 3rd edn Vol 13. Academic, New York, pp 89-173

Dioxygen Binding and Reduction

Oxygenases and Oxidases: Hypothesis

H. S. Mason[1]

I thank Professors Sund and Ullrich for their kind invitation to join this celebration of Otto Warburg's 100th birthday. It is now more than 50 years since he and Negelein established cytochrome a_3 as the "oxygen transporting" or autoxidizable component of the respiratory chain (Warburg and Negelein 1928), laying foundations for the modern structure of bioenergetics (Keilin 1966). Twenty-five years later, a new mode of dioxygen utilization was discovered (Mason et al. 1955; Hayaishi et al. 1955) and the number and variety of known biological dioxygen activation reactions greatly increased. The chemical biology of dioxygen is now very diverse and involves intercession of many dioxygen binding and activating proteins (Keevil and Mason 1978; Mason 1981).

In spite of this diversity, there is an underlying unity of dioxygen biochemistry in that major reaction types, prosthetic groups, and accessory reactions require only a small number of protein elements. These elements are *domains*, compactly folded stable regions of polypeptide chains associated with specific function (Edelman 1970; Liljas and Rossmann 1974; Rossmann et al. 1975, 1981; Wetlaufer 1973). The primary domains associated with dioxygen biochemistry are listed in Table 1 and the primary classes of enzymes activating dioxygen are in Table 2. We will show that these classes are characteristic assemblies of primary functional domains, as suggested in Fig. 1. Each such domain may have its own line of evolutionary descent. We propose that the diversity of dioxygen biochemistry can be well accounted for by mutation, duplication, transposition, and fusion of DNA coding for primary structural domains, as has been postulated for other proteins (e.g., Rossmann and Argos 1981; Doolittle 1981; Zuckerkandl 1975; Dayhoff et al. 1978; Rao and Rossmann 1973; Liljas and Rossmann 1974; Gilbert 1978, 1979; Blake 1978, 1979, 1983).

The concept of domains as genetically determined units of structure and function has come forward very powerfully in the last decade to illuminate the relationship between protein structure, function, and evolution (Edelman 1970; Wetlaufer 1973; Liljas and Rossmann 1974; Rossmann et al. 1975, 1981; Levitt and Chothia 1976; Schulz and Schirmer 1979; Schulz 1981; Crick 1979; Branden 1980; Rashin 1981, Richardsen 1981; Ghelis and Yon 1982; Dickerson and Geis 1983). The domain-gene or domain-exon axis appears to be a fundamental unit of evolution in that it may change spontaneously at the level of gene structure and be tested for fitness at the domain level. The extension of these new developments to the chemical biology of dioxygen (transport, transfer, and reduction) is the purpose of this communication.

1 Department of Biochemistry, Oregon Health Sciences University, Portland, Oregon 97201 USA

34. Colloquium - Mosbach 1983
Biological Oxidations

Table 1. Domain classes associated with dioxygen binding and activating proteins

1. *Dioxygen-binding domains* (O_2-transport, dioxygenation, monoxygenation, oxidase reduction

CU	Nonheme Fe	Fe porphyrin	Flavin
Mononuclear Cu	Mononuclear Fe	Fe protoporphyrin IX	FAD
Type I Cu		Covalently bound Fe porphyrine related to heme c	FMN
Type II Cu		Heme a	8a-Peptidyl flavin
Other		Heme a4	Biopterin
Binuclear Cu	Binuclear Fe	Heme d	
Type III Cu		Other (e.g., myelo-peroxidase)	

2. *Substrate binding domains*

3. *Electron donor binding domains,* for:

NADH	$FADH_2$	2-oxo-glutarate	Fe-S	cytochromes
NADPH	$FMNH_2$	ascorbate tetrahydrobiopterin	Fe (III) substrates	

4. *Electron transferring domains* in dioxygenases, monooxygenases, oxidases

5. *Complementary subunit binding domains*, yielding oliogomeric proteins

6. *Allosteric effector binding domains*, e.g., DPG, CO_2, H^+, other

Table 2. Classes and types of O_2 reactions catalyzed by O_2-binding proteins

I. *O_2 Transporters*: Reversible O_2 binding

II. *Dioxygenases*

1. Both O atoms inserted into acceptor, no reducing equivalents required
2. Reductive dioxygenases. 2-oxoglutarate as electron donor
3. Reductive dioxygenases. NADH or NADPH as electron donor
4. Reductive dioxygenase. Substrate as electron donor

III. *Monoxygenases*

1. Internal type: substrate as donor of electrons
2. External type: NADH or NADPH as electron donor
3. External type: FADH or FMNH as electron donor
4. External type: Fe-S as electron donor
5. External type: Tetrahydrobiopterin as electron donor
6. External type: ascorbate as electron donor

IV. *Oxidases*: O_2 Reduction to O_2^-, H_2O_2, or H_2O.

1. One equivalent donor producing O_2^-
2. One equivalent donor producing $2H_2O$
3. Two equivalent donor producing H_2O_2
4. Two two-equivalent donors producing $2H_2O$

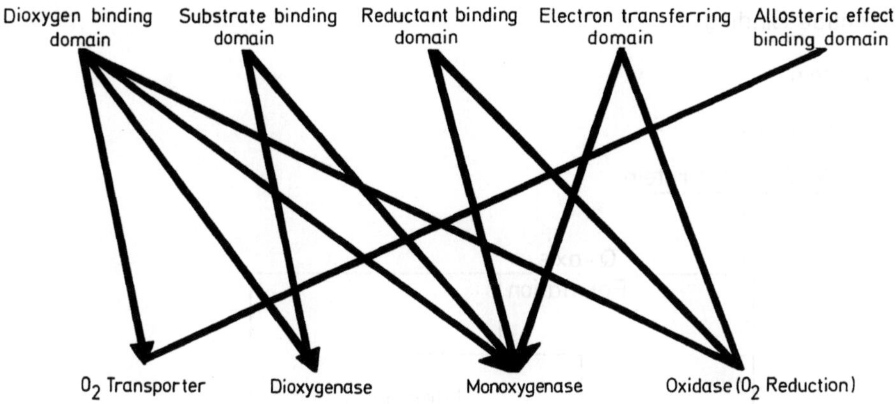

Fig. 1. Five domain classes: O_2-binding, substrate-binding, reductant-binding, electron transferring, and allosteric effector binding compose the principal classes of proteins which interact functionality with dioxygen, dioxygen-transport, dioxygenases, monooxygenases, and oxidases. The *lines* from domain classes to dioxygen-binding protein classes show the domains which must be assembled to produce the functions shown

Domains as Units of Protein Structure

Domains are defined as compactly folded stable segments of polypeptide within a polypeptide chain (Schulz and Schirmer 1979; Branden 1980; Richardson 1981; Rossmann and Argos 1981; Ghelis and Yon 1982). A polypeptide chain may fold into one or more such domains, each having the same or different function.

Rossmann et al. (1975) suggested that *proteins of similar function have similar tertiary structure*. This concept was based in part upon structural similarity of myoglobin to both the α- and β- chains of hemoglobin and conservation of cytochrome c structure over long evolutionary periods. The concept was extended to domains: i.e., *that domains with similar function should have similar structure*: "The structures of those parts of the dehydrogenase polypeptides whose function is to bind NAD^+ coenzymes are remarkably alike. The remainder of the polypeptide chains are required for substrate binding, catalysis, specificity, and formation of oligomeric structures. Thus, sophisticated dehydrogenases are mostly constructed out of a variety of structural domains which represent gene fusion of proteins of simple function". Similarity among NAD^+-binding domains was demonstrated among lactate dehydrogenase, liver alcohol dehydrogenase, s-malate dehydrogenases, and glyceraldehyde dehydrogenase. Further comment on NAD^+ and $NADP^+$ binding domains is found in the paper by Rossmann in this colloquium. Postulated evolutionary relationships among domains which bind nucleotide coenzymes are shown in Fig. 2.

The existence of domains is deduced from models of protein structure derived from X-ray diffraction crystallography by inspection and experience, or more positively by mathematical analysis of amino acid contiguity within the model protein structure. These analyses take the form of "neighborhood correlation plots" (Schulz and Schirmer 1979) or "interatomic distance matrices" (Nishikawa et al. 1972, 1974; Ooi and Nishikawa 1973), contour plots in which interatomic distances

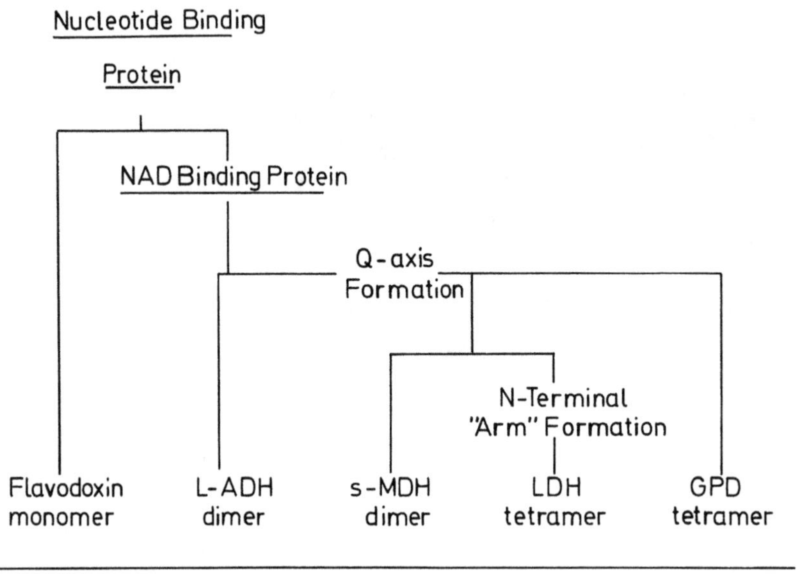

Fig. 2. Evolutionary development of the nucleotide binding domains. (Liljas and Rossmann 1974)

between C^{α} atoms are plotted as a matrix against amino acid sequence positions. A high degree of resolution is possible, e.g., the exact location of three and four domains in hemoproteins by Gō (1981).

Supporting evidence for domains has been obtained by selective proteolysis of polypeptide chains at "hinges" between domains when the hinge contains an appropriate peptide bond substrate for the protease being used. The split products may show characteristic structure or function and so identify the domain functionally; see for example, Craik et al. 1981 (hemoglobin); Porter 1959, and Edelman 1970 (immunoglobulins), Guiard and Lederer 1979 (flavocytochrome b_2, liver sulfite oxidase); van Holde and Miller 1982 (review of hemocyanin $(Cu)_2$ domains); Rossmann and Argos 1981 (review). Function can be demonstrated chemically or biochemically by specific high binding affinities toward prosthetic groups, cofactors, and substrates.

Guiard and Lederer (1979) compared flavocytochrome b_2, liver sulfite oxidase, and microsomal cytochrome b_5 using the selective proteolysis method, amino acid sequences, and spectral properties of heme-binding fragments. They concluded that the three heme-binding proteins are products of a divergent evolution from a common ancestor and that they must present a basically similar backbone called the "cytochrome b_5 fold". Their results also suggested that at some point in evolution several copies of an ancestor hemoprotein were formed in the cellular genome, that one copy was subsequently fused with the gene for flavin reduction in yeast cells (b_2) or a molybdoreductase (sulfite oxidase) in hepatic cells. Argos and Rossmann (1979) showed that a reasonable similarity also exists between the cytochrome c and globin folds and between them and the cytochrome b_5 fold (Fig. 3), but they comment that "the recognition of structural similarity must not be taken as sufficient evidence for divergent evolution from a common ancestor. Rather, it may be indicative of convergence to an energetically fa-

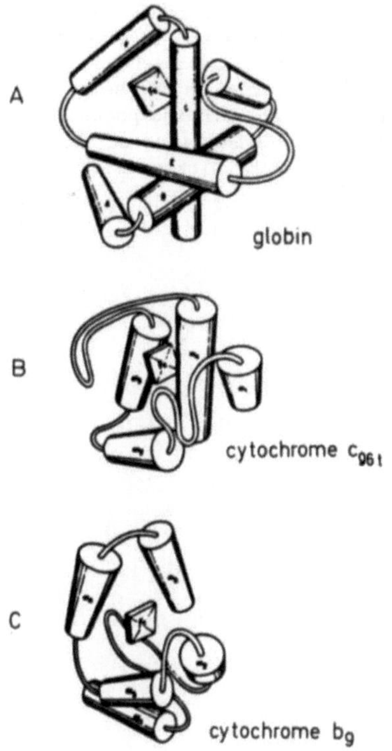

Fig. 3. Diagrammatic representation of the similar polypeptide topology in (a) hemoglobin beta; (b) cytochrome C_{551}, and (c) cytochrome b_5. (Argos and Rossmann 1979)

vorable protein fold suitable for particular functions". On the other hand, b-type and c-type cytochromes do not bind O_2 and almost certainly descend from ancestor proteins of the pre-aerobic period (Dickerson and Timkovich 1975; Almassy and Dickerson 1978) functioning in nonaerobic electron transfer systems. Their reduced forms are stable in the presence of O_2, but may be readily modified to O_2-binding and autoxidizable proteins. The genes for b- and c-type cytochromes were already ubiquitous during the evolution of globins and on these grounds divergent evolution of globins from pre-aerobic b-type cytochromes remains a good hypothesis. In any case, the decision whether or not similarity of amino acid sequences and tertiary structure among different proteins is due to chance or to common ancestry proves to be a very sophisticated exercise in probabilities (Doolittle 1981; Dayhoff, Barker, and Hunt 1983).

Occurrence of substantial amino acid sequence homology between two proteins which share a function, for example NAD^+-binding or heme-binding, suggests that they share a domain type. A well-studied example is the heme-binding domains of hemoglobins and myoglobins. These proteins are widely distributed in nature and all heme pockets in them have a similar configuration (Dayhoff 1972, 1973, 1976, 1978; Love et al. 1972; Thompson 1980; Dickerson and Geis 1983; Fig. 4). Nevertheless, the heme-binding domains of these globins may show only a few identically positioned specific amino acids, such as the proximal

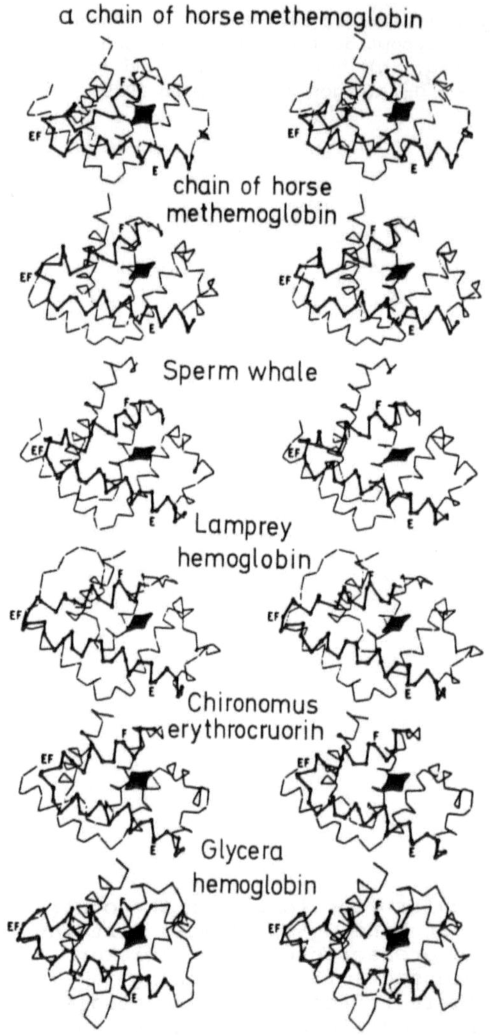

Fig. 4. Polypeptide backbone chain folding in vertebrate and invertebrate globins. The heme is represented by a *puckered black square* in the heme pocket and the heme-binding pocket by the *heavy black lines*. (Love et al. 1972, as modified by Dickerson and Geis 1983)

histidine residue as axial ligand to the heme Fe. Dickerson and Geis (1983) comment that amino acid homology is too restrictive a basis for establishing relationships from the functional point of view. Many positions may be substituted by functionally equivalent amino acid residues, although the conformation of the domain is essentially unchanged. Lesk and Chothia (1980) have examined with the globins the question, how do different amino acid sequences determine similar structures? They conclude that "the principal determinants of three-dimensional structure of these proteins are the residues involved in helix to helix and helix to heme packings. Half of these residues are buried within the protein. The variations in sequence keep the buried residues non-polar, but do not maintain their size, leading to changes in the geometry of the helix packings. These shifts are coupled so that the geometry of the residues forming the heme pocket is very similar in all globins". But it must be emphasized that variations in amino acid sequences may also be the origin of highly adaptive species specific modulations in function.

The tertiary structures of domains fall into clear-cut classes: (1) α-proteins which are mainly alpha-helical; (2) β-proteins which contain antiparallel beta strands; and (3) α/β proteins which have a central core consisting of a sheet of strands (Levitt and Chothia 1976; Sternberg et al. 1977; Branden 1980; Richardson 1981). Branden showed that substrate or cofactor binding occurs in crevices at the carboxyl end of the parallel strands of the sheet in α/β proteins and suggested that α/β structures are topologically favorable for forming binding crevices at the carboxyl end of the sheet. Rossmann and Argos (1981) reviewed the developments which led to protein structure taxonomies – a major forward step in protein chemistry – and extended previous classifications by indicating functional correlates. These showed surprising relationships among proteins hitherto considered unrelated, strengthening a sense of orderliness at the level of domain structure in the process of evolution yet to be explained. In a number of cases, domains repeat themselves along the polypeptide chain, implicating gene duplication. The O_2-transporting protein hemerythrin contains two symmetrical Fe-binding domains indicating gene duplication. The following are also candidates for gene duplication among the O_2 transporting, transferring, and reducing proteins: the $(Cu)_2$ site of hemocyanins and tyrosinase; the Type III $(Cu)_2$ site in laccase, ascorbate oxidase, and ceruloplasmin; the multiheme proteins (c_3); the diflavoreductases and oxidases; and the Cu_A cytochrome a Cu_B - cytochrome a_3 character of cytochrome c oxidases (see Addendum).

Mammalian cytochrome c oxidase contains two heme a and two Cu prosthetic groups among seven or more distinct subunits comprising the enzyme (and among two subunits comprising bacterial cytochrome c oxidase; Mason 1982). It has been reported that subunits I and II each carry a molecule of heme a (Mason 1982). It is a reasonable hypothesis that the two heme a binding domains originated by duplication from an ancestor protein carrying a single heme a. It would be reasonable to expect some evidence of it among subunits I and II just as Steffens et al. (1978) has observed homology between subunit II and the blue copper proteins. On the basis of this hypothesis, Dayhoff compared the sequences of human cytochrome c oxidase subunits (mol. wt. 57 000) and II (mol. wt. 24 500) by the RELATE and ALIGN procedures. She found 42 identities out of a possible 210 matches in the two fragments around subunit I residue 178 and subunit II residue 57 and she commented that "whatever the mechanism, either an evolutionary divergence or a transfer of a bit of genetic material in the distant past, these pieces are much more similar than you would expect by chance". This does not exclude the possibility that they *do* occur by chance nor the possibility that the homology has no relationship to heme a binding domains (Dayhoff et al. 1983a).

A classification of some Cu, Fe, heme, and flavin domains related to oxygenases and oxidases has been abstracted from the general Table 3 of Rossmann and Argos (1981) as Table 3 of this article. It demonstrates that the hypothesis that domains are involved in the structure and function of dioxygen-binding and activating proteins is likely to be generally applicable.

The relationship among these prosthetic groups (domains) and the kinds of interaction reaction with dioxygen that they catalyze is shown in Table 4 (Mason 1981). An important generalization is apparent: *every active site class (domain class) (Cu, Fe, heme, flavin) catalyzes all classes of substrate reactions with* O_2, except that no dioxygen-transport protein

Table 3. Taxonomy of polypeptide domains related to oxygenases and oxidases (adapted from Rossman and Argos 1981, Table 3)

I. *All alpha structures*

A. Heme binding proteins-Type A. Oxygen and electron carriers.
1. Globins: alpha chain of hemoglobin, beta chain of hemoglobin; single chain hemoglobins in worms, lamprey, leghemoglobin, myoglobin.
2. Cytochrome b_5 and related structures, such as domains in cytochrome b_2 and sulfite oxidases.
3. Cytochromes c, c_2, c_{550}, c_{551}, c_{555}.

B. Four-helical protein domains.
1. Oxygen carriers without a heme group, but using Fe: hemerythrin, myohemerythrin.
2. Heme-binding proteins-Type B. Electron carriers, cytochromes b_{562}, c'.

II. *All beta-structures*

A. Single sheets of antiparallel strands: second domain of p-hydroxygenzoate hydroxylase (FAD flavoprotein).

B. Greek key beta-barrels with 6-8 strands.
3. "Immunoglobulin domains": 3b. superoxide dismutase
3c. plastocyanin (Type I Cu)
3d. azurin (Type I Cu)

III. *Alpha/beta structures*

A. Domains that bind nucleotides toward the carboxyl end of a predominantly parallel b-pleated sheet.
1. NAD binding domains in dehydrogenases; LDH, MH, LADH, G3PD.
2. NADP-dependent dehydrogenases.
3. Flavin-binding domains; flavodoxin.

B. Structures based on a single primarily parallel beta-pleated sheet with active site toward the carboxyl end of the sheet, that do not bind nucleotides; thioredoxin, glutathione peroxidase.

C. Structures with eight successive 1 x crossovers that form a barrel: glycolate oxidase.

IV. *Small alpha + beta proteins*

A. Fe-S proteins: rubredoxin, both domains of ferredoxin, HPIP.

B. Multiheme cytochrome c_3.

Table 4. Distribution of prosthetic group classes among the reaction classes of the dioxygen-binding and activating proteins (Mason 1981)

Reaction class	Prosthetic group class						
	Cu	Cu_2	Fe	Fe_2	Heme	Flavin	Not known
A. Oxygen-transporting protein		1		1	5	0	
B. Dioxygenases	2	1	35		3	2	8
Reductive dioxygenases			(7)				
C. Mixed function oxidases	3	1	9		21	24	27
Internal source of $2e^-$	(2)	(1)	(1)		(0)	(4)	
External source of $2e^-$	(1)		(8)		(21)	(20)	
D. Electron transfer to O_2	12	3	4		9	33	16
Total	17	6	48	1	38	59	51

based on flavin as the prosthetic group has been found. (In principle it is possible because the law of spin conservation would be obeyed if the complex $HFl-O_2H$ were a triplet having two unpaired electrons.) Copper-domains and flavo-domains are favored for oxidases (electron or hydride ion transfer), and Fe, heme, and flavin-domains for monooxygenation and dioxygenation, including Fe for reductive dioxygenation, when 2-oxo-glutarate is the electron donor. However, the significance of this distribution is certainly skewed by the narrow choice of organisms studied by investigators in this field, e.g., *Pseudomonas* sp. is a popular source of oxygenases. In any case, the prosthetic group domains contain redox active systems capable of transferring electronic charge to dioxygen and so binding and activating it (Keevil and Mason 1978; Malmström 1982). The structures of the oxygenated-domains as far as known seem to conform to generalizations drawn by Vaska (1976) and Valentine (1973) that O_2 exhibits or approaches one or the other discrete states, superoxo or peroxo, when bound to a transition element (or reduced flavin). Since each prosthetic group class or domain is capable of catalyzing the same range of dioxygen reactions as the others do, *functional convergence* must have taken place if a single ancestor polypeptide was their evolutionary origin. Within a prosthetic group class, a whole range of reactions of dioxygen is catalyzed (transport, dioxygenation, monooxygenation, reduction). *Functional divergence* must have taken place.

Domain Structure and Function Determined by Gene Structure and Function

We have proposed that the diversity of dioxygen biochemistry can be accounted for by mutation, duplication, transposition, and fusion of a relatively small number of genes coding for primary structural domains in O_2-binding and activating proteins. This may occur in prokaryotes in which genes are whole or in eukaryotes in which genes are split into expressed and interspersed noncoding fragments (exons and introns, respectively). Substantial evidence has been obtained that exons code for domains in collagens, immunoglobulins, alpha-fetoproteins, hen and T4 lysozyme, and beta-hemoglobin (Blake 1978, 1979, 1983). The case of hemoglobin is particularly interesting in the context of O_2-binding and activating proteins because hemoglobin is related through its heme to O_2-activating enzymes (tryptophan dioxygenase, indoleamine dioxygenase, cytochromes P-450, cytochrome c oxidases, peroxidases, and catalase; Keevil and Mason 1978). Since the evolution of hemoglobin is relatively well-traced (Dayhoff 1972, 1978; Dickerson and Geis 1983), it should give important clues to the biological history of heme-containing oxidases and oxygenases.

Figure 5 illustrates the organization of a split gene similar to that for human beta-hemoglobin (Crick 1979). The gene has three exons marked 1, 2, and 3 and two introns marked A and B. There are no sequences in the mRNA (bottom lines) corresponding to those in the two introns. Gilbert proposed in 1979 that adaptively advantageous proteins might be produced by recombination of exons and introns. Blake suggested that exons might code for domains. They both recognized that the central exon of the hemoglobin gene (Fig. 6) coded for 16 of the 20 residues in contact with the heme of hemoglobin, i.e., the heme-binding pocket. Eaton (1980) then analyzed specific residue dependence of function in alpha- and beta-hemoglobin subunits (Table 5), based upon studies of abnormal hemoglobins and concluded that heme contacts are concentrated in the domain coded by the central exon; nearly all the contacts between the α_1 and β_2 subunits are coded by

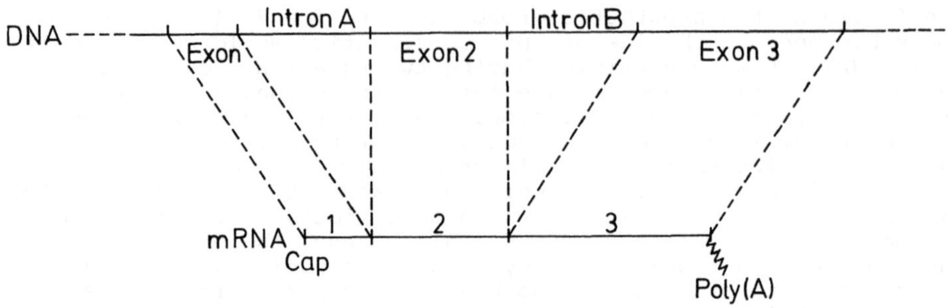

Fig. 5. A split gene similar to that for human beta-hemoglobin (Crick 1979). The *top horizontal line* represents a stretch of DNA in the genome the *bottom line* the mRNA produced from it. In this sequence, the gene has three exons marked 1, 2, and 3 and two introns (*intervening sequences*) lettered A and B. There are no sequences in the mRNA corresponding to those in the two introns

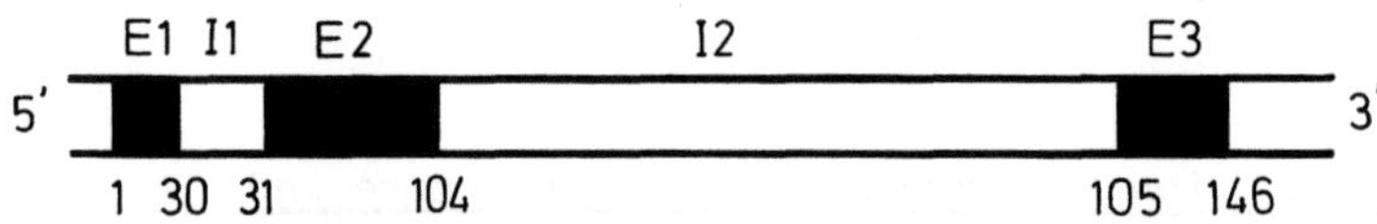

Fig. 6. Diagrammatic representation of the beta globin intron/exon structure. The black boxes E1, E2, E3 represent expressed DNA sequences (exons). The *white boxes* represent intervening sequences (introns). E2 represents the central exon that codes for heme-binding polypeptide, consisting of amino acid residues 31-104 (Craik, Buchman, and Beychok 1981)

Table 5. Summary of the locations on the gene for alpha and beta globin of amino acid residues supporting the functions of hemoglobin (Adapted from Eaton 1980)

Function	*Gene site*
Heme binding	Second exon
$\alpha_1\beta_2$ contacts	Second exon
$\alpha_1\beta_1$ contacts	Third exon
Bohr effect regulation of oxygenation	First and third exons
DPG regulation of oxygenation	First and third exons

the same exon, while those between α_1 and β_1 are coded by the third exon. Residues regulating oxygenation through the Bohr effect and the allosteric regulator diphosphoglycerate are coded predominantly by the first and third exons.

Subsequently, Craik et al. (1981) confirmed the location of the heme-binding function on the polypeptide coded by the central exon by cutting it out with proteolysis and studying its chemical and physical properties and its reconstitution with heme. The system of three recombined domains plus the corresponding complementary hemoglobin subunit reconstitutes the reversible O_2-binding function. The reconstituted heme-peptide alone did not have this property. The second exon of the hemoglobin gene thus codes for the heme-binding property, but

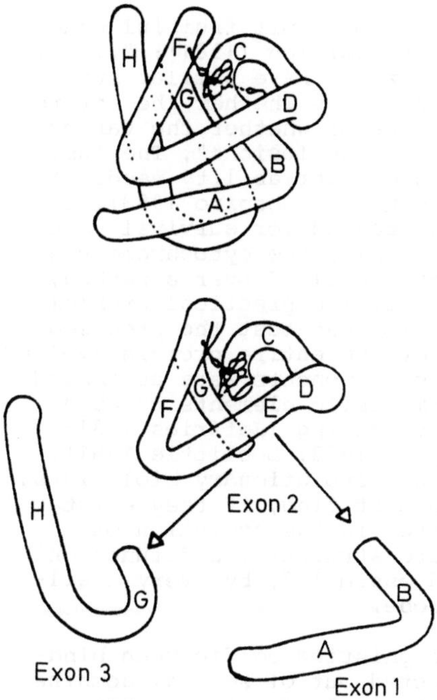

Fig. 7. Schematic drawing of a hemoglobin unit and the three exonic transcripts (Blake 1983)

only part of the O_2-binding property. We propose that homologous and ancestrally related genes may code for the heme-binding property of at least some of the heme-dependent oxygenases and oxidases for some contribution to catalytic function; but total function includes contributions arising from cooperative interactions with other domains, as illustrated in Fig. 7 and Table 5. This will apply to the Cu-, Fe-, and flavin-binding domains as well as the heme-binding domain of hemoglobin under discussion.

Testing Evolutionary Hypotheses for Oxygenases and Oxidases

We have now defined the extent of variety among the dioxygen transporting and activating proteins according to reaction type and prosthetic group (Tables 1, 2, 3, and 4). We have also shown that there is a reasonable basis for ascribing the reactive centers of dioxygen binding and activating proteins to a limited number of domains. And finally, we have shown that domains are coded for by specific gene pieces and speculated that during evolution, genetic variety was produced by mutation, transposition, recombination, and fusion of gene segments coding for domains functional in multidomain oxygenases and oxidases. We now consider possibilities for testing this hypothesis.

The actual mechanism for assembly, modification, and selection of sophisticated multidomain dioxygen binding and activating proteins during evolution are not well-understood. A role for mobile genetic elements is very likely, but Shapiro (1983), when discussing mobile genetic elements comments that "... their complex recombination activities and sometimes non-Mendelian behavior pose fundamental problems in understanding the organization of chromosomes and extra-chromosomal hereditary nucleic acids (viruses, phages, plasmids, etc.). They also open many mechanisms to the formation of novel genomic configurations.

However, our knowledge is still too immature to permit fruitful generalizations". Doolittle takes a more explicit position (1981): "... if the cofactor binding portions of a set of enzymes have similar structures or amino acid sequences, the question arises whether the entire enzyme protein is derived from a common source or whether the various components were assembled ... by a series of gene fusions". The fusion mechanism implies "a background noise of random variability needed to provide the raw material of evolutionary change"[2] (Shapiro 1983), in which every combination of domains would be tested for survival value. Multidomain oxygenases and oxidases, for example, the cytochrome P-450 systems and cytochrome c oxidases, would have evolved over a pathway of adaptively competent smaller intermediates. The practical problem is to find oxygenase and oxidase systems which can test the proposed alternative evolutionary mechanisms: whether the entire protein evolved from a common source or whether the various components were assembled as domains by a series of gene fusions from mobile elements capable of duplicating, jumping, diverging, and recombining (Jeffries 1981; Jeffries and Harris 1982; Lewin 1981; Shapiro 1983; Doolittle 1981). This is a province not only for molecular and evolutionary biologists, but also biochemists and biophysicists who must find, if they exist, the modern missing links – intermediate states in the evolution of multidomain proteins – and establish complete structural and function characterizations of domains, isolated and assembled, by every available contemporary physical and chemical probe.

If, to paraphrase Rossmann et al. (1975) sophisticated dioxygen binding and activating proteins may be constructed out of several domains (Fig. 1) by fusion of genes for proteins of simple function, then the component domains should show homologies of structure with domains for related function in other systems. An intense effort is being made now to characterize the structures of interesting oxygenases and oxidases (reviewed recently by Malmstrom 1980, 1982; Nozaki et al. 1982; King et al. 1982; Lamy and Lamy 1981; Dunford et al. 1981; Lee et al. 1981; Sigel 1981; Sato and Kato 1982, among others). Strong efforts are also being made to determine gene structures of prokaryotic and eukaryotic oxygenases and oxidases. The basis for genetic variability arising from mobile genetic elements (Shapiro 1983) is being actively pursued.

Test systems for the study of evolution of dioxygen-binding and activating proteins in terms of domains and gene function should have the following properties: (1) they should be of demonstrated ancient lineage and occur extensively in widely separated branches of the phylogenetic tree; (2) they should comprise multiple domains representing the major functions in oxygenases and oxidases; (3) they should represent the major classes of dioxygen-binding prosthetic groups, the major classes of reductive substrates, and the major classes of dioxygen reactions which are catalyzed by them.

Two such systems are cytochrome c oxidase and cytochrome P-450 mixed function oxidase. With respect to cytochrome oxidase, there is evidence that it shares domain structure with type I blue copper proteins (azurin, plastocyanin, stellacyanin), with type III binuclear copper pairs, and with the heme-binding domain of globin. The enzyme is widely distributed throughout the phylogenetic scale and has at least

---

2 But Shapiro continues with a different view: "In my opinion we will only integrate mobile elements fully into our picture of heredity when we have formulated entirely new mechanisms for cellular differentiation in both development and evolution ... Like scientists in other fields we geneticists now have come to terms with unanticipated levels of structure and organization".

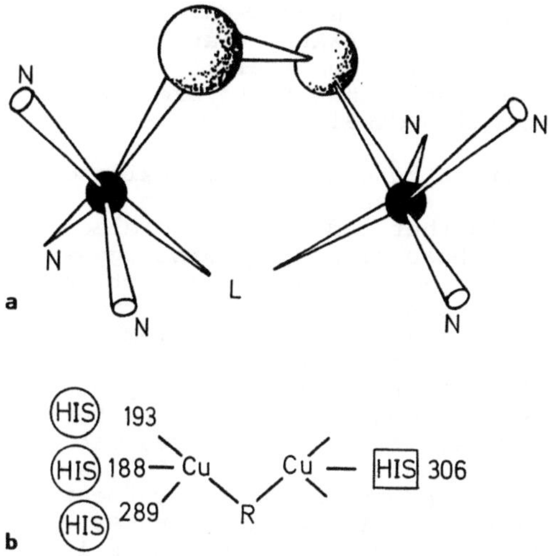

Fig.8a,b. Comparison of the active sites of neurospora tyrosinase (Pfiffner et al. 1981) and hemocyanin (van Holde and Miller 1982). (A) A schematic view of the active site of hemocyanin (not to scale). The *dark spheres* are the Cu ions, the *shaded spheres* represent the O_2^{-2}, *N* are the nitrogen ligands (presumably on histidines) and *L* is another bridging ligand (possibly tyrosine hydroxyl). (B) The proposed ligand environment of the active site of Neurospora Tyrosinase. *R* = bridging ligand; *His* = histidine, residue number indicated

5 and perhaps as many as 12 domains. One of them, cytochrome c binding, is demonstrably of ancient lineage. Mammalian cytochrome c oxidase comprises seven different polypeptides, more or less (cf. Winter et al. 1980). Three of them, subunits I, II, and III are coded by the mitochondrial genome in eukaryotes and the gene and amino acid sequences are known. Bacterial cytochrome oxidases on the other hand usually contain two subunits of molecular weight comparable to mammalian subunits I and II (57 000 and 25 400) (Yamanaka et al. 1979; Ludwig 1980). The oxidase contains two heme a and two Cu ions – cytochrome a, cytochrome a_3, Cu_A, and Cu_B – apparently contained entirely on subunits I and II (Winter et al. 1980; Thomson et al. 1982). The Cu_A on subunit II (EPR detectable) resembles Cu type I blue copper protein in its ligands and in its amino acid homology (Malmstrom 1980, 1982; Ryden and Lundgren 1976, 1979; Steffens and Buse 1978). The second, Cu_B, resembles stellacyanin (Powers et al. 1979, 1981; Chance 1981).

Type III Cu is an O_2-binding prosthetic group containing a binuclear Cu center. It occurs in hemocyanins, laccase, tyrosinase, ascorbate oxidase, and ceruloplasmin (representing dioxygen transport, monooxygenase, and 4-electron reducing oxidase function). The binuclear sites of hemocyanin and tyrosinase appear to be very similar, if not identical and almost certainly constitute a domain (Solomon et al. 1981; Pfiffner et al. 1981; Himmelwright et al. 1980; Schoot Uiterkamp and Mason 1973) (Fig. 8). Type I and III Cu centers have two or more imidazole N (histidine) ligands per Cu and there is some possibility of relationship through gene duplication. The analogous binuclear Fe_2 in hemerythrin, a dioxygen transporter, is symmetrical around the Fe-binding sites and is almost certainly a product of gene duplication (Doolittle 1981). Evidence for a relationship between type I Cu and type III has been reported by Reinhammar et al. (1980), Cline et al. (1983), and Malmstrom (this colloquium), in that a new rhombic EPR signal in partially reduced type III Cu of laccase represents one Cu of the binuclear pair. A similar EPR signal is detected from the EPR-indetectable Cu_B of reduced cytochrome c oxidase under turnover conditions. These investigators concluded that the ^{14}N ENDOR signal of this site and its EPR signal, corresponds extremely closely to that of the laccase type III signal, arising presumably from a stellacyanin type Cu uncoupled from cytochrome a_3.

Finally, Powers et al. (1981) concluded from extended X-ray absorption fine structure spectra of cytochrome c oxidase (EXAFS) that the first electron shell of carbonmonoxycytochrome a_3 is identical to that of carbonmonoxyhemoglobin. This makes it possible that there are similarities between the heme a and heme-binding domains.

It is also of considerable interest that in both cytochrome oxidase and laccase, the reduced bimetallic centers react rapidly with O_2 to form peroxide intermediates. These are further reduced by one electron to form paramagnetic intermediates whose structures are suggested to be $[Cu_B^{2+} - OH^-:..) = Fe_{a3}^{4+}]$ and $[Cu^{2+} - O^- - Cu^{2+}]$, respectively. The addition of one more electron then completes the dioxygen reduction in both cases (Malmstrom 1980, 1982, 1983, this colloquium). Thus, mechanistically, chemically, and physically, type I and type III Cu_2 domains participate in four electron reduction of dioxygen by one-equivalent reductants in diverse systems operating at bacterial, plant, and animal phylogenetic levels. It will be of greatest interest to compare the 3-dimensional structures of Cu_A and Cu_B domains with those of Type I and Type III Cu proteins and the cytochrome a and a_3 domains with those of other heme-containing dioxygenases and oxidases. We are attempting the isolation of the native subunits in the hope that the structures of the Cu and heme a binding domains will thereby become accessible. Preliminary results have been reported by Mason (1982).

A second example of a test system is cytochrome P-450 mixed function oxidase. Cytochrome P-450 is a dioxygen activating heme-thiolate protein occurring throughout the phylogenetic scale (mammals to bacteria and fungi). It acts on a wide range of substrates including steroids, fatty acids, prostaglandins, drugs, and many xenobiotics and is called a multi-substrate monooxygenase. Figure 9 depicts the composition of the P-450 monooxygenase systems which occur in mammalian microsomes, mitochondria, and bacteria. The underlying mechanics of dioxygen activation by them are probably identical, but the mechanism by which two reducing equivalents are delivered to the cytochrome P-450 varies, a very interesting opportunity for evolutionary studies.

Electron donor	*Electron transfer system*	*Electron acceptor*	*Organism*
NADPH	FAD FMN (NADPH-P-450 reductase)	$P\text{-}450^{3+}\text{-}SH$ $P\text{-}450^{2+}O_2\text{-}SH$	Microsomes
NADPH	FAD 2Fe-S/mol (NADPH-adrenodoxin reductase) (NADPH-hepatoredoxin reductase)	$P\text{-}450^{3+}\text{-}SH$ $P\text{-}450^{2+}O_2\text{-}SH$	Mitochondria (Adrenal cortex) (Liver mitochondria)
NADH	FAD 2Fe-S/mol (NADPH-putidaredoxin reductase)	$P\text{-}450^{3+}\text{-}SH$ $P\text{-}450^{2+}O_2\text{-}SH$	*Pseudomonas putida*

Fig. 9. Different P-450 systems in prokaryotes and eukaryotes

Liver microsomal cytochrome P-450 represents a fusion of at least four domains for the heme-thiolate protein (reductase-, heme-, O_2-, and substrate-binding) and seven for the reductase (NADPH-, FAD-, FMN-, and P-450 binding, and three domains transferring reducing equivalents) (Fig. 10).

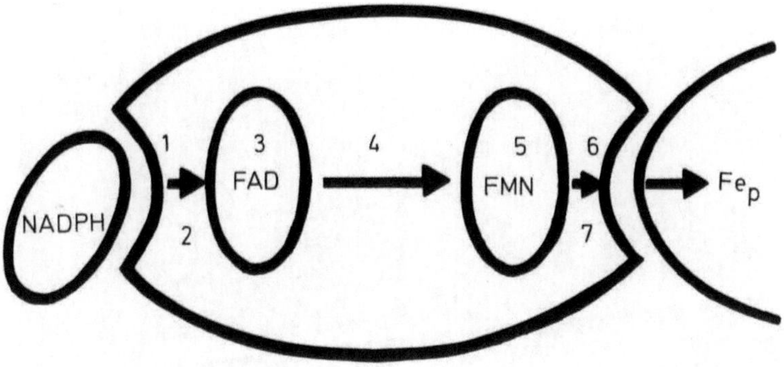

Fig. 10. The domains of NADPH-cytochrome P-450 reductase are shown: 1, 3, 5, and 6 represent NADPH-, FAD-, FMN-, and P-450 binding domains: 2, 4, and 7 represent domains transferring reducing equivalents from NADPH to P-450 (Mason 1981)

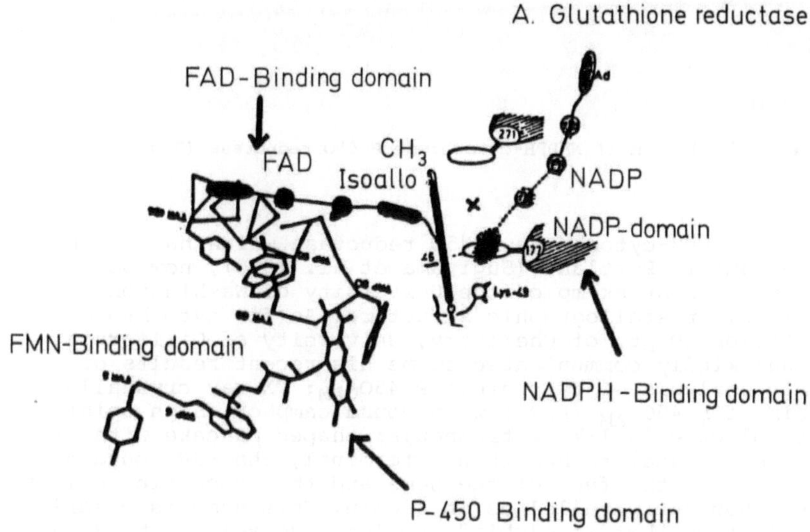

Fig. 11. Hypothetical construction of NADPH-cytochrome P-450 reductase, assuming NADPH, FAD, and FMN binding sites homologous to those of glutathione reductase and flavodoxin. Here electron transfer between $FADH_2$ and FMN is modulated by an aromatic amino acid (Sugiyama et al. 1983)

Structural relationships between these domains and their counterparts in other proteins are now being sought. Figure 11 illustrates a working hypothesis for the structure of the functional domains based upon the assumption that the NADPH and FAD-binding domains of glutathione reductase and the FMN-binding domains of flavodoxin will be similar to the corresponding domains of cytochrome P-450 reductase. Figure 12

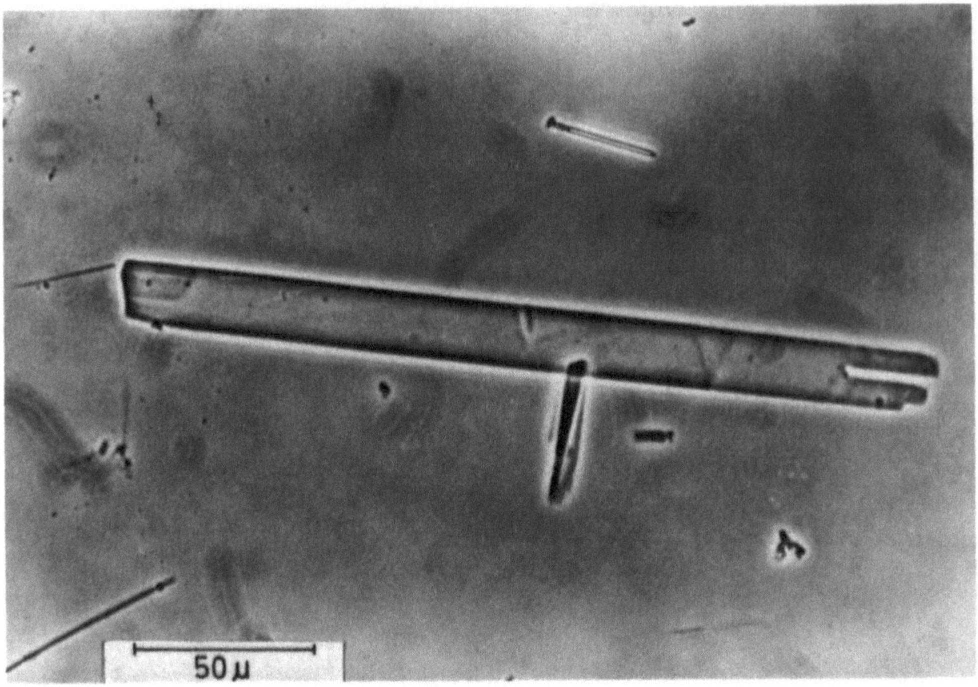

Fig. 12. Crystals of the T-form of NADPH-cytochrome P-450 reductase (Sugiyama et al. 1983)

shows crystals of NADPH-cytochrome P-450 reductase which have been obtained by Sugiyama in Portland (Sugiyama et al. 1983), now being examined by Sieker and Stenkamp of the University of Washington (Seattle) for X-ray crystallographic structure. As for cytochrome P-450 itself, Poulos (Dept. of Chemistry, University of California at San Diego) has kindly communicated to me his recent results of crystallographic analysis of cytochrome P-450_{CAM}: "X-ray crystallographic analysis of P-450_{CAM} (Fe^{3+}) with bound camphor (high spin) shows that the molecule is like a triangular shaped pancake with N-terminal cysteine (ligand) rather than C-terminal, the -SH containing peptide lying over the face of the heme and the bracketing helices parallel rather than antiparallel as in globin. This heme is clearly bound differently than that in globin". Poulos also solved the X-ray crystallographic structure of cytochrome c peroxidase and its complex with cytochrome c (Poulos et al. 1980; Poulos and Kraut 1980) and suggests that this peroxidase and probably all plant peroxidases belong to the globin family. Some amino acid sequence homology has already been reported between liver microsomal P-450 and P-450_{CAM} (Black et al. 1982; Hanui et al. 1982; Heinemann and Ozols 1982). Nevertheless the identification of the four domains of cytochrome P-450 itself is uncertain and the substrate-binding domains of cytochrome P-450s appear to be very variable, readily adapting to novel compounds. The mechanistic basis of this adaptation is of very great interest from the evolutionary point of view because the phenomenon appears to be evolution itself at the substrate-binding domain.

Conclusions

We have applied current evolutionary theory to dioxygen binding and activating proteins. We conclude: (1) that evolution of these proteins probably proceeded by assembly of multidomain polypeptides at the gene level, utilizing mobile genetic elements and other generators of genetic diversity; (2) that oxygenases and oxidases containing one O_2-binding prosthetic group of the four main types, evolved by divergent evolution from a common ancestor; and (3) that the lines of evolution from separate prosthetic group types converged to catalyze the same set of reaction types. Some approaches to testing these ideas were suggested, and their preliminary application to cytochrome c oxidase and cytochrome P-450 mixed function oxidase systems were described.

Acknowledgement. These studies were supported by grants from the American Cancer Society and the National Institutes of Health.

Addendum

We proposed that active sites containing two identical redox units could have arisen by gene duplication. After completing this manuscript, I found a paper by Ryden: Proc Natl Acad Sci 79, 6767-6771 (1982) in which evidence was obtained by comparing amino acid sequences of human ceruloplasmin and poplar plastocyanins that the portion of the ceruloplasmin molecule containing the binuclear Cu pair (Type III) could have evolved by a doubling of the plastocyanin structure.

Discussion

Buse: Could you tell us about your recent results concerning the binding of the coppers and the hemes to subunits I and II of the oxidase?

Mason: We have been working intensively on the separation of the native Cu- and heme a-binding subunits of cytochrome c oxidase and have been consistently obtaining subunits I and II each with copper and heme. However, the proof that these (or any) isolated subunits are native and essential for function of cytochrome c oxidase depends on their ability to be reconstituted to functional enzyme, as I have pointed out before. We have had inconsistent results in this respect, and the work is still being actively pursued.

References

Almassy RJ, Dickerson RE (1978) Proc Natl Acad Sci USA 75:2674-2678
Argos P, Rossmann MG (1979) Biochemistry 18:4951-4960
Black SD, Tarr GE, Coon MJ (1982) J Biol Chem 257:14616-14619
Blake CCF (1979) Nature 277:598
Blake CCF (1983) Trends Biochem Sci 8:11-13
Blake CCF (1978) Nature 273:267
Branden CI (1980) Quarterly Rev Biophys 13:317-338
Chance B (1981) Current Topics in Cell Regulation 18:343-360
Cline J, Reinhammar B, Jensen P, Venters R, Hoffmann BM (1983) J Biol Chem 258: 5124-5128

Craik CS, Buchman SR, Beychok S (1981) Nature 291:87-90
Crick F (1979) Science 204:264-271
Dayhoff MO (1972) Atlas of Protein Sequence and Structure, Vol 5, cf alignments 1 to 77. Nat Biomed Res Foundation, Washington DC
Dayhoff MO (1973) Atlas of Protein Sequence and Structure. Nat Biomed Res Foundation, Washington DC
Dayhoff MO (1976) Atlas of Protein Sequence and Structure, Vol 5, Supplement 2. Nat Biomed Res Foundation. Washington DC
Dayhoff MO (1978) Atlas of Protein Sequence and Structure, Vol 5, Supplement 3. Nat Biomed Res Foundation. Washington DC
Dayhoff MO, Barker WC, Hunt LT, Schwartz RM (1978) Atlas of Protein Sequence and Structure, Suppl 3, pp 9-24
Dayhoff MO, Barker WC, Hunt LT (1983) Methods in Enzymology 91:524-545
Dayhoff MO, Barker WC, Mason HS (1983a) Unpublished data
Dickerson R, Geis I (1983) Hemoglobin. Benjamin/Cummings Publishing, Menlo Park, California
Dickerson R, Timkovich R (1975) The Enzymes, 3rd edn, 11:397-547
Doolittle RF (1981) Science 214:149-159
Dunford HB et al (1981) The Biological Chemistry of Iron. D Reidel, Dordrecht, Holland
Eaton WA (1980) Nature 284:183-185
Edelman GM (1970) Biochemistry 9:3197-3204
Ghelis C, Yon J (1982) Protein Folding. Academic, New York
Gilbert W (1978) Nature 271:501-581
Gilbert W (1979) Eukaryotic Gene Regulation, ICN-UCLA Symposium on Molecular and Cellular Biology, Vol 14, pp 1-10
Gō M (1981) Nature 291:90-92
Guiard B, Lederer F (1979) J Mol Biol 135:639-650
Hanui M, Armes LG, Yasunobu KT, Shastry BA, Gunsalus IC (1982) J Biol Chem 257: 12664-12671
Hayaishi O, Katagiri M, Rothberg S (1955) J Am Chem Soc 77:5450-5451
Heinemann FS, Ozols J (1982) J Biol Chem 257:14988-14999
Himmelwright RS, Eikman NC, LuBien CD, Lerch K, Solomon EI (1980) J Am Chem Soc 102:7339
Jeffries AJ (1981) Genetic Engineering, Vol 2, pp 1-47
Jeffries AJ, Harris S (1982) Nature 296:9-10
Keevil T, Mason HS (1978) Methods in Enzymology LII, Part C:3-42
Keilin D (1966) The History of Cell Respiration and Cytochrome. Cambridge University Press
King TE, Mason HS, Morrison M (1982) Oxidases and Related Redox Systems. Pergamon Press, Oxford
Lamy J, Lamy J (1981) Invertebrate Oxygen-Binding Proteins. Marcel Dekker Inc, New York
Lee CP et al (1981) Mitochondria and Microsomes. Addison-Wesley, Reading, MA
Lesk AM, Chothia C (1980) J Mol Biol 136:225-270
Levitt M, Chothia C (1976) Nature 261:552-557
Lewin R (1981) Science 214:426-429
Liljas A, Rossmann MG (1974) Ann Rev Biochem 43:475-507
Love WE, Klock PA, Lattmann EE, Fadlan EA, Ward KB Jr, Hendrickson WA (1972) Cold Spring Harbor Symp Quant Biol 36:349-357
Ludwig B (1980) Biochim Biophys Acta 594:177-189
Malmström BG (1980) The Evolution of Protein Structure and Function, pp 87-96
Malmström BG (1982) Ann Rev Biochem 51:21-59
Mason HS (1981) Invertebrate Oxygen-Binding Proteins, pp 455-468
Mason HS (1982) The Biological Chemistry of Iron, NATO Advanced Study Institutes Series, Series c 89:459-474
Mason HS, Foulks WL, Peterson E (1955) J Am Chem Soc 77:2914
Nishikawa K, Ooi T, Isogai Y, Saito N (1972) J Phys Soc Japan 32:1331-1337
Nishikawa K, Momany FA, Scheraga HA (1974) Macromolecules 7:797-806
Nozaki M et al (1982) Oxygenases and Oxygen Metabolism. Academic, New York
Ooi T, Nishikawa K (1973) Conformation of Biological Molecules and Polymers, pp 173-187

Pfiffner E, Dietler C, Lerch K (1981) Invertebrate Oxygen-Binding Proteins, pp 541-552
Porter RR (1959) Biochem J 73:119-127
Poulos T, Kraut J (1980) J Biol Chem 255:8199-8205
Poulos TL, Freere ST, Alden RFA, Edwards SL, Skogland U, Takio K, Eriksson B, Xuong N-h, Yonetani T, Kraut J (1980) J Biol Chem 255:575-580
Powers L, Blumberg WE, Chance B, Barlow C, Leigh JS, Smith J, Yonetani T, Vik S, Peisach J (1979) J Biochim Biophys Acta 546:520-538
Powers L, Chance B, Ching Y, Angiolillo P (1981) Biophys J 34:465-498
Powers L, Chance B, Ching Y, Muhoberac B, Weintraub ST, Wharton DC (1982) FEBS Lett 138:245-248
Rao ST, Rossmann MG (1973) J Mol Biol 76:241-256
Rashin AA (1981) Nature 291:85-87
Reinhammar B, Malkin R, Jensen P, Karlsson B, Andreasson LE, Aasa R, Vanngård T, Malmström BG (1980) J Biol Chem 255:5000-5003
Richardson JS (1981) Adv Protein Structure 34:167-339
Rossmann MG (1981) Evolution of Glycolytic Enzymes, Phil Trans R Soc London B293:191-203
Rossmann MG, Argos P (1981) Ann Rev Biochem 50:497-532
Rossmann MG, Liljas A, Branden CI, Banaszak LJ (1975) The Enzymes, 3rd edn, pp 61-101
Ryden L, Lundgren JO (1976) Nature 261:344-346
Ryden L, Lundgren JO (1979) Biochimie 61:781-790
Sato R, Kato R (1982) Microsomes, Drug Oxidation, and Drug Toxicity. Japan Sci Soc, Tokyo
Schoot Uiterkamp AJM, Mason HS (1973) Proc Natl Acad Sci USA 70:993
Schulz GE (1981) Angew Chem Internat Ed Engl 20:143-151
Schulz GE, Schirmer RH (1979) Principles of Protein Structure. Springer, Berlin Heidelberg New York
Shapiro JA (1983) Mobile Genetic Elements. Academic, New York, pp xi-xvi
Sigel H (1981) Metal Ions in Biological Systems, Vol 12; Properties of Copper. Marcel Dekker, New York
Sigel H (1982) Metal Ions in Biological Systems, Vol 13; Copper Containing Enzymes. Marcel Dekker, New York
Solomon EI, Himmelwright RS, Eikman NC, LuBien CD, Schoeniger LO, Lerch K (1981) Invertebrate Oxygen-Binding Prcteins, pp 553-569
Steffens GJ, Buse G (1978) Hoppe-Seyler's Z Physiol Chem 84:1031
Sternberg JJE, Thornton JM (1977) J Mol Biol 110:285-296
Sugiyama T, Fischer B, Mason HS (1983) Federation Proc 42:2114
Thompson EOP (1980) The Evolution of Protein Structure and Function, pp 267-298
Thomson DA, Suarez-Villasan M, Ferguson-Miller S (1982) Biophys J 37:285
Vaska L (1976) Acc Chem Res 9:175
Valentine J (1973) Chem Rev 73:235
Van Holde KE, Miller KI (1982) Quart Rev Biophys 15:1-129
Warburg O, Negelein E (1928) Biochem Z 202:202-228
Wetlaufer DB, Ristow S (1973) Ann Rev Biochem 42:135-158
Winter DB, Bruyninckz WJ, Foulke FG, Grinich NP, Mason HS (1980) J Biol Chem 255: 11408-11414
Yamanaka T, Fukumori Y, Fujii K (1979) Cytochrome Oxidase, pp 399-407
Zuckerkandl E (1975) J Mol Evol 7:1-57
Zuckerkandl E (1978) Z Morph Anthrop 69:117-142

Reversible Dioxygen Binding[1]

K. Gersonde[2]

Introduction

For the transport of molecular oxygen by extracellular fluids, for diffusion enhancement, and for storage of O_2 within the cells (Hemmingsen 1963; Wittenberg et al. 1974) reversible dioxygen-metal adducts have to be formed (Antonini and Brunori 1971). Three classes of such adducts exist: Hemoglobins and myoglobins which provide mononuclear iron centers for binding of dioxygen (Antonini and Brunori 1971; Christomanos 1973), hemerythrins (Llinas 1973; Stenkamp and Jensen 1979; Stenkamp et al. 1981), and hemocyanins (Mangum 1980; Markl et al. 1982; van Holden and Miller 1982; Preaux and Gielens 1983), which utilize dinuclear iron and copper centers, respectively, for this purpose (see Table 1).

Table 1. Properties of biological dioxygen transport proteins

	Hemoglobin		Hemerythrin	Hemocyanin
	Hb	Mb		
Metal	Fe	Fe	Fe	Cu
Metal oxidation state of the deoxy-form	+2	+2	+2	+1
Spin state of the deoxy-form	4/2	4/2	4/2	0
Spin state of the oxy-form	0	0	0	0
Molar ratio, metal:O_2	1:1	1:1	2:1	2:1
p_{50}[a] (at 20°C and pH 8) (mmHg)	2.5	0.7	4.0	4.3
Number of subunits	4	1	8	Variable
function	Transport	Diffusion enhancement; storage	Storage	Transport

[a] p_{50} is the O_2 partial pressure in mmHg at which 50% of the binding sites are oxigenated

Hemerythrins occur in certain species of four invertebrate phyla: *Spinunculids*, *Polychaetes*, *Pripulids*, and *Brachiopods*. Each subunit of these octameric proteins contains two iron atoms and binds one molecule of O_2. In the deoxy-form, the iron centers are in the Fe(+2) oxidation state, whereas in the oxy-form, they are in the Fe(+3) oxidation state with the O_2 moiety being a peroxo-ligand, i.e., O_2^{2-} (see Fig. 1). The

1 Dedicated to the memory of Professor Eraldo Antonini

2 Abteilung Physiologische Chemie, Rheinisch-Westfälische Technische Hochschule (RWTH), D-5100 Aachen, FRG

34. Colloquium-Mosbach 1983
Biological Oxidations

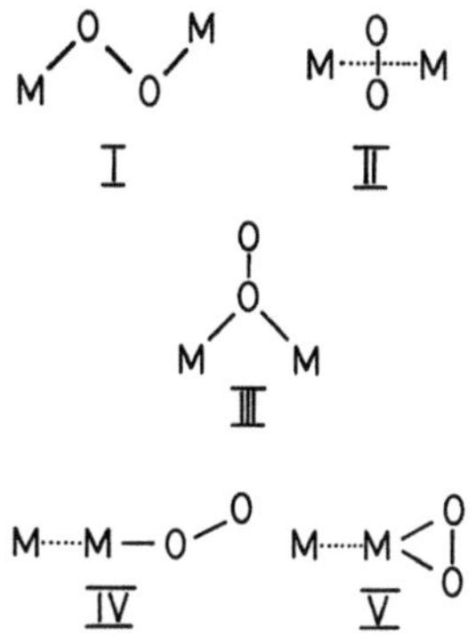

Fig. 1. Possible structures of the metal-dioxygen unit in hemerythrins and hemocyanins

two iron atoms, which are directly linked to protein side groups, exhibit different ligand environments and therefore are electronically nonequivalent (Kurtz et al. 1976; Kurtz et al. 1977). Furthermore, resonance Raman studies demonstrate that the two oxygen atoms of the peroxo-ligand are not equivalent consistent with structures III and IV shown in Fig. 1 (Dunn et al. 1975).

Hemocyanins appear in some species of invertebrate phyla: *Arthropoda* and *Mollusca*. The two copper atoms are directly bound to the protein side chains. In the deoxy-form, the copper centers are in the Cu(+1) oxidation state, whereas in the oxy-form they change to the Cu(+2) oxidation state with the O_2 moiety being a peroxo-ligand, i.e., O_2^{2-}. Resonance Raman studies provide evidence for equivalent oxygen atoms, which are bound in a nonplanar μ-dioxygen fashion as indicated by structure I shown in Fig. 1 (Kirchner and Loew 1977).

Hemoglobins and myoglobins are dioxygen-binding pigments being present in blood, lymph, or tissues of numerous invertebrates and vertebrates. This class of oxygen-transporting proteins, binds one O_2 molecule per heme prosthetic group. The heme iron is chelated to four nitrogen atoms of the protoporphyrin IX dianion, which provides the so-called planar ligands (Perutz 1980; Perutz 1976). Among the hemoglobins, only one exception exists which does not possess the protoporphyrin IX, i.e., the chlorocruorin, found in some species of polychaete worms of the phylum *Annelida*. In chlorocruorins, the vinyl group at position 2 of the protoporphyrin IX is replaced by a formyl group (see Fig. 2), thus, resulting in chlorocruoroporphyrin or Spirographis porphyrin (Asakura and Sono 1974). The base imidazole provided by the invariant "proximal" histidine and dioxygen as the exogenous ligand occupy the axial positions of the heme iron. In the deoxy-form, the iron is in the five-coordinate high-spin (S=2)Fe(+2) oxidation state, whereas in the oxy-form heme iron is in the six-coordinate low-spin Fe(+3) oxidation state. The heme system is embedded in a pocket formed by the E and F helices of the polypeptide chain. The heme is held in place by a large number of nonbinding interactions and the single covalent attachment to the already mentioned proximal histidine F8 (Perutz et al. 1968; Bolton and Perutz 1970). Myoglobins, leghemoglobins (Christahl et al. 1981; Armstrong et al. 1980), and some insect hemoglobins (Gersonde et al. 1972, 1982) are monomeric, i.e., they are composed of only one polypeptide chain. Most vertebrate hemoglobins are tetrameric with pairwise equal polypeptide chains ($\alpha_2\beta_2$). The tertiary structures of all hemoglobin protomers are in general quite similar (Huber et al. 1971).

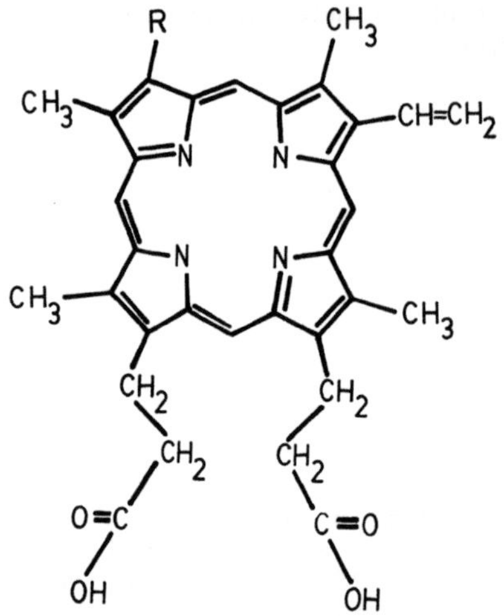

Fig. 2. Structures of natural porphyrins. R, $-CH=CH_2$ (protoporphyrin IX); R = -CHO (chlorocruoroporphyrin)

Terminology and Definitions

The generic term "dioxygen" designates diatomic oxygen in all states and forms of O_2, i.e., O_2, O_2^-, O_2^{2-} are species of dioxygen. The only common feature is an oxygen-oxygen covalent bond regardless of whether dioxygen is free, part of another complex, or carries an electronic charge.

"Dioxygen-metal" adducts, i.e., O_2-M, are defined as having an additional bond between dioxygen and metal atom without any further specification of structure and bonding.

"Molecular oxygen", i.e., O_2, means free or isolated neutral O_2. This term refers usually to the ground-state electronic configuration ($^3\Sigma_g^-$).

"Oxygenation and deoxygenation" indicate reversible addition and subtraction of molecular oxygen or dioxygen. Thus, these metal complexes can exist in the deoxy- and oxy-form, respectively.

"Superoxide anions", i.e., O_2^-, which are covalently bound to metal centers, lead to superoxo-covalently bound dioxygen and to the formation of superoxo-metal complexes.

"Peroxide anions", i.e., O_2^{2-}, which are covalently bound to metal centers lead to peroxo-covalently bound dioxygen and to the formation of peroxo-metal complexes.

"Oxide anions", i.e., O^{2-}, which are covalently bound to metal centers lead to covalently bound mono-atomic oxygen and to the formation of oxo-metal complexes.

The ground-state electronic structure ($^3\Sigma_g^-$) of the molecular oxygen is given by equations (1) and (2) (see Fig. 3).

$$2\ O \rightarrow O_2 \qquad (1)$$

$$(1s)^2(2s)^2(2p)^4 \rightarrow (KK)(2s\sigma_g)^2(2g\sigma_u^*)^2(2p\sigma_g)^2(2p\pi_u)^4(2p\pi_g^*)^1(2p\pi_g^*)^1 \qquad (2)$$

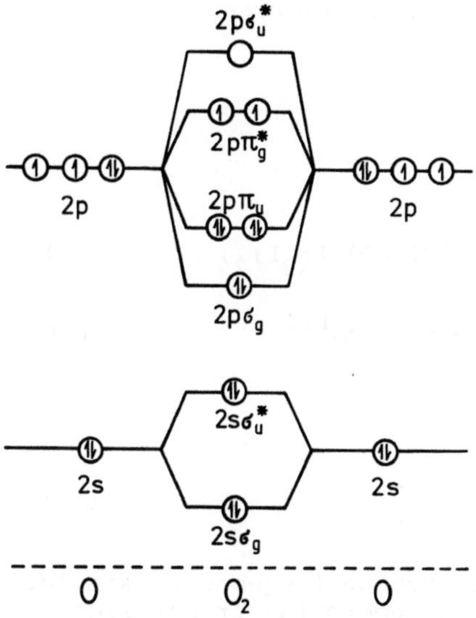

Fig. 3. Molecular orbital energy level description for molecular oxygen

Table 2. Oxidation and reduction of molecular oxygen. Relationship between bond order, O - O distance, and O - O stretching frequency

$$O_2^+ \xleftarrow{-e} O_2^{\pm 0} \xrightarrow{+e} O_2^- \xrightarrow{+e} O_2^{2-} \xrightarrow{+2e} 2\, O^{2-}$$

Bond order	2.5	2.0	1.5	1.0	0
O - O (Å)	1.12	1.21	1.33	1.49	
ν_{O_2} (cm^{-1})	1905	1580	1097	802	

The electrons added to the dioxygen enter the partially vacant antibonding orbitals ($2p\pi_g^*$) causing a decrease of the O - O dissociation energy (see Table 2). Therefore, an increase of the interatomic distance and a decrease in vibrational frequency can be observed. The addition of one or two electrons to the neutral dioxygen molecule results in a formation of superoxide (O_2^-) and peroxide (O_2^{2-}) anions, respectively (Vaska 1976).

Nature of Reversible Dioxygen-Metal Adducts

Dioxygen attached to a metal center can exhibit two discrete geometries, i.e., it forms superoxo- (see I, in Fig. 4) or peroxo- (see II in Fig. 4) complexes. In model compounds, the coordinated O_2 strongly tends to reach a particular state of geometry in either class of the above mentioned groupings. However, in O_2-binding proteins with apolar constraints on the exogenous ligand or with polar interactions from amino acid side groups intermediate geometries are possible.

In addition, binding of dioxygen to a metal complex often involves formal oxidation of the metal center and concomitant reduction of the coordinate dioxygen described in equations (3) - (5) (Vaska 1976; Ellis and Pratt 1973; Caughey et al. 1975).

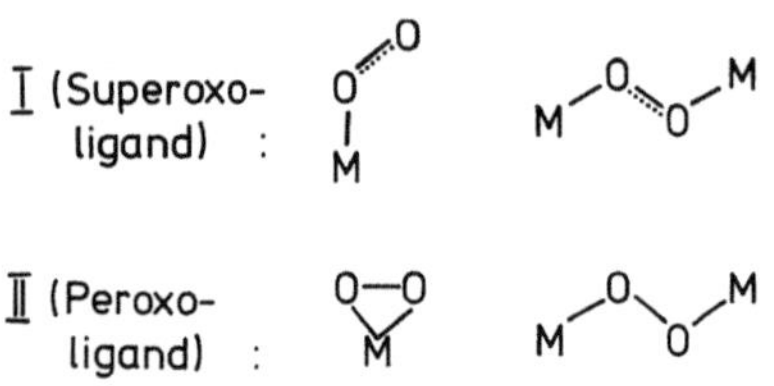

Fig. 4. Dioxygen-metal groupings

$$\text{trans-}(Ir^{I}Cl(CO)(Ph_3P)_2) + O_2 \rightleftharpoons (O_2^{2-} \cdot Ir^{III}Cl(CO)(Ph_3P)_2) \quad (3)$$

$$Co^{II} + O_2 \longrightarrow O_2^{-} \cdot Co^{III} \longleftarrow O_2^{-} + Co^{III} \quad (4)$$

$$Fe^{II}(Hb) + O_2 \longrightarrow O_2^{-} \cdot Fe^{III}(Hb) \quad (5)$$

$$O_2^{-} \cdot Fe^{III}(Hb) + CN^{-} \longrightarrow CN^{-} \cdot Fe^{III}(Hb) + O_2^{-} \quad (5a)$$

The dioxygen-metal complexes have O - O stretching frequencies similar to those obtained from compounds containing superoxide or peroxide anions suggesting a substantial transfer of electron density from the metal center to the coordinated dioxygen (Jones et al. 1979).

The bonding of dioxygen in oxy-hemoglobin or oxy-myoglobin is commonly assumed to be of the superoxo-like type. To explain the diamagnetism of oxy-hemoglobin, Pauling proposed in 1936 an even number of electrons about dioxygen represented by resonance hybrid structures I and II in Fig. 5 (Pauling and Coryell 1936). Pauling already supposed that "the oxygen molecule undergoes a profound change in electronic structure on combination with hemoglobins". Weiss (1964) proposed for the bonding of O_2 in oxy-hemoglobin an odd number of electrons on the coordinated dioxygen represented by structures III and IV in Fig. 5. In this case, the unpaired spins on Fe^{III} and O_2^{-} are paramagnetically coupled resulting in a nonmagnetic ground-state elec-

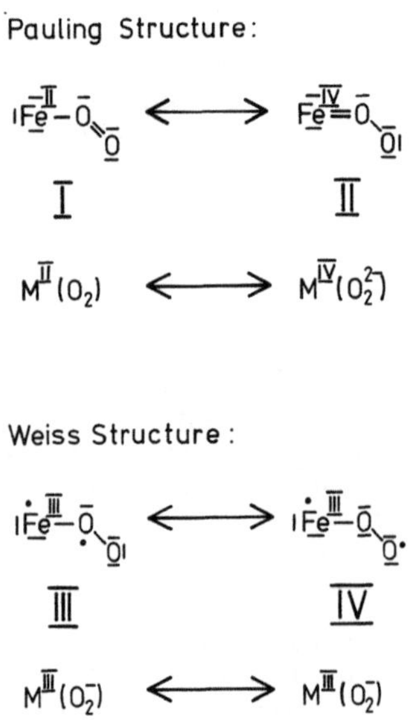

Fig. 5. Electronic structures of dioxygen-iron complexes in hemoglobin and myoglobin

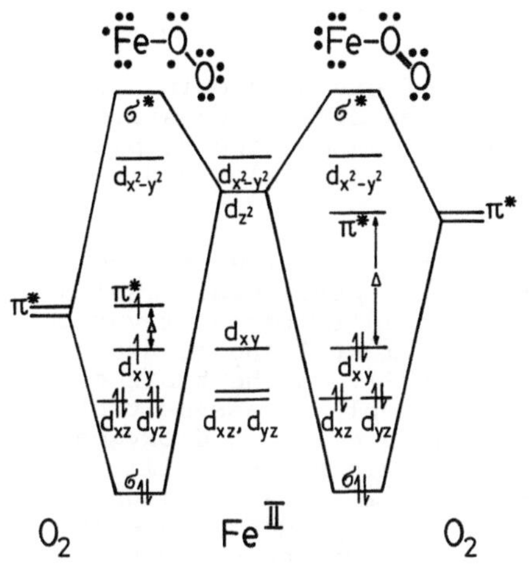

Fig. 6. Qualitative molecular orbital scheme for the bonding in dioxygen-iron complexes

tronic structure and a low lying paramagnetic excited state with S=1 (see Fig. 6). On the basis of a qualitative MO scheme for the dioxygen-iron adduct shown in Fig. 6 the magnetic behavior of oxy-hemoglobin can be better understood and it is possible to differentiate between the two above mentioned electronic structures of dioxygen. Which of the two possible electronic configurations is preferred depends on the energy separation, Δ, between the nonbonding d_{xy} orbitals and antibonding π^* orbitals. If Δ is greater than the electron-pairing energy, the spin-paired configuration (Pauling model) occurs. With small Δ values, the two unpaired electrons occur in the d_{xy} and $\pi^*_{O_2}$ orbitals (corresponding to the Weiss model).

ESR (Chien and Dickinson 1972; Getz et al. 1975; Dickinson and Chien 1980; Christahl et al. 1981; Gersonde et al. 1982) and magnetic susceptibility measurements (Floriani and Calderazzo 1969) on cobalt-dioxygen adducts have clearly demonstrated the presence of one unpaired electron with >80% spin density residing on the dioxygen. Thus, the dioxygen-cobalt adduct is represented by $Co^{III} \cdot O_2^-$. Bonding between Co and O_2 arises primarily from the overlap of a dioxygen $\pi^*_{O_2}$ orbital with the d_{z^2} orbital of the cobalt. The $Co\text{-}O_2\pi$-overlap is negligible and the unpaired electron resides in the $\pi^*_{O_2}$.

Unfortunately, the situation was not as clear for the dioxygen-iron adducts. Most of the experimental evidence, which has been collected for the dioxygen-iron adduct, was interpreted in terms of the $Fe^{III} \cdot O_2^-$ formalism (Jones et al. 1979):

1) The infrared stretching frequencies of the coordinated dioxygen moiety were similar to that of O_2^- (Maxwell et al. 1974; Rimai et al. 1975; Collman et al. 1976).
2) Similarity exists between optical spectra of oxy- and hydroxy-methemoglobin (Wittenberg et al. 1967; Antonini and Brunori 1971).
3) Large quadrupole splittings in the ^{57}Fe Mößbauer resonance spectra were observed (Trautwein 1974; Spartalian and Lang 1975).
4) Release of free superoxide anions O_2^- from the interaction of oxy-hemoglobin with anions like N_3^- was observed (Wallace et al. 1974).
5) The residual relatively high paramagnetism at low temperature (Hoenig and Gersonde 1976; Cerdonio et al. 1977, 1978) was inter-

preted as due to a charge-transfer configuration as a low-lying excited state.

6) Polarized single crystal optical studies of oxy-myoglobin helped to distinguish between the two possibilities of excited state contributions to O_2-binding and seemed to rule out the triplet-triplet model of dioxygen-iron adduct (Makinen et al. 1978; Churg and Makinen 1978).

However, theoretical calculations do not uniformly provide or predict the electronic ground state of the oxy-complex. Experimental results mentioned under 1) - 4) can be finally interpreted in terms of various models. Recently a critical discussion of experimental and quantum chemical data has been presented by Makinen (1979). On the basis of the results mentioned under 5) and 6) we have simply to assume that low lying excited states must contribute to the binding of dioxygen in oxy hemoglobins. We have determined the magnetic susceptibility of oxyhemoglobin over a wide range of temperatures (2-100 K) in magnetic fields up to 0.07 T and found a weak temperature-dependent magnetism which was absent in the respective carbonmonoxy hemoglobin (Hoenig and Gersonde 1976). With a weight of 0.41 for the triplet state and the finding of an excitation energy of about 17 cm^{-1} (25 K) between singlet ground and triplet excited spin state (see Fig. 7), we proposed the charge transfer model of oxy-hemoglobin (Hoenig and Gersonde 1976). For clarification, it should be stated in agreement with Makinen (1979) and on the basis of the polarized single crystal optical studies that at least in the case of dioxygen-iron adducts the charge-transfer complex is due to anisotropic covalent bonding interaction yielding substantial delocalization of the iron and O_2 orbitals with charge-transfer character.

Also peroxo-like metal complexes have been prepared with the group VIII metals. For example, dioxygen-porphinato manganese complexes can be formed in noncoordinating organic solvents at low temperature which show the so-called Griffith-type of O_2-bonding (Weschler et al. 1975; Hoffman et al. 1976). The first report of the preparation of a peroxo-like complex by reaction with molecular oxygen and Ir^{I} complexes (see Eq. 3 and Fig. 4, structure II) has been given by Vaska (1963). The oxygenation which is reversible under ambient conditions emerges

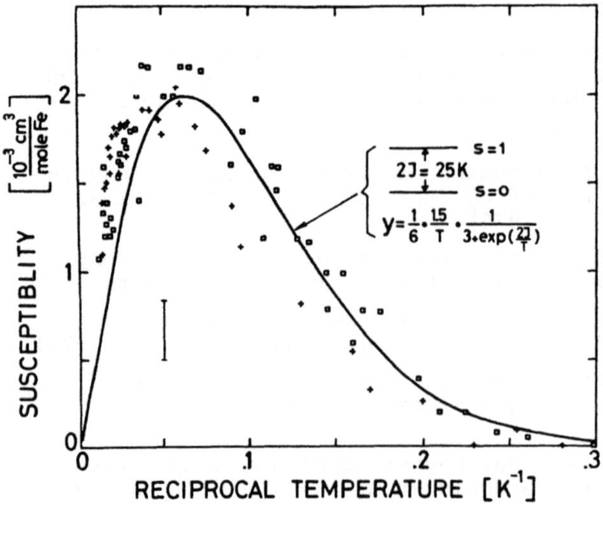

Fig. 7. Temperature dependence of the magnetic susceptibility of human-oxy hemoglobin. The *full line* is the calculated molar susceptibility for a spin singlet-triplet level sequence scaled by a factor 0.17. (Hoenig and Gersonde 1976)

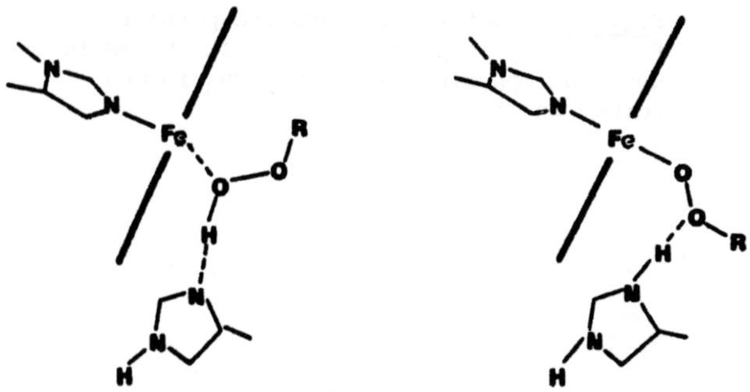

Fig. 8. Two types of distal histidine-dioxygen interactions in horse radish peroxidase compound I. (Schonbaum et al. 1979)

as an oxidation of the univalent (spin-paired d^8) and four-coordinate iridium complex to a tervalent (spin-paired d^6) and six-coordinate iridium peroxo-complex. It has been demonstrated that such peroxo-complexes with short O - O bond lengths (1.3 - 1.4 Å) are reversible, whereas those with larger O - O distances (1.5 - 1.6 Å) are irreversible. The O - O distance gradually increases with increasing electron donation and hence the basicity or charge density of the metal center controls the particular functions, i.e., reversible O_2-binding or O_2 activation. The interaction of solvent molecules with the bound dioxygen can also modulate the charge transfer. In proteins for example, hydrogen-bonding to dioxygen increases this charge transfer and leads to further lengthening of the dioxygen bond. Dioxygen-imidazole interaction in hemoglobins and oxidases can involve the oxygen atom proximal or distal to iron, resulting in peroxo-like and superoxo-like structures, respectively (see Fig. 8). If this particular interaction changes with pH in the same protein, differences in O_2 affinity (Christahl et al. 1981) or activation products (Schonbaum et al. 1979) will be the result.

Microallosteric Effects in Hemoglobins and Model Systems

During the evolution from primitive to more sophisticated O_2 transport proteins, the autoxidation of the heme group had to be avoided and the O_2 affinity of the adduct had to be lowered by the protein environment. Furthermore, the individual binding sites in multi-site hemoglobins had to be modulated by an allosteric mechanism. The allostery of hemoglobins is not only important for the adaptation of the organism to various physiological and environmental conditions, but is absolutely necessary for the control of the O_2 flux into or out of red blood cells under given rheological conditions.

"Microallosteric" effects on O_2-binding are due to conformation changes in the protein subunit itself or in the nearest surrounding of the dioxygen-metal moiety and have been mimicked and extensively studied for iron and cobalt porphyrin compounds as well as for monomeric hemoglobins. Microallosteric effects are important for stabilizing the reversible O_2 adducts and for modulating the O_2 affinity.

Porphyrin Effect

The porphyrin effect can be of electronic or of stereochemical nature. It has been studied by spectroscopy and O_2-binding experiments on monomeric hemoglobins reconstituted with synthetic unnatural porphyrins (see Fig. 9). Principally, an increase in basicity of the in-plane li-

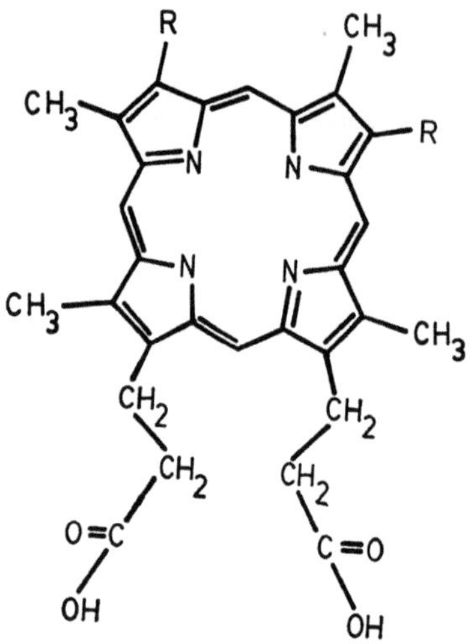

Fig. 9. Structure of synthetic porphyrins. R, $-CH=CH_2$ in protoporphyrin IX; R, -H in deuteroporphyrin IX, R, $-CH_2-CH_3$ in mesoporphyrin IX

Table 3. Optical absorption maxima of cobalt proto and meso insect hemoglobin IV at pH 6.5 and 22°C

Ligation state	Protoporphyrin IX CTT IV	Hb	Mb	Mesoporphyrin IX CII IV	Hb	Mb
Oxy	570	571	577	560	561	560
	538	538	539	528	530	529
	423	422	426	414	412	414
Deoxy	554	552	558	544	542	542
	403	402	406	395	392	395

For comparison, data from human hemoglobin (Hb) and sperm whale myoglobin (Mb) taken from Yonetani et al. 1974

gands, i.e., the pyrrole nitrogens, leads to an increase in O_2 affinity as shown for the monomeric insect hemoglobin CTT IV (Gersonde et al. 1982). If one replaces proto by meso heme, the latter has a less electron-withdrawing potential, the visible absorption spectra of both forms, cobalt and iron hemoglobins, become blue-shifted (see Table 3) and in the ESR spectra the ^{59}Co axial component of the hyperfine coupling (I=7/2) for the monomeric deoxy cobalt hemoglobin CTT IV increases, whereas the $^{14}N_\varepsilon$ hyperfine coupling (I=1) remains fairly constant (Yonetani et al. 1974; Gersonde et al. 1982).

The O_2 affinities for monomeric iron and cobalt hemoglobins CTT IV and myoglobin increase at temperatures >25°C with the order: proto < meso < deutero. At temperatures <25°C, the order of affinities changes to: proto < deutero < meso. Only the low temperature order of O_2 affinities is exactly consistent with the above mentioned rule of negative inductive effects of the groups in position 2 and 4 of the porphyrin ring. In particular, proteins and under special experimental conditions the heme-protein interactions, which introduce steric strains, can override the favorable electronic effect.

At high temperature, vinyl and ethyl heme substituents, also for a heme in the protein pocket, change their rotational position from the favored out-of-plane to the in-plane position, resulting in an increase of the electron-withdrawing effect and modifying the basi-

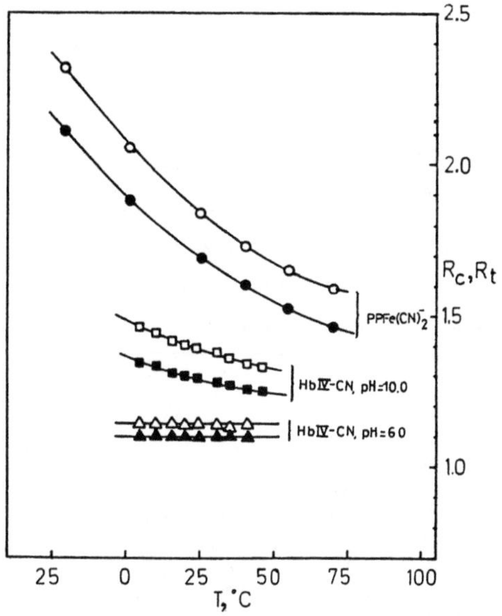

Fig. 10. Average rotational position of a vinyl group in a free heme and heme of the insect hemoglobin CTT IV. For definition of R_c and R_t see La Mar et al. 1978b. PPFe(CN)$_2^-$ (○ and ●), cyano-met CTT IV at pH 10.0 in its r state (□ and ■), and cyano-met CTT IV at pH 6.0 in its t state (Δ and ▲). The *open markers* are for R_c, and the *darkened markers* are for R_t

city of the pyrrole nitrogens. Therefore, steric effects change the relative O_2 affinities. It has been demonstrated by ^1H NMR investigations that at least one of the vinyl groups in CTT IV changes its rotational position depending on temperature and tertiary conformation (see Fig. 10; La Mar et al. 1978b). The in-plane position of the vinyl group is preferred at high temperature and in the t conformation (with low O_2 affinity). In the t state, the temperature dependence of the rotational probability is very small, whereas in the r state with its more flexible protein structure, the typical temperature dependence of a vinyl group with high degree of rotational freedom can be observed.

Although this vinyl group in CTT IV quantitatively reflects the Bohr effect, substitution of proto by meso and deutero heme has no influence on the magnitude of the Bohr effect (La Mar et al. 1978a; Gersonde et al. 1982). Therefore, we must assume that the porphyrin itself modulates the reversible O_2-binding, but does not control it, although it is sensitive to the allostery of the protein.

Metal Effect

The metal effect in hemoglobins can be demonstrated by comparing native with the respective cobalt-substituted proteins. The replacement of $3d^6$ Fe^{II} by a $3d^7$ Co^{II} center leads to low-spin complexes, where a considerable spin density exists in the d_{z^2} orbital. Therefore, cobalt hemoglobins or cobalt porphyrin model complexes show much smaller O_2 affinities than the respective iron complexes (see Table 4; Gersonde et al. 1982). The cobalt(II) complexes as strong σ donors lead to two or three orders of magnitude larger k_{off} values compared with the respective dioxygen-iron adducts, whereas k_{on} changes relatively little with substitution. The k_{off} values differ remarkably, when comparing different hemoglobins, indicating the particular influence of the protein moiety. The p_{50} values (half-saturation pressures) and k_{off} values increase in the following sequence for the iron form: legHb < Mb < CTT IV < HbA and for the cobalt form: legHb < Mb < HbA < CTT IV. The ratios $p_{50}(Co)/p_{50}(Fe)$ decrease in the

Table 4. O_2-binding equilibrium parameters of iron- and cobalt-substituted leghemoglobin (legHb), myoglobin (Mb), Chironomus hemoglobin IV (CTT IV), and human hemoglobin A (HbA) at 25°C and pH 7.0. The Hill coefficient is n = 1 and for HbA given in brackets

Parameters	Central atom	legHb	Mb		CTT IV		HbA	
		Proto	Proto	Meso	Proto	Meso	Proto	meso
p_{50} (mmHg)	Iron	0.05	0.90	0.54	1.33	0.32	12.2(2.9)	2.0(2.0)
p_{50} (mmHg)	Cobalt	22.2	50.0	50.0	309.0	165.0	125.0(2.2)	40.0(1.2)
$p_{50}(Co)/p_{50}(Fe)$		444.0	55.6	92.6	232.2	515.6	10.3	20.0

For references see Gersonde et al.(1982)

order: legHb > CTT IV > Mb > HbA. From the comparison of these sequences, the conclusion can be drawn that substitution for cobalt affects k_{off} and p_{50} of CTT IV much more than that of the other hemoglobins. This specific behavior in O_2-binding of CTT IV is due to the lack of the distal histidine-dioxygen interaction which will be discussed separately. Cobalt substitution has no influence on the allostery of the monomeric hemoglobins (Gersonde et al. 1982).

cis-Effect

The cis-effect in hemoglobin is defined here as the direct interaction of protein side groups with the bound dioxygen. This definition should not lead to confusion with the cis-effect described for isolated transition metal porphyrin complexes where it is a substituent effect (see also review by Buchler et al. 1978). In the cobalt forms of leghemoglobin and myoglobin hydrogen-bonding of the distal histidine to dioxygen stabilizes the $Co^{III}O_2^-$ moiety and diminishes the k_{off} value (Christahl et al. 1981). The distal histidine hinders the dioxygen to escape from its binding site. This distal histidine-dioxygen interaction can be demonstrated by ESR spectra which are pH-dependent for the oxy-form of cobalt leghemoglobin. By comparison with the oxy-form of cobalt CTT IV, the latter has no distal histidine, conclusions on the bond strengths between cobalt and the axial ligands, geometries of the Co-O-O grouping, and affinity states can be drawn (Gersonde et al. 1982).

In oxy-cobalt legHb the O_2 moiety exists at 77 K in two pH-dependent stereoelectronic structures characterized by distinct ^{59}Co, $^{14}N_\varepsilon$, and $^{14}N_{pyrrole}$ hyperfine tensors. Both structures differ with regard to the back-transfer of spin from dioxygen to cobalt, which is by about 40% larger in that spectral species which preferentially exists at low pH. The geometries of the Co-O-O grouping are also pH-dependent. At low pH, more equivalent oxygen atoms indicate a peroxo-like (olefin type) structure, whereas at high pH, a bond angle of about 120° indicates a superoxo-like (ozonoid type) structure. The high-pH form of leghemoglobin shows greater spin density on the dioxygen as reflected by the smaller hyperfine coupling at ^{59}Co, $^{14}N_\varepsilon$, and ^{14}N pyrroles. ESR spectroscopic parameters are correlated with functional states of some monomeric hemoglobins in Table 5.

The pH-dependent change of the ESR hyperfine pattern of oxy-leghemoglobin can be demonstrated for each form, low-pH (protonated), and high-pH (deprotonated form) by typical proton titration curves with a unity Hill coefficient and an inflection point at pH 6.4, indicative

Table 5. Correlation of ESR parameters with functional properties of monomeric oxy-cobalt hemoglobins

pH	ESR parameters	Hemoglobins leg Hb[a]	Mb[a]	CTT IV[b]
High pH	$g_{\parallel}$	2.078	2.080	2.078
	$a_{\parallel}$ (^{59}Co) (mT)	1.55	1.67	1.81
	$a_{\parallel}$ ($^{14}N_{\varepsilon}$) (mT)	0.44	n.d.	n.d.
	$a_{\parallel}$ ($^{14}N_{pyrrole}$) (mT)	0.20	n.d.	n.d.
	O_2 affinity state	low	low	high
	conformation state	t	t	r
	Co-O-O grouping	superoxo-like	superoxo-like	superoxo-like
	distal histidine-dioxygen interaction	no	no	no
	spin transfer Co → O_2	large	large	intermediate
	Co-O_2 bond strength	weak	weak	strong
	Co-N_{ε} bond strength	strong	strong	weak
Low pH	$g_{\parallel}$	2.092	2.080	2.077
	$a_{\parallel}$ (^{59}Co) (mT)	2.10	2.40	1.78
	$a_{\parallel}$ ($^{14}N_{\varepsilon}$) (mT)	0.65	n.d.	n.d.
	$a_{\parallel}$ ($^{14}N_{pyrrole}$)	0.24	n.d.	n.d.
	O_2 affinity state	high	high	low
	conformation state	r'	r'	t
	Co-O-O grouping	peroxo-like	peroxo-like	superoxo-like
	distal-histidine-dioxygen interaction	yes	yes	no
	spin transfer Co → O_2	small	small	large
	Co-O_2 bond strength	strong	strong	weak
	Co-N_{ε} bond strength	weak	weak	strong

[a]With acid Bohr effect; [b]with alkaline Bohr effect. For references see Gersonde et al. (1982)

for a distal histidine titration. The protonated form of the distal histidine is hydrogen-bonded to dioxygen stabilizing or inducing formation of the peroxo-like structure. Deprotonation at high pH leads to a change in the Co-O-O grouping with more superoxo-like structures. At higher temperature, however, only one bond geometry is observed because of fast relaxation between these two substates which are frozen at low temperatures.

In the case of CTT IV, the distal histidine is lacking. Consistent with the ESR data obtained for leghemoglobin and myoglobin, the axial hyperfine constant is close to that found for leghemoglobin with a deprotonated distal histidine. Thus, the Co-O-O grouping in CTT IV is superoxo-like, but nevertheless, pH-dependent. This pH-effect correlates with the alkaline Bohr effect, which has another origin and is not linked to the distal histidine dissociation.

trans-Effect

The trans-effect is linked to the axial ligand trans to the dioxygen, which is in hemoglobins, the proximal histidine. The geometry of oxy-hemoglobin is under intimate control of the relative energy of the d_{z^2} orbital, which is in turn controlled by the distance between the metal and the 5th ligand. Whereas all the other effects mentioned before have no influence on the allosteric control of the dioxygen-binding, the proximal histidine has. The distance between proximal histidine and central atom as well as the geometry of the metal-ligand grouping can be modified by conformation changes. Monomeric cobalt-substituted insect hemoglobins, which do not possess a distal histidine, but show a pH-dependent O_2 affinity change, are ideal systems to demonstrate the trans-effect between the axial ligands. Furthermore, in cobalt porphyrin model compounds, the dioxygen binding can be modified by adding nitrogen bases differing in basicity.

The ESR feature of the deoxy-form of cobalt CTT IV is not changed by pH reflecting an invariant electronic structure of the cobalt atom. The following questions arise: Is the deoxy state, i.e., the pentacoordinated state, frozen in only one conformation state or does the pH-induced conformation transition exist in the deoxy-state, but cannot affect the axial imidazole-cobalt bond and the electronic structure? By measuring the hyperfine-shifted resonances of the heme and the proximal imidazole protons, it has been shown that the imidazole exchangeable NH(1) proton does not shift with pH, whereas the methyl protons at the heme periphery are sensitive to pH-induced conformation changes (La Mar et al. 1981). Thus, we must assume that deoxy-CTT IV shows the pH-induced conformation transitions, which cannot be transmitted to the central atom and its axial ligands. The pH-independent on-rate constants for O_2-binding (Sick and Gersonde 1983) and the pH-independent temperature dependence of the quadrupole splitting of iron in deoxy-CTT IV (Parak and Gersonde 1983) are further evidence for this assumption.

The ESR spectra of oxy-cobalt CTT IV are pH-dependent (Gersonde et al. 1982). The pH-induced transition from the low-pH to the high-pH spectral form correlates with the transition from the low-affinity to the high-affinity state of this hemoglobin, reflecting the alkaline Bohr effect. Because distal histidine interaction is missing in CTT IV, the increase in the ^{59}Co hyperfine coupling in the low-affinity form is due to the trans-effect of the proximal histidine. In both conformation states, dioxygen is bound in the mode of superoxo-like bonding, however, with more bending in the high-pH form (high-affinity state) (see Table 5).

A direct correlation between the pK_a of structurally related axial bases and O_2 uptake have been observed for Co(PPIXDME)·(B) (see Fig. 11) (Jones et al. 1979; Stynes et al. 1973). Other synthetic porphyrins

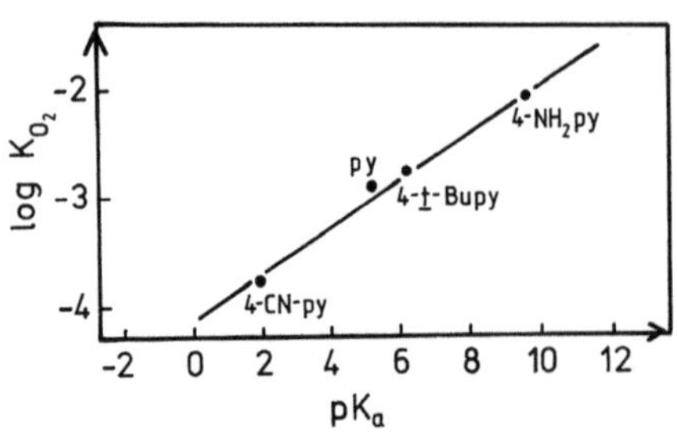

Fig. 11. Relation between log K_{O_2} for O_2·Co(PPIXDME)·B and pK_a for the axial base (B) in toluene at -45°C. (Data from Stynes et al. 1973)

Table 6. Energies for Fe-N_ε and O_2-Fe bonds in O_2-hem-imidazole complexes calculated by iterative extended Hueckel theory

Distance (pm)		Energy (eV)
Fe-N_ε	Fe-O_2	
196.0[a]	170.0	-2624.831
182.0	175.0	-2624.795
182.0	180.0	-2624.798
182.0[b]	185.0	-2624.799
182.0	190.0	-2624.787
210.0	170.0	-2624.809
210.0	165.0	-2624.817
210.0[c]	160.0	-2624.818
210.0	155.0	-2624.810

[a]Energy minimum; [b]energy minimum for t or low-affinity state; [c]energy minimum for r or high-affinity state

yield approximately the same results. With P-donor ligands again a straight line, as shown in Fig. 11, is obtained. This log K_{O_2} versus pK_a plot is steeper indicating differences in the π-bonding between base and Co^{II} center (Takayagi et al. 1975). Deviations from these simple linear correlations indicate additional sterical effects.

The trans-effect as a possible trigger of allostery in monomeric hemoglobins or in model systems can be demonstrated by semi-empirical calculations (Iterative Extended Hueckel Theory; Fleischhauer et al. 1983). As a model O_2-Fe-imidazole porphyrin has been used. The minimum in energy has been found for a Fe-N_ε distance of 196.0 pm and a corresponding Fe-O_2 distance of 170.0 pm (see Table 6). Decrease and increase of the iron-imidazole bond distances due to protein relaxation or constraints lead to energy minima, which are correlated with an increase and a decrease of the Fe-O_2 distance, respectively (see Table 6). The t or low-affinity state model is characterized by bond lengths of 182.0 pm and 185.0 pm for Fe-N_ε and Fe-O_2, respectively; the r or high-affinity state is characterized by bond lengths of 210.0 pm and 160.0 pm for Fe-N_ε and Fe-O_2, respectively.

Ligand-Effect

The ligand effect leads to differences in the allosteric response of the hemoglobins. If one compares ligands, like NO, CO, and O_2 with regard to the allosteric interaction energy, it seems to be dioxygen which shows the largest interaction energy. Therefore, the conformation of hemoglobins becomes most sensitive for allosteric interactions when O_2 is bound. Furthermore, the allosteric ligand effects are largest in hemoglobins without distal histidine. In the case of the monomeric insect hemoglobin CTT IV, the Bohr effect for different ligands decreases in the following sequence: $O_2 > CO > NO$ (Trittelvitz et al. 1973; Sick and Gersonde 1974).

Heme Disorder and Structural Heterogeneity

Heme disorder has been described 1978 for the first time for monomeric insect hemoglobins (La Mar et al. 1978a). The heme group is incorporated into the protein pocket by two ways due to the rotation of the heme about the α,γ -meso ayis (see Fig. 12; La Mar et al. 1978, 1980, 1981; Sick et al. 1980). Thus, a chemically homogeneous monomeric hemoglobin becomes structurally heterogeneous because of the nonsymmetric structure of the protoheme and because of the specific interactions of the heme substituents with the protein environment in both disorder states. The two heme disorder states interconverting at pH 7.0 and 25°C in the dimeric insect hemoglobin CTT II with a rate of 0.29 h^{-1} can be demonstrated by proton MR (Gersonde et al. 1983). Both heme disorder states of the hemoglobin differ with regard to the magnitude of the Bohr effect (La Mar et al. 1978a) and we have also experimental evidence that the O_2 affinities for both forms are different. If in future evidence for heme disorder in the tetrameric human hemoglobin is obtained, the consequence will be that the structural heterogeneity, due to the existence of the α and the β chains, is multiplied resulting in a much more complex nature of the cooperative O_2-binding curve. Then the allosteric models for hemoglobin have to be reevaluated and the O_2-binding curves have to be remeasured under new experimental conditions and reanalyzed.

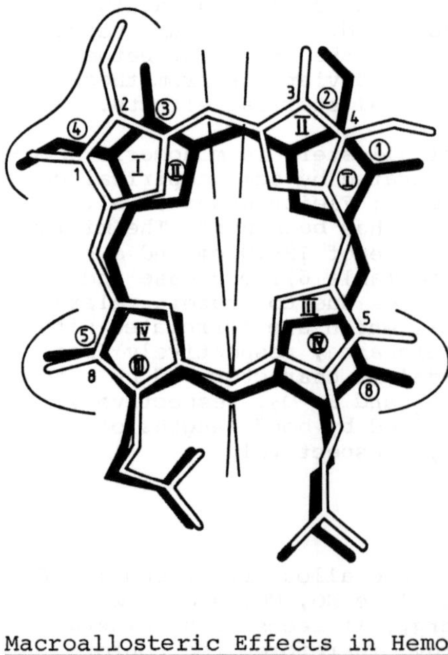

Fig. 12. Porphyrin-protein interactions in the protein pocket of monomeric and dimeric insect hemoglobins. The *full* and *open lines* and *numbers* represent the two heme disorder components, respectively. The two protein forms differ due to rotation of the heme about the α-γ-meso axis

Macroallosteric Effects in Hemoglobins

The allosteric control of the O_2-binding in tetrameric hemoglobins is based on the cooperativity and on a variety of heterotropic effects (2,3-bisphosphoglycerate, Cl^-, CO_2, Bohr effect). As extensive reviews are available (Perutz 1976, 1979; Kilmartin and Rossi-Bernardi 1973), no further detailed descriptions are given here. Most of the heterotropic interactions result in "right-shifting" of the O_2-binding curve which leads to an increase of the O_2 release capacity in a given O_2 gradient (Gersonde and Nicolau 1979; Gersonde and Nicolau

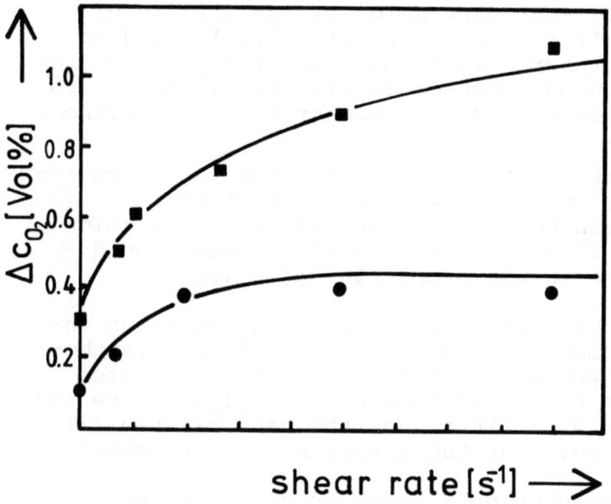

Fig. 13. Influence of the allostery of hemoglobin on the O_2 efflux from red blood cells under shearing conditions. ■ ■ ■, p_{50} = 23.7 mmHg; ● ● ●, p_{50} = 6.5 mmHg; blood flow, 5 ml/min

1982). O_2-binding with high exchange rates of the ligand is an absolute necessity to maintain and modulate an enhanced O_2 flux from red blood cells into the surrounding plasma. This purpose of the allosteric hemoglobin became even more important when nature started to use compact O_2 containers, like red blood cells. This molecular advantage of hemoglobin, however, can only be used, if an intracellular hemoglobin solution is rapidly transported by the tank tread-like motion of the red blood cell membrane (Zander and Schmid-Schönbein 1973; Mottaghy et al. 1982). Under shearing forces, red blood cells with a high fraction of low-affinity hemoglobin exhibit a higher O_2 flux than those with a high fraction of high-affinity hemoglobin (see Fig. 13; Mottaghy and Gersonde 1983). This can be demonstrated with red blood cells which are characterized by right-shifting of the hemoglobin-O_2-dissociation curve, i.e., high p_{50} values, due to an increase in the 2,3-bisphosphoglycerate level in the cells. The O_2 efflux increases if red blood cells are sheared with increasing rates (Mottaghy et al. 1982), however, more prominent if the p_{50} of red blood cells is high (Mottaghy and Gersonde 1983).

Thus, the understanding of all aspects of the reversible O_2-binding in hemoglobins and its allosteric control on the molecular level is just a starting point for the elucidation of the more complex interactions in a physiological system of O_2 transport.

Disscusion

Jungermann: Nature has provided a number of interesting hemoglobin variants in patients with hemoglobinopathies. Have you ever looked at these?

Gersonde: We have investigated hemoglobin M Iwate in more detail, which has only two binding sites (i.e., on the β-chains) and a quaternary structure which is frozen in the T state with low dioxygen affinity. With this model of hemoglobin, allosteric phenomena and O_2-binding properties linked to tertiary structural changes have been analyzed in more detail.

Karlson: I was quite impressed by your discussion of the fitting of protoporphyrin IX versus III. It is the first time that I heard an attempt to explain why protoporphyrin IX or more generally porphyrins derived biosynthetically from uroporphyrinogen III are "superior" to those of uroporphyrinogen I. However, your argument has not convinced me for two reasons:

(1) If the two carboxylate ions are the points of attachment, you can do the inversion also with protoporphyrin III instead of IX.
(2) More generally, proteins can be modelled around any small molecule. We know this from about 2000 enzymes. Therefore, one should look for some effects intricate in the prophyrin molecule.

Now comes my question. Uroporphyrinogen III is synthesized by an intricate rearrangement. The isomer I is formed much easier. Nevertheless, evolution has selected Type III. Was this molecule "invented" first, and the hemo*proteins* had to be modelled accordingly, or are there indications that some proteins were first, and the heme of Type III fitted better? I know, of course, that the answer must be speculative.

Gersonde: The asymmetric protoporphyrin IX seems to have the advantage of allowing a modulation of the globin structure via heme interconversion, i.e., the rotation of the heme about the α,β-meso axis results in different globin-heme interactions and thus in different functions of the two forms of hemoglobin. Heme interconversion may be another mechanism of changing hemoglobin functions. As the interconversion is very slow, hemoglobin may not exist in an equilibrium state during its life time. The interconversion happens also with symmetric porphyrins. However, the two forms are degenerate due to the identical globin-heme interactions in the two forms. As nearly all hemoglobins use protoporphyrin IX, it seems to be the globin which is modified during evolution to allow specific globin-heme interactions.

References

Antonini E, Brunori M (1971) Hemoglobin, Myoglobin in their Reactions with Ligands. North-Holland Publishing Company, Amsterdam
Armstrong RS, Irwin MJ, Wright PE (1980) Biochem Biophys Res Commun 95:682-689
Asakura T, Sono M (1974) J Biol Chem 249:7087-7093
Bolton W, Perutz MF (1970) Nature 228:551-552
Buchler JW, Kokisch W, Smith PD (1978) Structure and Bonding 34:79-134
Caughey WS, Barlow CH, Maxwell JC, Volpe JA, Wallace WJ (1975) Ann NY Acad Sci 244:1-9
Cerdonio M, Congiu-Castellano A, Moguo F, Pispisa B, Romani GL, Vitale S (1977) Proc Natl Acad Sci USA 74:398-400
Cerdonio M, Congiu-Castellano A, Calabrese L, Merante S, Pispisa B, Vitale S (1978) Proc Natl Acad Sci USA 75:4916-4919
Chien JCW, Dickinson LC (1972) Proc Natl Acad Sci USA 69:2783-2787
Christahl M, Raap A, Gersonde K (1981) Biophys Struct Mech 7:171-186
Christomanos AA (1973) Molekulare Biologie der Hämoglobine. Bd I-III. Scientific Greg Parisianos, Athen
Churg AG, Makinen MW (1978) J Chem Phys 68:1913-1925
Collman JP, Brauman JI, Halbert TR, Suslick KS (1976) Proc Natl Acad Sci USA 73: 3333-3337
Dickinson LC, Chien JCW (1980) Proc Natl Acad Sci USA 77:1235-1239
Dunn JB, Shriver DF, Klotz IM (1975) Biochemistry 14:2689-2695
Ellis J, Pratt JN (1973) J Chem Soc Chem Commun 781-782
Fleischhauer J, Meier U, Gersonde K (1983) Unpublished results
Floriani C, Calderazzo F (1969) J Chem Soc A:946-953

Gersonde K, Nicolau C (1979) Blut 39:1-7
Gersonde K, Nicolau C (1982) In: Ho C (ed) Hemoglobin and Oxygen Binding. Elsevier North Holland, Amsterdam, pp 277-282
Gersonde K. Sick H, Wollmer A, Buse G (1972) Eur J Biochem 25:181-189
Gersonde K, Twilfer H, Overkap M (1982) Biophys Struct Mech 8:189-211
Gersonde K, Chatterjee C, La Mar GN (1983) Unpublished results
Getz D, Melemud E, Silver BC, Dori Z (1975) J Am Chem Soc 97:3846-3847
Hemmingsen EA (1963) Comp Biochem Physiol 10:239-244
Hoenig HE, Gersonde K (1976) International Conference on Superconducting Quantum Devices in Berlin (West), Abstracts D3, pp 64-67. In: Lübbing H (ed) Superconducting Quantum Interference Devices and their Application. Walter de Gruyter, Berlin 1977, pp 250-254
Hoffman BM, Weschler CJ, Basolo F (1976) J Am Chem Soc 98:5473-5482
Holden van KE, Miller KJ (1982) Quart Rev Biophys 15:1-129
Huber R, Epp O, Steigemann W, Formanek H (1971) Eur J Biochem 19:42-50
Jones RD, Summerville DA, Basolo F (1979) Chem Rev 79:139-179
Kilmartin JV, Rossi-Bernardi L (1973) Physiol Rev 53:836-890
Kirchner RF, Loew GH (1977) J Am Chem Soc 99:4639-4647
Kurtz DM Jr, Shriver DF, Klotz IM (1976) J Am Chem Soc 98:5033-5035
Kurtz DM Jr, Shriver DF, Klotz IM (1977) Coord Chem Rev 24:145-178
La Mar GN, Overkamp M, Sick H, Gersonda K (1978a) Biochemistry 17:352-351
La Mar GN, Viscio DB, Gersonde K, Sick H (1978b) Biochemistry 17:361-367
La Mar GN, Smith KM, Gersonde K, Sick H, Overkamp M (1980) J Biol Chem 255:66-70
La Mar GN, Anderson RR, Budd DL, Smith KM, Langry KC, Gersonde K, Sick H (1981) Biochemistry 20:4429-4436
Llinas M (1973) Structure and Bonding 17:135-220
Makinen MW (1979) In: Biochemical and Clinical Aspects of Oxygen. Academic, New York, pp 143-155
Makinen MW, Churg AK, Glick HA (1978) Proc Natl Acad Sci USA 75:2291-2295
Mangum CP (1980) Amer Zool 20:19-38
Markl J, Decker H, Linzen B (1982) Hoppe Seyler's Z Physiol Chem 363:73-87
Maxwell JC, Volpe JA, Barlow CH, Caughey WS (1974) Biochem Biophys Res Commun 58:166-171
Mottaghy K, Gersonde K (1983) Unpublished results
Mottaghy K, Haest CWM, Schleuter HJ (1982) Chem Eng Commun 15:157-167
Parak F, Gersonde K (1983) Unpublished results
Pauling L, Coryell CD (1936) Proc Natl Acad Sci USA 22:210-216
Perutz MF, Muirhead A, Cox JM, Goaman LCG (1968) Nature 219:131-139
Perutz MF (1976) Brit Med Bull 32:195-212
Perutz MF (1979) Ann Rev Biochem 48:327-386
Perutz MF (1980) Proc R Soc Lond B 208:135-162
Préaux G, Gielens C (1983) In: Lontie R (ed) Copper Proteins and Copper Enzymes, Vol II. CRC Boca Raton, in press
Rimai L, Salmeen I, Petering DH (1975) Biochemistry 14:378-382
Schonbaum GR, Houtchens RA, Caughey WS (1979) Biochemical and Clinical Aspects of Oxygen. Academic, New York, pp 195-211
Sick H, Gersonde K (1974) Eur J Biochem 45:313-320
Sick H, Gersonde K (1983) Unpublished results
Sick H, Gersonde K, La Mar GN (1980) Hoppe-Seyler's Z Physiol Chem 361:333-334
Spartalian K, Lang G (1975) J Chem Phys 63:5375-5382
Stenkamp RE, Jensen LH (1979) Advances in Inorgan Biochem 1:219-233
Stenkamp RE, Siecker LC, Jensen LH, Sanders-Loehr J (1981) Nature 291:263-264
Stynes DV, Stynes HC, James BR, Ibers JA (1973) J Am Chem Soc 95:1796-1801
Takayagi T, Yamamoto H, Kwan T (1975) Bull Chem Soc Jpn 48:2618-2622
Trautwein AX (1974) Structure and Bonding 20:101-167
Trittelvitz E, Sick H, Gersonde K, Rüterjans H (1973) Eur J Biochem 35:122-125
Vaska L (1963) Science 140:809-810
Vaska L (1976) Accounts Chem Res 9:175-183
Wallace WJ, Maxwell JC, Caughey WS (1974) Biochem Biophys Res Commun 57:1104-1110
Weiss JJ (1964) Nature (London) 202:83-84, 182-183
Weschler CJ, Hoffman BM, Basolo F (1975) J Am Chem Soc 97:5278-5280

Wittenberg JB, Wittenberg BA, Peisach J, Blumberg WE (1967) Proc Natl Acad Sci USA 67:1846-1853
Wittenberg JB, Bergersen FJ, Appleby CA, Turner GL (1974) J Biol Chem 249:4057-4066
Yonetani T, Yamamoto H, Woodrow III GV (1974) J Biol Chem 249:682-690
Zander R, Schmid-Schönbein H (1973) Respiration Physiology 19:279-289

Cytochrome *c* Oxidase and Related Enzymes

B. G. Malmström[1]

Introduction

Historic Notes

In 1929 Otto Warburg, in whose honor this Mosbach Colloquium is held, reported a most impressive experimental achievement with the *Atmungsferment*, nowadays known as cytochrome *c* oxidase. He showed that the action spectrum for the photochemical reversal of the CO inhibition of respiration displayed the characteristics of a heme protein (Warburg and Negelein 1929). Despite the clear implication of this observation, David Keilin long remained unconvinced that the oxidase is a cytochrome and as late as 1938, he advanced several arguments for it being a copper-protein compound (Keilin and Hartree 1938). We now know that the views of these two pioneer investigators were, in fact, both correct: the functional unit of cytochrome oxidase contains two heme A prosthetic groups and two ions of copper (Malmström 1982).

The Asymmetric Nature of Cytochrome Oxidase and Related Enzymes

Cytochrome oxidase is structurally and functionally asymmetric (Malmström 1982). The protein environment imparts different properties to the two heme A groups, which are consequently distinguished as cytochrome a and cytochrome a_3. The two copper ions are also bound in different ways to the protein and designated Cu_A (also known as Cu_d or Cu_a) and Cu_B (also known as Cu_u or Cu_{a3}). In the oxidized forms of the enzyme, the metals are present as Fe^{3+} and Cu^{2+}, respectively. It is, therefore, surprising that only two of the metal components, namely, cytochrome a^{3+} and Cu_A^{2+}, are detectable by EPR, as Fe^{3+} and Cu^{2+} are both paramagnetic. The fact that cytochrome a_3^{3+} and Cu_B^{2+} are EPR-undetectable can be explained on the basis of a structural model, in which these components form a bimetallic (binucelar) site with an antiferromagnetic coupling between the two metal ions (Van Gelder and Beinert 1969; Tweedle et al. 1978).

The two substrates of cytochrome oxidase, reduced cytochrome *c* and dioxygen, react at separate sites, which are available from different sides of the inner mitochondrial membrane (Wikström et al. 1981). Cytochrome *c* rapidly donates electrons, one at a time, to the EPR-detectable centers, cytochrome a and Cu_A (Antalis and Palmer 1982). The two electrons are then transferred intramolecularly to the bimetallic center, cytochrome a_3 - Cu_B, which is the dioxygen-reducing site (Clore et al. 1980).

The so-called blue oxidases, i.e., ascorbate oxidase, ceruloplasmin, and laccase, like cytochrome oxidase, catalyze the four-electron re-

1 Institutionen för biokemi och biofysik, Göteborgs universitet and Chalmers tekniska högskola, S-412 96 Göteborg, Sweden

34. Colloquium - Mosbach 1983
Biological Oxidations

duction of dioxygen to two molecules of water. They show great structural and functional similarities to cytochrome oxidase (Malmström 1982; Reinhammar 1983), despite the fact that they contain copper only as prosthetic metal ions. Laccase, from the tree *Rhus vernicifera* and the fungus *Polyporus versicolor*, is the best characterized member of the group. It contains four copper ions in its functional unit. In the oxidized enzyme, two of these are EPR-detectable Cu^{2+} ions, designated type 1 and type 2, respectively, which, like cytochrome *a* and Cu_A in cytochrome oxidase, are the primary acceptors of electrons from the reducing substrate (Andréasson and Reinhammar 1976). The two EPR-undetectable ions form an exchange-coupled $Cu^{2+}-Cu^{2+}$ pair (Petersson et al. 1978; Dooley et al. 1978), called type 3, and this is the dioxygen-reducing site, like the cytochrome a_3-Cu_B unit in the oxidase. The reduction of the type 3 copper pair occurs by an intramolecular transfer of two electrons from the reduced type 1 and 2 sites (Andréasson and Reinhammar 1979).

Scope of the Chapter

In this communication, I will review some recent advances in our knowledge of the structure of the metal centers in cytochrome oxidase and in laccase. First, I will discuss how a protein-forced distorted site geometry of the type 1 center in laccase and other blue copper proteins is responsible for their high reduction potentials and facile electron-transfer kinetics. I will also make a comparison of the type 1 site in laccase with the Cu_A center in cytochrome oxidase.

The second major topic will be the nature of the bimetallic sites in both cytochrome oxidase and laccase. In particular, I will present recent evidence that the coordination environment of Cu_B in cytochrome oxidase is closely related to the type 3 center in laccase. Finally, I will discuss the mechanism of dioxygen reduction at the bimetallic centers. Emphasis will be placed on some new results, which allow a detailed proposal for the structure of the paramagnetic intermediate in the cytochrome oxidase reaction. In addition, a catalytic reaction cycle for cytochrome oxidase will be presented.

A Rack Mechanism for Blue Copper Centers

The Unique Properties of Blue Copper

Blue copper proteins have unique optical and EPR spectra, as illustrated in Fig. 1. When these proteins were first isolated, colored Cu^{2+} complexes with extinction coefficients of the order of 10^3 M^{-1} cm^{-1} were not known, which made coordination chemists suggest that the strong absorption bands were due to charge-transfer transitions involving Cu^{1+} (Orgel 1958). Together with Vänngård, I showed, however, with the aid of EPR, that ceruloplasmin and laccase contain Cu^{2+} (Malmström and Vänngård 1960). Furthermore, we found that the EPR spectra showed exceptionally small hyperfine splitting constants compared to a large range of other Cu^{2+} complexes, including natural as well as artificial Cu^{2+} proteins, such as erythrocuprein (superoxide dismutase) and Cu^{2+}-carbonic anhydrase (cf. Fig. 1). We suggested that the unique spectral properties could be related to a coordination environment forced upon the Cu^{2+} ion by the protein conformation. This interpretation was in accordance with the rack hypothesis introduced by Lumry and Eyring (1954) to explain alterations in the reactivity of amino acid side chains and metal-ion prosthetic groups observed in the active sites of enzymes and other biologically

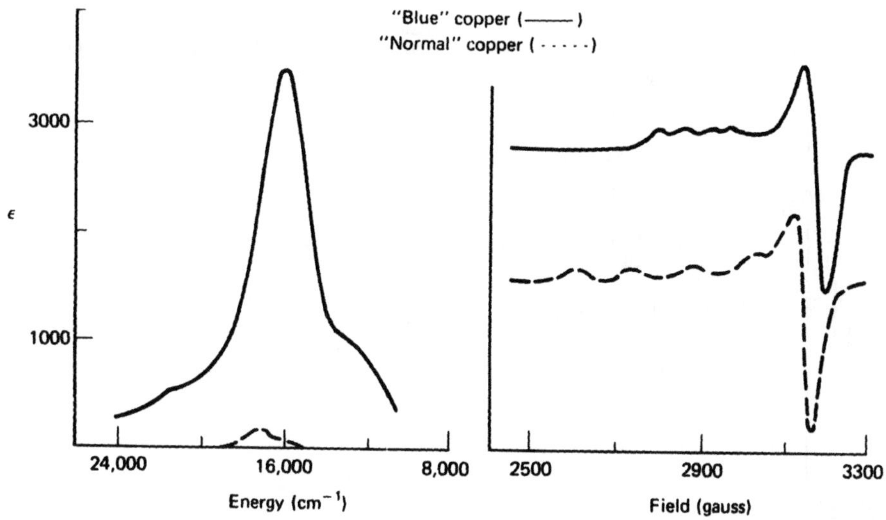

Fig. 1. Comparison of the optical and EPR spectra of a blue copper site and a normal tetragonal copper center. (Gray and Solomon 1981)

active proteins. The hypothesis stated that key functional groups are distorted by the overall protein conformation and this introduces a strain leading to anomalous properties.

Experimental developments in the decades following our original interpretation (Malmström and Vänngård 1960) have now made it possible to perform a more concrete and quantitative analysis of the rack mechanism as a basis for the unique coordination properties of blue copper proteins. Three types of new information, in particular, provide cornerstones of the extended hypothesis. First, the three-dimensional structure of two single-site blue proteins, plastocyanin and azurin, has been determined (Freeman 1981). This has unabmiguously identified the amino acid side chains coordinated to the Cu atom and revealed the coordination geometry around that atom. Second, extensive spectroscopic investigations based on CD, MCD, and EPR measurements have allowed a complete description of the electronic structures of blue copper centers (Gray and Solomon 1981). This showed that the *d-d* transition energies of blue proteins are relatively low. Third, the thermodynamic parameters for electron transfer to blue copper centers have been determined in several cases (Taniguchi et al. 1980; Taniguchi et al. 1982). In general, the variation observed in reduction potentials appears to be dominated by changes in reaction enthalpy.

Protein-Forced Ligand Fields in Blue Copper Proteins

Together with Gray, I have estimated the minimum protein conformational energy that is required to force Cu^{2+} to adopt the special structure of a blue site (Gray and Malmström 1983). We also discussed how this distorted structure is related to the spectroscopic properties and reduction potentials of the proteins. The starting point for our analysis was the blue center in plastocyanin, whose structure in the vicinity of the copper is known in detail (Freeman 1981).

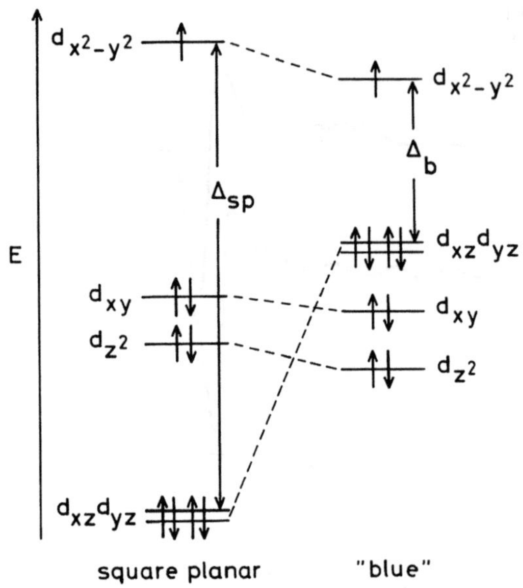

Fig. 2. Energy diagram for the *d* orbitals of Cu^{2+} in a square planar site and a blue copper center

The Cu atom in plastocyanin is coordinated by the side chains of His-37, Cys-84, His-87, and Met-92. The coordination geometry is that of a considerably distorted tetrahedron. Usually Cu^{2+} complexes have square planar or tetragonal geometries. The distortion from the preferred geometry results in a considerable change in the energy scheme for the *d* orbitals of the metal ion, as illustrated in Fig. 2 and this is reflected in the relatively low *d*-*d* transition energies observed. The d_{xz} and d_{yz} orbitals undergo the largest change in energy in going from square planar coordination to the geometry of a blue site, whereas the energies of the d_{xy} and $d_{z}2$ orbitals are relatively unaffected. A minimal model for the calculation of the ligand-field (LF) destabilization of the CuN_2SS^* site need consequently only include the $Cu(x^2-y^2)$ and Cu(xz,yz) antibonding levels. The difference $(\Delta_{sp}-\Delta_b)$ (Fig. 2) is at least 9000 cm^{-1} (Solomon et al. 1980), which means that the minimum blue site electronic destabilization is 2/3 (9000 cm^{-1}) = 6000 cm^{-1}.

The observation that the plastocyanin binding site is not distorted significantly upon incorporation of Cu^{2+} in the apoprotein (Freeman 1981) clearly shows that the energy for the electronic destabilization is provided by the protein conformation in accordance with the rack mechanism. The lowest energy of a protein conformation that would accommodate a planar CuN_2SS^* unit must be over 70 kJ mol^{-1} (~6000 cm^{-1}) above that of the observed conformation. This is well above the energy rhat would be released if Cu^{2+} could choose its own preferred geometry, with the consequence that the metal ion is forced to accept the site geometry that the protein presents. This protein-forced LF-destabilization is undoubtedly related to the high reduction potentials and facile electron transfer kinetics observed for blue copper proteins (Reinhammar 1983).

The model of a protein-forced distorted site geometry of blue copper centers can also provide an explanation for the puzzling observation that the energy of the $\sigma S \rightarrow Cu(x^2-y^2)$ charge-transfer transition does

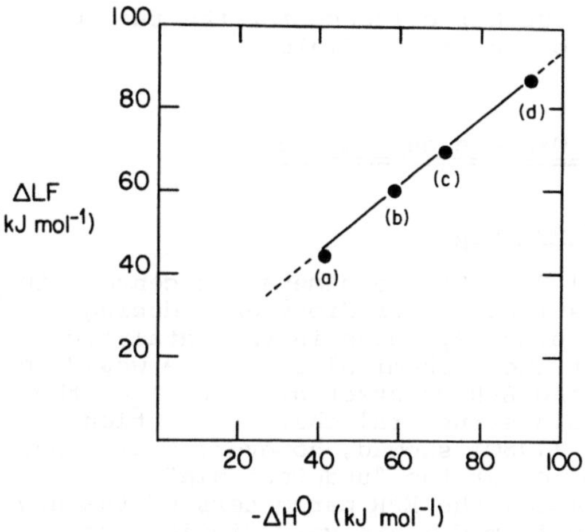

Fig. 3. Plot of ΔLF versus $-\Delta H^o$ for stellacyanin (*a*), plastocyanin (*b*), azurin (*c*), and fungal laccase (*d*). (Gray and Malmström 1983)

not vary significantly from protein to protein (Gray and Solomon 1981), whereas the standard chemical reduction potential does (Taniguchi et al. 1980; Reinhammar 1983). It is suggested (Gray and Malmström 1983) that small differences in the ligand-Cu^{2+} interaction, for example, in the length of copper-thioether bond, introduce a variation in π back bonding (Cu → L). This creates a higher ligand field for strong π interactions and at the same time, it confers a special stabilization to the Cu^{1+} state, leading to an increased reduction potential. Our electronic structural model is supported by an observed linear correlation of increasing ΔLF with decreasing electron-transfer reaction enthalphy ($-\Delta H^o$), as shown in Fig. 3. This correlation can be ascribed to the fact that both ΔLF and ΔH^o are related to the degree of antibonding in the Cu(xz,yz) level (cf. Fig. 2). Variations in Cu^{2+}-ligand binding interactions in the various blue copper sites are also evidenced by resonance-Raman spectra (Woodruff et al. 1983). The intense peaks in the 400-cm^{-1} region shift to higher energies as the ligand fields strengthen. It is thus apparent that the protein-forced LF model can provide a rationale for the common unique properties of blue copper centers and at the same time explain the variations between different blue sites.

Finally in this section, I will briefly discuss to what extent Cu_A in cytochrome oxidase can be considered a blue site. At first, this may appear unlikely, as the EPR parameters of Cu_A^{2+} put it in a class all by itself (Malmström 1982; Chan et al. 1983). Amino acid sequence homologies between subunit II and blue copper proteins (Buse et al. 1982), on the other hand, suggest that cytochrome oxidase does possess a blue site. In addition, there is evidence that Cu_A is bound to subunit II (Winter et al. 1980). The cytochrome oxidase subunit, unlike the blue proteins, has, however, *two* conserved cysteine residues in its sequence. Chemical modification (Darley-Usmar et al. 1981) and EXAFS measurements (Scott et al. 1981) suggest that these residues are ligands to Cu_A. Consequently, it has been proposed that the Cu_A site has a CuN_2S_2 coordination, compared to the CuN_2SS^* sites in blue proteins and Stevens et al. (1982) have provided additional evidence for such a model. In elegant EPR and ENDOR experiments with isotopically substituted yeast cytochrome oxidase, they could unambiguously identify cysteine and histidine residues as ligands to Cu_A. The data also indicate that substantial spin density is delocalized onto a cysteine sulfur to render the site $Cu^{1+} - S^{\cdot}$ in the oxidized enzyme. These results then provide further evidence for the tuning of copper site properties by protein-forced ligand fields.

The Bimetallic Sites in Cytochrome Oxidase and Laccase

A Comparison Between Type 3 Copper and Cu_B

Earlier it was mentioned (Section 1.2) that the bimetallic centers in cytochrome oxidase and laccase constitute their dioxygen-reducing sites. Consequently, these centers play key roles in the catalytic mechanisms, so that knowledge about their chemical structure would be of great interest. Unfortunately, the EPR-undetectable nature of the sites has long posed an obstacle to a structural characterization. A few years ago, Reinhammar et al. (1980) showed, however, that under some conditions, it is possible to render the "undetectable" Cu^{2+} centers detectable by EPR. Furthermore, the EPR parameters of the new Cu^{2+} signals from type 3 copper and from Cu_B are very similar, as shown in Table 1, which also includes data for two other proteins having Cu^{2+} in bimetallic sites (half-met hemocyanin, superoxide dismutase). This suggests that Cu^{2+} found in bimetallic structures in several different proteins has similar coordination sites. The coordination is neither of type 1 nor of type 2 character. This means that Cu_B in cytochrome oxidase is not a blue copper.

Table 1. EPR parameters for type 3 and Cu_B signals (Reinhammar et al. 1980)

Protein	g_x	g_y	g_z	A_z (mT)
Fungal laccase	2.025	2.148	2.268	12.5
Tree laccase	2.05	2.150	2.305	7.8
Cytochrome oxidase	2.052	2.109	2.278	10.2
Half-met hemocyanin	2.05	2.12	2.302	11.6
Superoxide dismutase	2.025	2.103	2.257	13.2

Table 2. Ligand hyperfine couplings in ENDOR spectra of type 3 Cu^{2+} in tree laccase and Cu_B^{2+} in cytochrome oxidase (Cline et al. 1983)

Protein	Peak (α)	Hyperfine coupling	$A^N(\alpha)$ (MHz)
		A_x	A_z
Laccase	1	50	40
	2	45	36
	3	36	30
Cytochrome oxidase	1		40
	2		35
	3		28

Reinhammar, in collaboration with Hoffman's group, has now provided additional evidence for similar structures of the type 3 and Cu_B sites (Cline et al. 1983). With tree laccase, a type 3 EPR signal was produced by partial reduction of the type 2 copper-depleted enzyme (Reinhammar 1982), after which ^{14}N, ^{1}H, and $^{63,65}Cu$ ENDOR data were collected. The results indicate that type 3 Cu^{2+} is coordinated to three imidazole residues with H_2O or OH^- as the fourth ligand. In Table 2, the ^{14}N ENDOR data for type 3 copper are compared with results for

Cu_B and it is apparent that the metal sites in the two proteins must have closely related structures. There is no indication from proton resonances for a cysteine side chain as a ligand to Cu_B.

The bimetallic sites in cytochrome oxidase and tree laccase also display similarities in their reactions with nitric oxide (Brudvig et al. 1980; Martin et al. 1981). In the presence of reductant, the sites in both enzymes are oxidized by NO to their half-oxidized states. The oxidized enzymes, on the other hand, are reduced by NO and then reoxidized to form catalytic cycles in which NO_2^- and N_2O are produced.

In view of the spectroscopic and chemical comparisons just presented, it is interesting that the type 3 copper-containing protein, ceruloplasmin, shows sequence homologies with cytochrome oxidase as well as with superoxide dismutase in a region different from the blue site (Takahashi et al. 1983). It would appear that multisite metalloproteins, such as cytochrome oxidase and blue oxidases, have been synthesized by a combination of individual polypeptide stretches, each one formed by divergent evolution of genes for ancestral proteins (Malmström 1980).

The Mechanism of Dioxygen Reduction

Dioxygen is quite an inert molecule from a kinetic point of view. One reason for its kinetic stability is related to the reduction potentials for the successive steps in a stepwise four-electron reduction (Malmström 1982). The potential for the O_2/O_2^- couple is such that the first one-electron step is associated with a considerable energy barrier when the electron donor has a high, positive reduction potential, which is the case with both cytochrome oxidase and laccase. This is undoubtedly one reason for the dioxygen-reducing sites being bimetallic centers in these enzymes. In their two-electron reduced form, they can then reduce dioxygen directly to the level of peroxide, in this way by-passing the energetically unfavorable formation of superoxide. With cytochrome oxidase, it has been demonstrated that the dioxygen reaction with the fully reduced enzyme involves three consecutive intermediates, the first one of which has the cytochrome a_3-Cu_B-O_2 unit formally at the level of peroxide (Clore et al. 1980). The second intermediate is paramagnetic (Karlsson et al. 1981), which means that the peroxide must be further reduced in separate one-electron steps. Laccase has also been shown to form a paramagnetic oxygen intermediate, which has dioxygen formally at the three-electron reduced level (Aasa et al. 1976).

The ready formation and stabilization of peroxide in the reaction of dioxygen with reduced bimetallic complexes is well-established in simple systems (Jones et al. 1979), but the further reduction of the peroxide intermediate to the paramagnetic intermediate has no good inorganic analogs. Consequently, an important clue to the mechanism of dioxygen reduction should center around the chemical nature of intermediate II (the paramagnetic intermediate) in cytochrome oxidase. An analysis of its unusual EPR spectrum, including the effect on the spectrum of using dioxygen enriched in ^{17}O has led to a detailed structural proposal: $Cu_B^{2+}-{}^-OH{-}{-}{-}O=Fe_{a_3}^{IV}$ (Hansson et al. 1982). A brief recapitulation of the arguments for this structure will be given.

The presence of four hyperfine lines in the EPR spectrum of the intermediate demonstrates that Cu_B is in the Cu^{2+} state. The unusual relaxation properties (Karlsson et al. 1981) show that the nearby $[Fe_{a_3}O]^{2+}$ unit must also be paramagnetic and a ferryl ion structure seems most

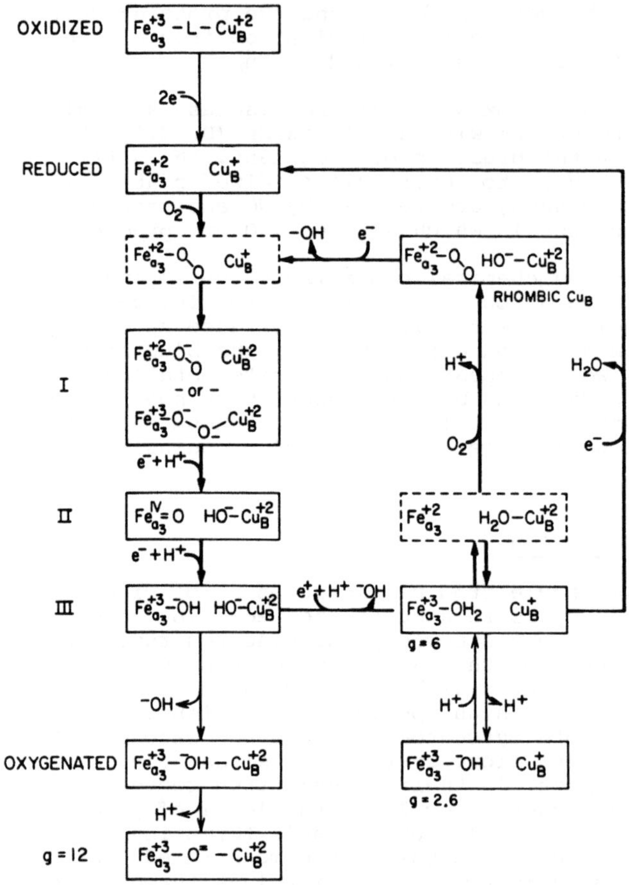

Fig. 4. Proposed mechanism for the reactions of dioxygen with the cytochrome a_3 - Cu_B site in cytochrome oxidase. (Blair et al. 1983)

likely. The broadening effect of ^{17}O means that an oxygen atom derived from dioxygen is coordinated to Cu_B^{2+}. This same oxygen atom cannot be bound also to Fe_{a_3}, as a μ-oxo bridge should result in a strong exchange interaction; an analysis by perturbation theory has shown an interaction energy around 10 cm^{-1} (Hansson et al. 1982). A hydrogen-bonded structure can, on the other hand, lead to an interaction of the observed order of magnitude.

Unfortunately, the proposed structure is not a unique interpretation of the experimental findings. If the electron, which is added to intermediate I (the peroxide intermediate), reduces $Fe_{a_3}^{3+}$ and at the same time, a proton is added to the peroxide, then a compound of the following structure would be formed: $Cu_B^{2+} - O^-_{\,H} - O - Fe_{a_3}^{2+}$. In this $Fe_{a_3}^{2+}$ would be expected to be high-spin ($S = 2$), which could also account for the relaxation properties. The observed zero-field splitting parameters (Hansson, unpublished experiments) are, however, the same as for Fe^{4+} in compounds I and II of horseradish peroxidase and in the myoglobin-H_2O_2 complex, which supports the ferryl ion alternative.

On the basis of spectroscopic observations during turnover, together with the information on intermediates just reviewed, Wilson et al. (1982) proposed a catalytic cylce for cytochrome oxidase. Additional experimental results have already necessitated a modification and extension of our scheme. In my opinion, the best up-to-data reaction cycle is that formulated by Blair et al. (1983), which is given in Fig. 4. The structures shown for intermediates II and III differ slightly from the proposals of Wilson et al. (1982), who assumed that the oxygen atom associated with Cu_B had been released in the form of water. Our ^{17}O results (Hansson et al. 1982), however, support the structures in Fig. 4.

The scheme in Fig. 4 also includes the other species that have been observed by EPR, such as those labelled g = 2.6, g = 6, g = 12, rhombic Cu_B, and oxygenated (Brudvig et al. 1981; Malmström 1982). None of these enzyme forms can be trapped during turnover (Wilson et al. 1982), showing that their steady-state concentrations are low. It is particularly striking that the odd-electron species, intermediate II and the g = 6 form are not detectable. This may suggest that successive one-electron steps are tightly coupled, so that in a kinetic sense, the four electron reduction occurs in two-electron steps, as I proposed over a decade ago (Malmström 1970).

Acknowledgements. I wish to thank Drs. R Aasa, SI Chan, HB Gray, B Reinhammar, and T Vänngård for stimulating collaboration and many useful discussions. My own investigations have been supported by grants from Statens naturvetenskapliga forskningsråd.

Discussion

Buse: I would like to make a comment on the structure of cytochrome *c* oxidase.

From electron microscopic studies with membrane crystals beef heart cytochrome *c* oxidase is known to be a Y-shaped complex with the two arms imbedded in the inner mitochondrial membrane (the M_1 and M_2 domains) and the stem with half of the molecular mass protruding at the cytoplasmic side (the C-domain) (1) Recently, this structure has been confirmed using our preparation of the beef heart oxidase (2,3). It is thus assured that the 2 copper, 2 heme a, proton pumping enzyme complex purified to 10 nmol heme a/mg protein consists of the protein components described in the Table.

Protein chemical data on beef heart cytochrome *c* oxidase (4)

Polypeptide	Mr	Residues	Synthesis	Stoichiometry	Hydrophobic-domains	N-terminal sequences
I	56993[a]	514	Mit.	1	>2, Core	Formyl-Met-Phe-Ile-Asn-
II	26049	227	Mit.	1	2	Formyl-Met-Ala-Tyr-Pro-
III	29918[a]	261	Mit.	1	>2, Core	(Met)Thr-His-Gln-
IV	17153	147	Cyt.	1	1	Ala-His-Gly-Ser-
V	12436	109	Cyt.	1	none	Ser-His-Gly-Ser-
VIa	10670	98	Cyt.	1	none	Ala-Ser-Gly-Gly-
VIb	9419	85	Cyt.	1	none	Ala-Ser-Ala-Ala-
VIc	8480	84	Cyt.	1	1	Ser-Thr-Ala-Leu-
VII	10068	73	Cyt.	1	none	Acetyl-Ala-Glu-Asp-Ile-
VIIIa	5441	47	Cyt.	1	1	Ser-His-Tyr-Glu-
VIIIb	4962	46	Cyt.	2	1	Ile-Thr-Ala-Lys-
VIIIc	6244	56	Cyt.	1	1	Phe-Glu-Asn-Arg-

Complete primary structures for all the 12 components have now been obtained and show the functional monomer to contain altogether 1793 amino acids with Mr close to 200 000. The sequence analysis shows the presence of at least 11 hydrophobic membrane penetrating sequence stretches.

The data represent the complex pattern of a mammalian type cytochrome *c* oxidase, while at the other end of the evolutionary scale, the 2 copper, 2 heme a, proton pumping cytochrome oxidase of *Paracoccus denitrificans* consist of only 2 subunits (5). Our recent investigations of the primary structures of this polypeptide show a surprisingly close homology of these subunits to subunits I and II (Table) of the eukaryotic oxidases. Taken together, these findings imply the following conclusions:

1) The mitochondrial subunits I and II fo the eukaryotic oxidases bind the coppers and hemes a and contain the coupling site for e^- and H^+ translocation.
2) While subunit II is the substrate binding and e^- accepting site of the enzyme subunit, I may contain the O_2 activating center, i.e., display the functions, which Warburg termed "Atmungsferment".
3) The folding pattern of subunit II implies that all the metal centers are localized in the C-domain of the oxidase.
4) The mitochondrial subunit III probably functions as a proton channel from the m-side of the inner membrane to the C-domain of the oxidase.
5) There can be no doubt about the symbiotic origin of mitochondria.

References

1. Fuller SD, Capaldi RA, Henderson R (1979) J Mol Biol 134:305-327
2. Steffens GJ, Buse G (1976) Hoppe-Seyler's Z Physiol Chem 357:1125-1137
3. Fuller SD, Capaldi R, Henderson R (1982) Biochem 21:2525-2529
4. Buse G, Steffens GJ, Steffens GCM, Sacher R, Erdweg M (1982) Primary Structure and Function of Cytochrome c oxidase Subunits. In: Chien H (ed) "Electron Transport and Oxygen Utilization" (Elsevier North-Holland, Amsterdam) pp 157-163
5. Ludwig B, Schatz G (1980) Proc Natl Acad Sci (USA) 77:196-200

Buse: Is the very first e^- acceptor of the oxidase the heme a or could it be the copper A?

Malmström: This is a question that is still being debated. Capaldi has recently reported that cytochrome c binds very closely to Cu_A and recent work of Palmer shows that Cu_A and cytochrome a are in very rapid equilibrium; one cannot really distinguish which is first. One says that Cu_A and cytochrome a are reduced at the same time.

Hess: Please comment on the interaction of Cu_A with SH-groups in the near neighborhood.

Malmström: My suggestion of two interacting groups was taken from the results of Sunney Chan. His ENDOR data shows that there is definitely thiol coordination, but how many he cannot say. Of course, there has been a suggestion from Britton Chance that there is a sulfhydryl group sitting on Cu_B, but the ENDOR results of Brian Hoffman show there is no such thing on Cu_B, although he can find it on Cu_A. My suggestion is that there are two cysteines on Cu_A and none on Cu_B.

Mason: Please comment on your observation that the EPR spectrum of your Cu_B form is similar to that of half methemocyanin.

Malmström: It is a little misleading when I say the EPR spectra are similar, because if you look at all the EPR parameters, you find that they are not by any means identical. If a plot of $g_{\parallel}$ against $A_{\parallel}$ is made, these signals do not fall in the domains in which either Type I or Type II copper falls. If you take all these Type III or Cu_B signals, you find that they form a new domain, which is intermediate between Type I and Type II. If you look at just the spectra, you see they are not identical, but if you make such a plot you see they are all related.

References

Aasa R, Brändén R, Deinum J, Malmström BG, Reinhammar B, Vänngård T (1976) A ^{17}O-effect on the EPR spectrum of the intermediate in the dioxygen-laccase reaction. Biochem Biophys Res Comm 70:1204-1209

Andréasson L-E, Reinhammar B (1976) Kinetic studies of *Rhus vernicifera* laccase - Role of the metal centers in electron transfer. Biochim Biophys Acta 445: 579-597

Andréasson L-E, Reinhammar B (1979) The mechanism of electron transfer in laccase-catalysed reactions. Biochim Biophys Acta 568:145-156

Antalis TM, Palmer G (1982) Kinetic characterization of the interaction between cytochrome oxidase and cytochrome *c*. J Biol Chem 257:6194-6206

Blair DF, Martin CT, Gelles J, Wang H, Brudvig GW, Stevens TH, Chan Si (1983) The metal centers of cytochrome *c* oxidase: structures and interactions. Chemica Scripta 21:43-53

Brudvig GW, Stevens TH, Chan SI (1980) Reactions of nitric oxide with cytochrome *c* oxidase. Biochemistry 19:5275-5285

Brudvig GW, Stevens TH, Morse RH, Chan SI (1981) Conformations of oxidized cytochrome *c* oxidase. Biochemistry 20:3912-3921

Buse G, Steffens GJ, Steffens GCM, Sacher R, Erdweg M (1982) Primary structure and function of cytochrome *c* oxidase subunits. In: Ho C (ed) Electron transport and oxygen utilization. Elsevier Biomedical, New York, p 157

Chan SI, Martin CT, Wang H, Brudvig GW, Stevens TH (1983) The structure of the metal centers in cytochrome *c* oxidase. In: Bertini I, Drago RS, Luchinat C (eds) The coordination chemistry of metalloenzymes. D Reidel, Dordrecht, p 313

Cline J, Reinhammar B, Jensen P, Venters R, Hoffman BM (1983) ENDOR from the type 3 copper center of tree laccase and Cu_B of cytochrome *c* oxidase. J Biol Chem, 258:5124-5128

Clore GM, Andréasson L-E, Karlsson B, Aasa R, Malmström BG (1980) Characterization of the low-temperature intermediates of the reaction of fully reduced soluble cytochrome oxidase with oxygen by electron-paramagnetic-resonance and optical spectroscopy. Biochem J 185:139-154

Darley-Usmar VM, Capaldi RA, Wilson MT (1981) Identification of cysteines in subunit II as ligands to the redox centers of bovine cytochrome *c* oxidase. Biochem Biophys Res Commun 103:1223-1230

Dooley DM, Scott RA, Ellinghaus J, Solomon EI, Gray HB (1978) Magnetic susceptibility studies of laccase and oxyhemocyanin. Proc Natl Acad Sci USA 75:3019-3022

Freeman HC (1981) Electron transfer in 'blue' copper proteins. In: Laurent JP (ed) Coordination chemistry - 21. Pergamon, Oxford, p 29

Gray HB, Malmström BG (1983) On the relationship between protein-forced ligand fields and the properties of blue copper centers. Comments Inorg Chem, 2:203-209

Gray HB, Solomon EI (1981) Electronic structures of blue copper centers in proteins. In: Spiro TG (ed) Copper proteins. John Wiley, New York, p 2

Hansson Ö, Karlsson B, Aasa R, Vänngård T, Malmström BG (1982) The structure of the paramagnetic oxygen intermediate in the cytochrome *c* oxidase reaction. EMBO J 1:1295-1297

Jones RD, Summerville DA, Basolo F (1979) Synthetic oxygen carriers related to biological systems. Chem Rev 79:139-179

Karlsson B, Aasa R, Vänngård T, Malmström BG (1981) An EPR-detectable intermediate in the cytochrome oxidase-dioxygen reaction. FEBS Lett 131:186-188

Keilin D, Hartree EF (1938) Cytochrome *a* and cytochrome oxidase. Nature 141:870-871
Lumry R, Eyring H (1954) Conformation changes of proteins. J Phys Chem 58:110-120
Malmström BG (1970) Co-operation of electron-accepting sites in oxygen reduction by oxidases. Biochem J 117:15-16P
Malmström BG (1980) Metalloenzymes. In: Sigman DS, Brazier MA (eds) The evolution of protein structure and function. Academic, New York, p 87
Malmström BG (1982) Enzymology of oxygen. Annu Rev Biochem 51:21-59
Malmström BG, Vänngård T (1960) Electron spin resonance of copper proteins and some model complexes. J Mol Biol 2:118-124
Martin CT, Morse RH, Kanne RM, Gray HB, Malmström BG, Chan SI (1981) Reactions of nitric oxide with tree and fungal laccase. Biochemistry 20:5147-5155
Orgel LE (1958) Enzyme-metal-substrate complexes as coordination compounds. In: Crook EM (ed) Metals and enzyme activity. The University Pree, Cambridge, p 8
Petersson L, Ångström J, Ehrenberg A (1978) Magnetic susceptibility of laccase and ceruloplasmin. Biochim Biophys Acta 526:311-317
Reinhammar B (1982) Optical properties and a new epr signal from type 3 copper in metal-depleted *Rhus vernicifera* laccase. J Inorg Biochem, 18:113-121
Reinhammar B (1983) Metal coordination and mechanism of blue copper-containing oxidases. In: Bertini I, Drago RS, Luchinat C (eds) The coordination chemistry of metalloenzymes. D Reidel, Dordrecht, p 177
Reinhammar B, Malkin R, Jensen P, Karlsson B, Andréasson L-E, Aasa R, Vänngård T, Malmström BG (1980) A new copper (II) electron paramagnetic resonance signal in two laccases and in cytochrome *c* oxidase. J Biol Chem 255:5000-5003
Scott RA, Cramer SP, Shaw RW, Beinert H, Gray HB (1981) Extended X-ray absorption fine structure of copper in cytochrome *c* oxidase: Direct evidence for copper-sulfur ligation. Proc Natl Acad Sci USA 78:664-667
Solomon EI, Hare JW, Dooley DM, Dawson JH, Stephens PJ, Gray HB (1980) Spectroscopic studies of stellacyanin, plastocyanin, and azurin. Electronic structure of the blue copper sites. J Am Chem Soc 102:168-178
Stevens TH, Martin CT, Wang H, Brudvig GW, Scholes CP, Chan SI (1982) The nature of Cu_A in cytochrome *c* oxidase. J Biol Chem 257:12106-12113
Takahashi N, Bauman RA, Ortel TL, Dwulet FE, Wang C-C, Putnam FW (1983) Internal triplication in the structure of human ceruloplasmin. Proc Natl Acad Sci USA 80:115-119
Taniguchi VT, Malmström BG, Anson FC, Gray HB (1982) Temperature dependence of the reduction potential of blue copper in fungal laccase. Proc Natl Acad Sci USA 79:3387-3389
Taniguchi VT, Sailasuta-Scott N, Anson FC, Gray HB (1980) Thermodynamics of metalloprotein electron transfer reactions. Pure and appl Chem 52:2275-2281
Tweedle FM, Wilson LJ, García-Iñiguez L, Babcock GT, Palmer G (1978) Electronic state of heme in cytochrome oxidase III. The magnetic susceptibility of beef heart cytochrome oxidase and some of its derivatives from 7-200 K. Direct evidence for an antiferromagnetically coupled Fe(III)/Cu(II) pair. J Biol Chem 253:8065-8071
Van Gelder BF, Beinert H (1969) Studies of the heme components of cytochrome *c* oxidase by EPR spectroscopy. Biochim Biophys Acta 189:1-24
Warburg O, Negelein E (1929) Über das Absorptionsspektrum des Atmungsferment. Biochem Z 214:64-100
Wikström M, Krab K, Saraste M (1981) Cytochrome oxidase - a synthesis. Academic, London
Wilson MT, Jensen P, Aasa R, Malmström BG, Vänngård T (1982) An investigation by e.p.r. and optical spectroscopy of cytochrome oxidase during turnover. Biochem J 203:483-492
Winter DB, Bruyninckx WJ, Foulke FG, Grinich NP, Mason HS (1980) Location of heme a on subunits I and II and copper on subunit II of cytochrome *c* oxidase. J Biol Chem 255:11408-11414
Woodruff WH, Norton K, Swanson BI, Fry HA, Malmström BG, Pecht I, Blair DF, Cho W, Campbell GW, Lum V, Miskowski VM, Chan SI, Gray HB (1983) Cryo-vibrational spectroscopy of blue copper proteins. Inorg Chim Acta, 79:51-52

Oxygenases

Lipoxygenases from Plant and Animal Origin

J. F. G. Vliegenthart, G. A. Veldink, J. Verhagen, and S. Slappendel[1]

Introduction

Lipoxygenases (linoleate:oxygen oxidoreductase EC 1.13.11.12) catalyze the dioxygenation of polyunsaturated fatty acids which posses a 1,4-*cis,cis*-pentadiene system. Under optimum conditions regarding pH, temperature, and concentration of the reactants, the products are optically active *cis, trans* conjugated hydroperoxides. For example, the products of the enzymic dioxygenation of linoleic acid are
13-S(L)-hydroperoxy-9-*cis*,11-*trans*- (13-S-HPOD),
9-R(L)-hydroperoxy-10-*trans*,12-*cis*- (9-R-HPOD) or
9-S(D)-hydroperoxy-10-*trans*,12-*cis*-octadecadienoic acid (9-S-HPOD)
(Scheme 1).

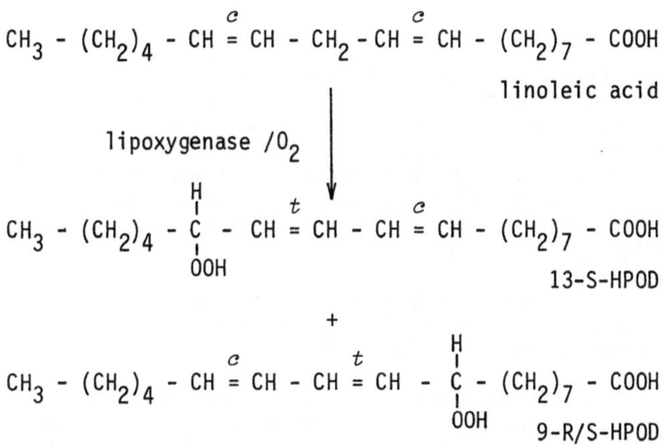

Scheme 1. The dioxygenation reaction of linoleic acid catalyzed by lipoxygenase

The enzyme is widespread in the plant kingdom, especially in legume seeds and cereals. The existence of lipoxygenase activity has also been demonstrated in a number of animal and human systems, e.g. blood cells. Interestingly, the end product of mammalian enzymes is often a hydroxy-polyene with *cis,trans*-conjugated double bonds rather than a hydroperoxy-polyene.

Besides the dioxygenation reaction, several lipoxygenases are capable of metabolizing the primary product, i.e., hydroperoxide. A variety of secondary products, including dimers and oxodienoic acids, can be

1 Department of Bio-organic Chemistry, State University of Utrecht, Croesestraat 79, NL-3522 AD Utrecht, The Netherlands

34. Colloquium - Mosbach 1983
Biological Oxidations

obtained in this way. In animal and human systems the primary products are precursors for compounds which may possess physiological activity, like prostaglandins, thromboxanes and leukotrienes.

In this paper, we present recent developments in the biochemistry and biophysics of plant and mammalian lipoxygenases and related enzymes like cyclooxygenase. An emphasis is given to soybean lipoxygenase, because this enzyme has most extensively been investigated.

Plant Lipoxygenases

Variations in substrate and product specificity, pH-optimum, and stability have been observed for lipoxygenases from different plants. For soybeans, four more or less different isoenzymes have been described. Lipoxygenase-1 from soybeans is the most extensively investigated isoenzyme, due to its high stability and easy accessibility. It was already crystallized as early as 1947. For soybean lipoxygenase-1, several isolation methods have been reported (Vliegenthart and Veldink 1982).

Soybean Lipoxygenase-1

Lipoxygenase-1 consists of a single polypeptide chain and has a molecular weight of approx. 100 000. Theorell et al. (1947) reported an iron content of 0.3 mol per mol enzyme, but considered iron as an impurity. Therefore, for a long time lipoxygenase was regarded as an unique nonmetal dioxygenase. However, in 1973 several research groups (e.g., Roza and Francke 1973; Chan 1973) reported the occurrence of 1 mol iron per mol enzyme. The iron is not part of a heme group, but is directly bound to the polypeptide backbone.

A specific activity of 280 μmol $O_2 \cdot min^{-1} \cdot mg^{-1}$ has been reported as determined by a polarographic method. This activity corresponds to 4.67 $\mu kat \cdot mg^{-1}$ or to a turnover number of 467 s^{-1}.

The amino acid composition of lipoxygenase-1 has been published by several investigators, but the sequence has not yet been established. Information on the amino acids located in or near the active site of the enzyme has been obtained from various studies. On the basis of fluorescence experiments (Finazzi-Agrò et al. 1975), it has been proposed that soybean lipoxygenase-1 has a large hydrophobic active site, which contains tryptophan residues. A histidine residue has been suggested to have a functional role in the enzymic dioxygenation of unsaturated fatty acids (Yamamoto et al. 1970). Furthermore, it has been reported (Spaapen et al. 1980) that lipoxygenase-1 contains five free sulfhydryl groups and no disulfide bridges. Three sulfhydryl groups react readily with methylmercuric halides leading to significant changes in the catalytic properties of the enzyme.

The optical spectrum of the native enzyme shows only a protein absorption band with a maximum at 280 nm and at high enzyme concentration shoulders are discernable at around 330 nm and 400 nm (Spaapen et al. 1979).

Reactions Catalyzed by Soybean Lipoxygenase-1

1. The Dioxygenation Reaction. The main product of the dioxygenation of linoleic acid catalyzed by soybean lipoxygenase-1 at pH 9.0 is 13-S(L)-hydroperoxy-9-*cis*,11-*trans*-octadecadienoic acid (13-S-HPOD) (Van Os et al. 1979). The kinetics of this reaction have been studied extensively. Using a prochiral ^{3}H-labeling of the methylene group of the pentadiene system, it has been demonstrated that an antarafacial relationship exists between hydrogen abstraction and dioxygen insertion (Hamberg and Samuelsson 1967; Egmond et al. 1972). From studies with substrates ^{2}H-labeled at the n-8 position[2], it could be concluded that abstraction of hydrogen is the rate-limiting step of the reaction (Egmond et al. 1973).

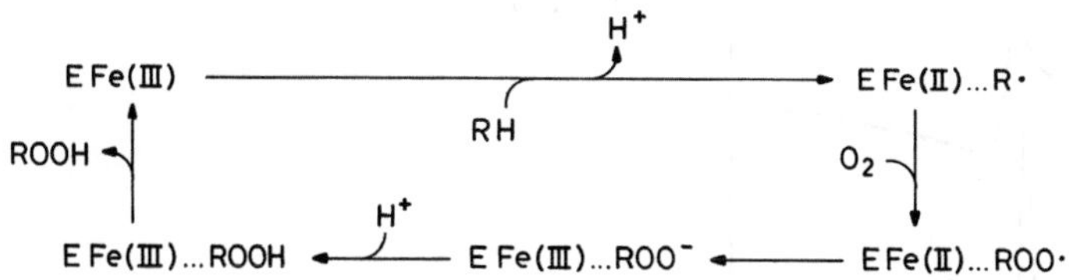

Scheme 2. The aerobic reaction of lipoxygenase-1

Based on qualitative EPR results, a scheme with a key role for iron in the dioxygenation reaction has been proposed (Scheme 2; De Groot et al. 1975). The native enzyme is almost EPR-silent (Fig. 1) and iron is thought to be in the Fe(II) state. A yellow enzyme form is obtained upon addition of one molar equivalent of 13-S-HPOD to native lipoxygenase-1. Yellow lipoxygenase-1 shows an EPR spectrum with resonances around *g* 6 characteristic for high-spin Fe(III) (Fig. 1). This oxidized enzyme form can be reduced by linoleic acid. Anaerobic addition of this substrate leads to an enzyme form which shows no EPR spectrum as reported for the native enzyme. In an aerobic system, the dioxygenation proceeds and the formation of the product hydroperoxide is coupled with a reoxidation of the iron in lipoxygenase and the enzyme becomes available for a new cycle.

2. The Anaerobic Reaction. When dioxygen is depleted during the dioxygenation reaction and both linoleic acid and product (13-S-HPOD) are available, lipoxygenase-1 starts to catalyze an anaerobic reaction (Scheme 3; Garssen et al. 1971, 1972). The yellow Fe(III) enzyme is reduced by linoleic acid and the reduced enzyme form is oxidized by the product 13-S-HPOD. The radicals formed give rise to a variety of products, including dimers and oxodienoic acids. From steady-state kinetics of the anaerobic reaction (Verhagen et al. 1978), it is concluded that lipoxygenase has one active site, which alternatively binds the two substrates of the anaerobic reaction (linoleic acid and 13-S-HPOD).

3. The Conversion of Linoleic Acid Hydroperoxide. In the mechanism of the anaerobic reaction (Scheme 3), the enzymic conversion of 13-S-HPOD in the presence of linoleic acid is shown. However, lipoxygenase also catalyzes the conversion of the product 13-S-HPOD in the absence of

---

2 Different systems are in use for the indication of the position of atoms in polyunsaturated fatty acids. For plant lipoxygenases, the position is indicated counting from the methyl group. For mammalian lipoxygenases, the positions in the molecule are given by numbers counting from the carboxylic group

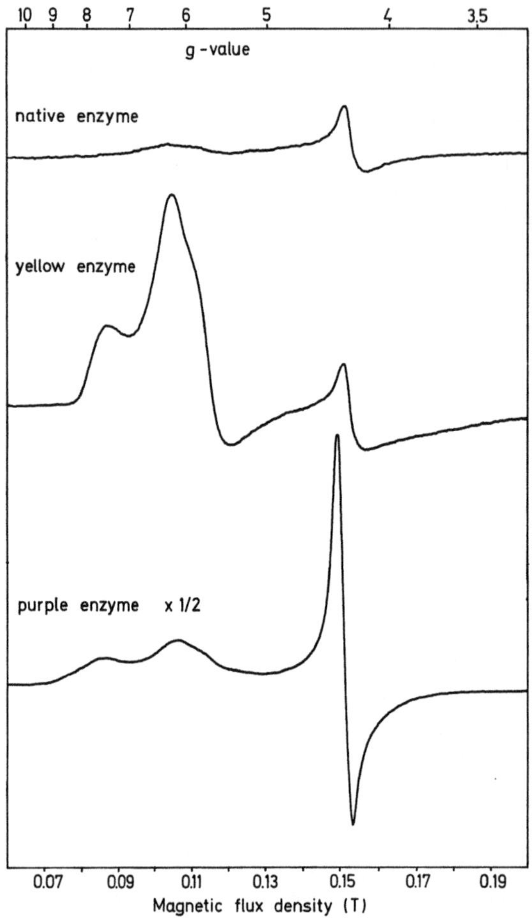

Fig. 1. EPR spectra of various enzyme forms of soybean lipoxygenase-1. *Native enzyme*: 0.54 m*M* in 0.1 *M* sodium borate buffer, pH 9.0. *Yellow enzyme*: 1 molar equivalent of 13-S-HPOD was added to the native enzyme solution. *Purple enzyme*: 3 molar equivalents of 13-S-HPOD were added to the yellow enzyme solution. Microwave frequency, 9.12 GHz; microwave power, 2 m*W*; temperature 15 K. The g_z parts of the spectra are presented in Fig. 3

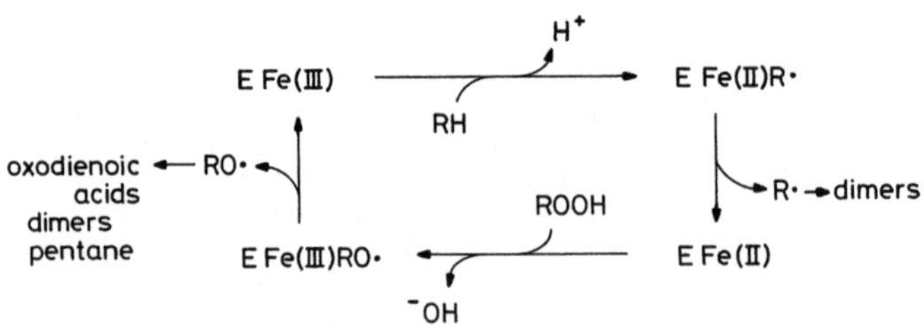

Scheme 3. The anaerobic reaction of lipoxygenase-1

linoleic acid. The conversion of 9-R- and 13-S-HPOD catalyzed by soybean lipoxygenase-1 has been investigated both under aerobic and anaerobic conditions (Verhagen et al. 1977, 1979). These reactions are much slower than the anaerobic reaction in the presence of linoleic acid. Under anaerobic conditions,
13-oxo-9-*cis*,11-*trans*-octadecadienoic acid,
13-oxo-9-*cis*(*trans*),11-*trans*-tridecadienoic acid, and
11-hydroxy-12:13-epoxy-9-*cis*-octadeceoic acid are the main products. In the aerobic conversion of 13-S-HPOD, the reaction rate is at least 4 times slower and no 13-oxo-9-*cis*,11-*trans*-tridecadienoic acid is

formed. The amount of dioxygen consumed is 0.7 mol per mol 13-S-HPOD leading to a large amount of products probably from oxidative polymerization. Two tentative schemes for the action of lipoxygenase on hydroperoxides have been presented, but the detailed mechanism is as yet unknown (Verhagen 1978).

4. Double Dioxygenation Reaction. The reactions described sofar concern the interaction between lipoxygenase and linoleic acid as well as products derived from linoleic acid. In polyunsaturated fatty acids with an 1,4,7-octatriene or an 1,4,7,10-undecatetraene system, like linolenic and arachidonic acid, lipoxygenase has the capacity of incorporating two molecules of dioxygen per substrate molecule (Bild et al. 1977). The first step is the "normal" dioxygen insertion at the n-6 position, followed by a second dioxygenation leading to a dihydroperoxy compound, which has either two conjugated *cis,trans* diene systems or a conjugated *trans,cis,trans* triene system, depending on the substrate and the position of insertion of the second dioxygen molecule (Scheme 4). The second dioxygenation step proceeds with a reaction rate which is several orders of magnitude smaller than that of the first step (Van Os et al. 1981).

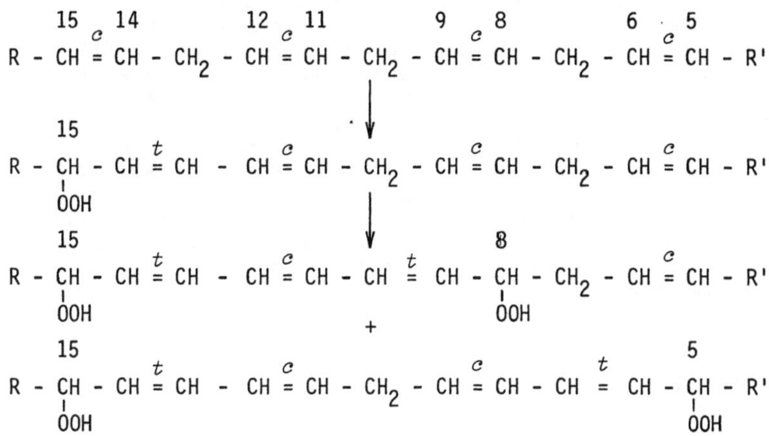

Scheme 4. Dioxygenation of arachidonic acid and 15-hydroperoxy-arachidonic acid

Magnetic and Spectroscopic Studies on Soybean Lipoxygenase-1

Lipoxygenase is a metalloprotein which contains one mol nonheme iron per mol protein. In general, the metal environment and the reaction mechanism of metalloenzymes can adequately be studied by magnetic and spectroscopic methods.

1. Description of the EPR Signals Around g 6. Previously, the functional role of iron has been studied by qualitative EPR spectroscopy (De Groot et al. 1975a). The EPR signals around g 6 of yellow lipoxygenase result from contributions of high-spin Fe(III) species with more or less axial symmetry (g 6.2 - 7.4). The components contributing to the signal are enzyme species differing only in structure of the environment of iron, because the signals could not be attributed to either different enzyme - product complexes or to protonated and nonprotonated enzyme forms (Slappendel et al. 1982a). A shift to an axial type of EPR spectrum is observed after storage of the enzyme in the yellow form at 4°C. Storage of the native enzyme in a sodium-borate buffer solution (pH 9.0) instead of in a sodium-acetate buffer solution (pH 5.0) brought to 25% saturation with solid ammonium sulfate at 4°C also leads to changes of the metal environment, which become apparent in a similar axial type of EPR spectrum after oxidation.

2. Determination of the Amount of EPR-Visible Iron. For further proof of the main reaction scheme (Scheme 2), it is necessary to quantitate the amount of iron visible in the EPR spectra. Because the signals around *g* 6 (Fig. 1) stem from only one of the three Kramers' doublets, the population of the doublets had to be determined. This requires knowledge of the energy differences between the doublets, which are described by the zero-field splitting constants (*D*) of the different components building up the signal around *g* 6 of lipoxygenase. The *D*-values are determined by two methods: (1) temperature dependence studies of the signal intensity and (2) by establishing *g*-shift upon increasing the microwave frequency (Slappendel et al. 1980). The ranges of *D* for the axial and rhombic species are found to be 1.5 - 3.0 K and 1.8 - 4.4 K, respectively. The absence of a rigid coordination sphere for iron in lipoxygenase like porphyrin in heme proteins might be the reason for the observed width of the ranges of the *D*-values and the variation in the shape of the EPR signal around *g* 6. Heme proteins with iron in the high-spin Fe(III) state also give EPR signals around *g* 6, however, these proteins have much higher *D*-values (approx. 14 K). The intensity of the EPR signals of lipoxygenase has been determined by both double integration and simulation methods. The complex EPR signal around *g* 6 of yellow lipoxygenase could only be simulated by using three components, which differ in degree of axiality (Fig. 2 and Table 1) (Slappendel et al. 1981). The amounts of iron visible in the EPR spectra have been calculated from the spectral intensity using the *D*-values described above and as references a Cu^{2+} solution in $NaClO_4$, pH 2, and a solution of sero-transferrin, an iron-transporting protein, which shows an EPR signal around *g* 4.3 (Table 2). A considerable amount of iron is EPR-visible in the yellow enzyme form which indicates that the qualitative EPR results earlier described dealt with major enzyme forms and this study strongly supports the mechanism of the dioxygenation presented in Scheme 2.

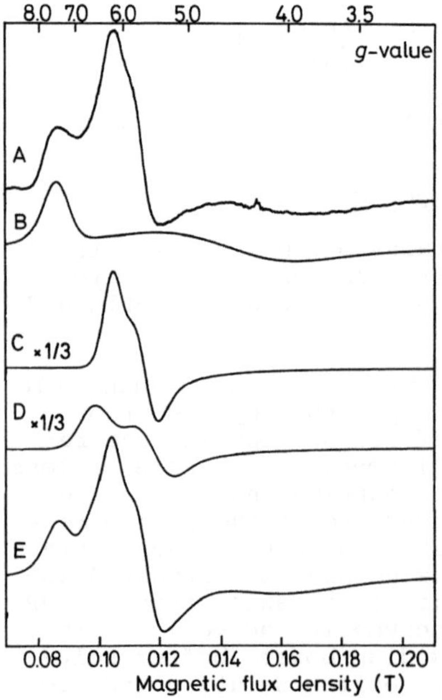

Fig. 2. Experimental and simulated spectra of yellow Fe(III)-lipoxygenase-1. *A* Spectrum of yellow lipoxygenase. The signal at *g* 4.3, which stems for at least 90% from an impurity in the EPR cavity and quartz dewar, is eliminated by subtraction of the spectrum of native enzyme from the spectrum of yellow enzyme. *B*, *C*, and *D* Simulated spectra giving the sum spectrum *E*. Spectrum *C* and *D* are recorded with a gain making their total integrated intensities one-third of the intensity of spectrum *B*. Simulation data is given in Table 1

Table 1. Simulation data of yellow lipoxygenase[a]

Component	g value			Line width (mT)			Relative weight in spectrum E
	x	*y*	*z*	*x*	*y*	*z*	
1 (B)	7.35	4.55	1.88	7	35	25	1.0
2 (C)	6.20	5.60	2.00	5.5	6.5	25	0.4
3 (D)	6.55	5.45	1.98	9	10	25	0.5

[a]The letters refer to the spectra presented in Fig.2

Table 2. EPR-visible iron in lipoxygenase-1[a]

Enzyme form	Signal around *g* 6 Simulation	Signal at *g* 4.3 Simulation	Integration
Native	<1	0.1	
Yellow	80	0.6	
Purple	76	8	13

[a]The amount of iron visible in the EPR spectra presented in Fig. 1 is given as a percentage of the total iron content. Corrections have been made for a signal at *g* 4.3 stemming from an impurity in the EPR cavity and quartz dewar (90% of the signal shown for native lipoxygenase)

3. Purple Lipoxygenase. Besides yellow Fe(III)-lipoxygenase, a second colored enzyme form has been described by De Groot et al. (1975b). Upon addition of a molar excess of 13-S-HPOD to native or yellow lipoxygenase, a purple enzyme form is obtained, which shows in the EPR spectrum a more rhombic type of signal around *g* 6 than yellow enzyme and in addition a signal at *g* 4.3 (Fig. 1). The line shape of the signal aroung *g* 6 has strongly changed as compared with yellow lipoxygenase, but the amount of iron visible in this part of the spectrum has only slightly been diminished (Table 2). The EPR signal at *g* 4.3 is typical for high-spin Fe(III) iron in a ligand field of rhombic symmetry. In the case of lipoxygenase, it has been attributed to an enzyme – 13-S-HPOD complex. Recently, it has been shown that the intensities of the signals around *g* 6 and at *g* 4.3 are linearly correlated with the absorbance of yellow lipoxygenase at 370 nm and the absorbance of purple lipoxygenase at 570 nm, respectively (Slappendel et al. 1983). The amount of iron visible in the EPR signal at *g* 4.3 is only approx. 10% of the total iron content (Table 2). This has led to a reinterpretation of the EPR results concerning the purple enzyme form. Addition of 3 to 4 molar equivalents of 13-S-HPOD to native or yellow lipoxygenase results in an almost complete formation of the enzyme – 13-S-HPOD complex because the affinity of 13-S-HPOD for lipoxygenase is large (K_{aff} = 10 μ*M*; Egmond et al. 1977). This means that only part of the enzyme – 13-S-HPOD complex (approx. 10% cf. Table 2) has a conformation of the environment of iron which leads to an EPR signal at *g* 4.3 and an absorption at 570 nm (Scheme 5). The molar absorption coefficient ε_{570} is 10^4 $M^{-1} \cdot cm^{-1}$ and the absorption is proposed to originate from charge transfer transitions between iron and amino acids.

4. EPR Spectrum Around g 2. In the EPR spectra of the various forms of soybean lipoxygenase-1, a signal around *g* 2 can be discerned (Fig. 3). Several different contributions to this signal can be distinguished, including those from contaminants. The latter ones can be analyzed by studying the native enzyme, which is in principle EPR-silent. A

E - Fe(III) + 13-L-LOOH $\rightleftharpoons$ E - Fe(III)·· 13-L-LOOH $\longleftrightarrow$ E - Fe·· 13-L-LOOH

YELLOW	YELLOW	PURPLE
EPR AROUND G 6	EPR AROUND G 6	EPR AT G 4.3
	(LESS AXIAL THAN E - Fe(III))	

Scheme 5. Interaction of lipoxygenase-1 and 13-hydroperoxy-linoleic acid (13-L-LOOH)

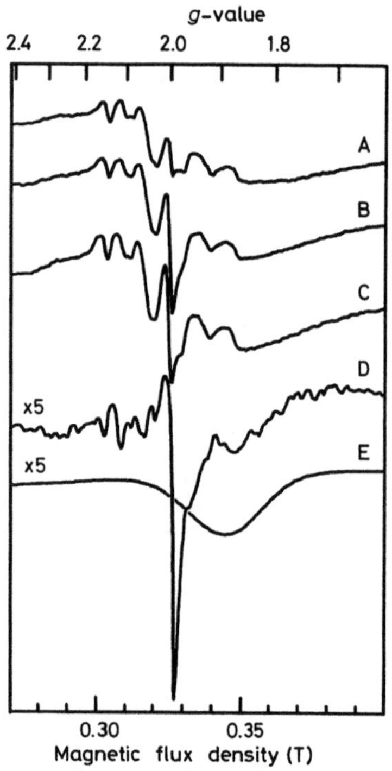

Fig. 3. EPR signals around g 2 of soybean lipoxygenase-1. *A* Native enzyme; *B* yellow enzyme; *C* purple enuyme; *D* spectrum obtained by subtraction of spectrum *A* from *B*. *E* Simulated spectrum of the g_z component of the high-spin Fe(III). Spectra *D* and *E* are presented with a gain 5 times higher than spectra *A* - *C*. For experimental conditions and simulation data see Fig. 1 and Table 1

contamination of manganese (0.07 mol per mol enzyme) cause a signal with a hyperfine splitting in six lines (I = 5/2) (Fig. 3A-C). Contributions from copper (0.006 mol per mol enzyme) and from other impurities were not detectable. Besides the manganese-signal, the signal around g 2 of yellow and purple lipoxygenase consists of two parts:

a. The high-field component (g_z) of the high-spin Fe(III) signal with g_x and g_y components aroung g 6. The g_z component of the high-spin Fe(III) enzyme species having the least axial symmetry is discernable at g 1.9 in the spectrum obtained by subtraction of the spectrum of native enzyme from that of yellow (Fig. 3D) or from that of the purple enzyme. A simulated spectrum of the g_z component of this high-spin Fe(III) enzyme species is given in Fig.3E. The g_z parts of the two more axial species coincide with the signals at g 2.

b. A radical type of signal (Fig. 3B-C). This signal is better recognizable after subtraction of the manganese signal (Fig. 3D). In the different schemes describing lipoxygenase catalysis (Scheme 2 and 3) radicals are present in either an enzyme-bound or in a free-radical form. The dissociation of the enzyme - radical complexes in the aerobic catalytic cycle is thought to be insignificant, since the high degree of optical purity of the hydroperoxides formed in the lipoxygenase-1-catalyzed dioxygenation indicates that the enzyme controls the way of dioxygen insertion. In the anaerobic reaction of lipoxygenase, the dissociation of the enzyme - radical complex is more pronounced than in the aerobic reaction and part of the radicals are thought to be free radicals which react with each other to form a variety of dimers. Knowledge on the structure of these free radicals in solution could be obtained by a spin-trapping method (De Groot et al. 1973). The water soluble spin-trap 2-methyl-2-nitrosopropanol has been used in the anaerobic reaction. From incubation with linoleic acid, specifically ^2H-labeled in the pentadiene system, the trapped radical could be identified as a linoleyl radical. Spin adducts of oxygen-centered radicals are thought to decompose too rapidly at ambient temperature to be observed by EPR spectroscopy. For direct detection of radicals at low temperature (<77 K), rapid-freeze equipment has been used. Exploratory studies with this technique have given experimental evidence for the formation of alkoxy radicals during the incubation of lipoxygenase with 13-S-HPOD.

5. EPR Spectra of the Protein Chain. The EPR spectra of lipoxygenase described sofar contain signals from iron and radicals. However, EPR signals arising from the protein chain of lipoxygenase are observed after storage of a sample of yellow lipoxygenase-1 for 6 months at 77 K (Fig. 4). The signals stem from the protein moiety of the enzyme, because the conversion products of 13-S-HPOD give a different type of EPR signals. According to Schaich and Karel (1976), signals as shown in Fig. 4 could stem from either cysteine or amino acids with nitrogen-containing side chains, but assignment to one particular amino acid can not be made. Either radicals from the conversion products of 13-S-HPOD or iron may be involved in the reaction leading to this new radical signal.

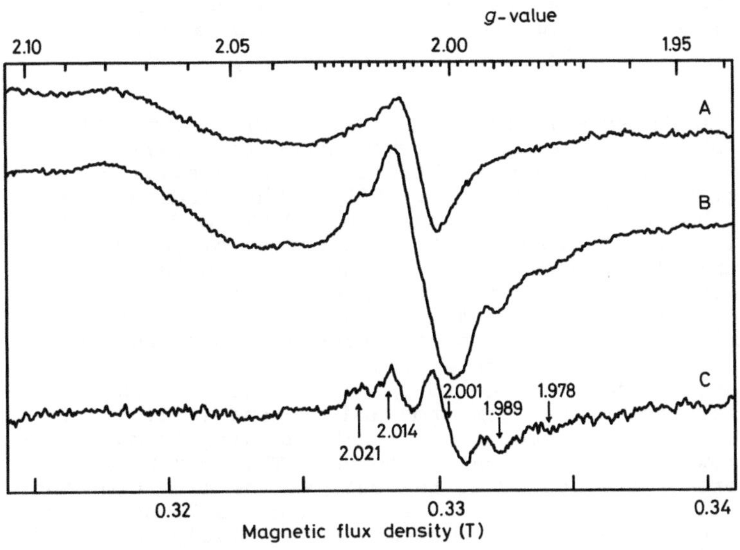

Fig. 4. EPR spectra around *g* 2 of yellow lipoxygenase. *A* Yellow enzyme; *B* yellow enzyme after storage at 77 K over 6 months; *C* difference spectrum of spectra *A* and *B*. For complete base-line correction, it was necessary to multiply spectrum *A* with a factor 1.5. Microwave frequency 9.251 GHz; microwave power 2 *mW*; temperature 15 K

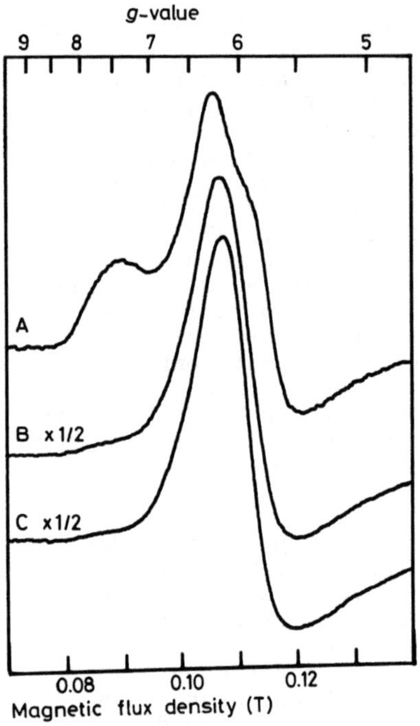

Fig. 5. EPR line-shape changes of yellow lipoxygenase-1. Yellow lipoxygenase (*A*) was incubated with 10 molar equivalents KCN (*B*) and 30 molar equivalents ethanol (*C*). Microwave frequency 9.256 GHz; temperature 15 K

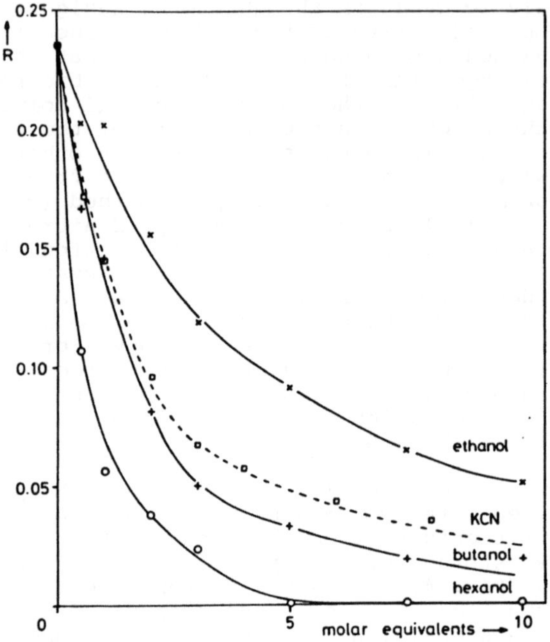

Fig. 6. Titration of yellow lipoxygenase-1 with alcohols and cyanide. On the *ordinate*, the ratio of the rhombic and axial components (*R*) is given. The amplitude of the rhombic part measured at *g* 7.5 is corrected for the contribution of the axial part at this *g*-value

6. The Binding of Alcohols. Interestingly, it has been found that small amounts of ethanol (30 molar equivalents) when added to a yellow Fe(III)-lipoxygenase solution induce a shift to a nearly axial type of EPR spectrum (Fig. 5; Slappendel et al. 1982a). The binding of alcohols to the enzyme has been further studied by ^{1}H-NMR and EPR spectroscopy (Slappendel et al. 1982b,c). Titration curves of yellow lipoxygenase with ethanol, butanol-1, and hexanol-1 are given in Fig. 6, where changes in the EPR spectra are presented as the ratio (*R*) of the amplitudes of the rhombic (at *g* 7.4) and axial (at *g* 6.0 - 6.2) components. ^{1}H-NMR spectra of solutions of butanol-1, to which either native or yellow lipoxygenase has been added, show a line-broadening of the proton resonances, which is due to proton relaxation enhancement from magnetic interaction between iron and protons (Fig. 7). In comparing the same proton resonances, the line-broadening is more pronounced for the yellow than for the native enzyme. This can be explained by one or more of the following reasons: a. the line-broadening is determined by the spin quantum number *S*, which is 2 (see next section) and 5/2 for native and yellow enzyme, respectively; b. a difference in affinity of the alcohol for binding to the enzyme forms; and c. a difference in relaxation times (rotational correlation time, exchange life time, and electron spin relaxation time). For the different proton resonances of the same enzyme sample (native or yellow enzyme added to a butanol-1/D_2O solution), the line-broadening increases going from the methyl protons to the protons bound to carbon

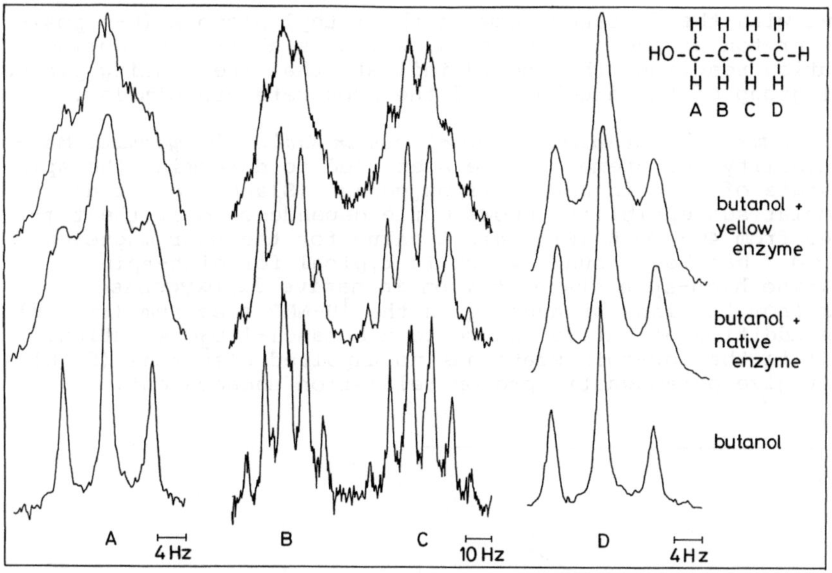

Fig. 7. NMR spectra of butanol-1 showing the effect of paramagnetic iron in native and yellow lipoxygenase on the proton resonances of butanol-1. *A* 8 m*M* butanol-1; *B* and *C* enzyme: 26 μ*M* and butanol-1 22.8 m*M*. Buffer: 0.1 *M* sodium borate/D_2O, pH 9.0; temperature 297 K

Table 3. ^{1}H-NMR study of the binding of various alcohols to yellow lipoxygenase-1[a]

Alcohol	Observed line-broadening (Hz)		K_a (m*M*)	Distance to iron (A)	
	Methyl protons	Protons bound to the carbon atom 1		H_3C-	$-CH_2OH$
Ethanol	1.5	2.0	260	5.1	4.7
Butanol-1	5.0	11	30	7.2	6.0
Hexanol-1	8	21	3	6.3	5.3

[a]The observed line-broadening of the methyl protons and the protons bound to carbon atom 1 of the alcohol is given after addition of yellow lipoxygenase-1 to an alcohol/D_2O solution; concentrations: alcohol 8 m*M*; enzyme 0.03 m*M*. The affinity constant K_a has been derived from titration curves of yellow enzyme with the alcohols. The distances between iron and methyl protons and protons bound to carbon atom 1, respectively, have been calculated with the Solomon-Bloembergen equation

atom 1. This differential line-broadening implies no difference in mobility of the protons, because from a competition experiment with cyanide, which also gives a shift to an axial type of EPR spectrum (Fig. 5 and 6), it is concluded that the alcohols do not bind via the hydroxyl group. Quantitative results on the binding of various alcohols to lipoxygenase have been obtained from titrations and calculations using the Solomon-Bloembergen equation (Table 3). The affinity of the various unbranched alcohols for binding to yellow lipoxygenase increases with increasing carbon chain length (see also Fig. 6). The branched alcohol t-butanol gives neither a shift to an axial type of EPR spectrum nor a line-broadening in NMR upon addition to a yellow lipoxygenase solution. Remarkably, the calculated distance between the methyl protons of the various alcohols and iron is in good agreement with the distance between the methyl protons (n-1 position) and the proton at the n-8 position, which is abstracted during the dioxygenation reaction. This could indicate that the binding places of the methyl group of the alcohols and the substrate are similar.

7. Determination of the Spin and Valence State of Iron in Native Lipoxygenase. Magnetic susceptibility measurements have been used to determine the spin and valence state of iron in native lipoxygenase (Slappendel et al. 1982d). The molar susceptibility shows Curie dependence over the temperature range from 40-200 K (Fig. 8). A value for the Bohr magneton equal to 5.2 ± 0.3 has been found, which is typical for high-spin Fe(II) (S=2). The high-spin state of iron in native lipoxygenase is also clear from the line-broadening in the ^{1}H-NMR spectrum (Fig. 6) observed upon addition of native enzyme to a butanol-1/D_2O solution, because Fe(II) in the low-spin state has no unpaired electrons (S = 0) and should not give paramagnetic proton relaxation enhancement.

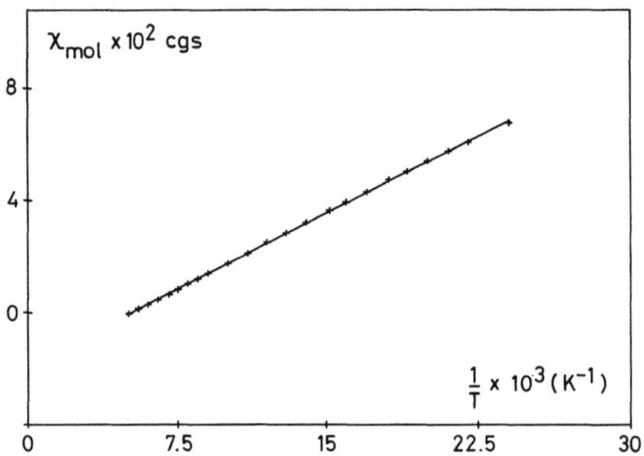

Fig. 8. The temperature dependent contribution of the magnetic susceptibility of native lipoxygenase-1. Enzyme: 1.19 m*M* in 0.1 *M* sodium borate buffer, pH 9.0

Physiological Role of Plant Lipoxygenases

Although lipoxygenases are widespread in the plant kingdom and occur in relatively large amounts, especially in legume seeds and cereals, little is known about the physiological role of these enzymes. Hypotheses include a role of the enzymes in various stages of germination and growth, in the process of wound healing via traumatic acid, and in the production of volatile compounds which render legumes and fruits their typical flavors. These functions are possibly also coupled with lipoperoxidase activity of the enzyme.

Mammalian Lipoxygenases and Cyclooxygenase

Since 1974, lipoxygenases have also been isolated from mammalian origin, predominantly from various blood cells. As mentioned before, these mammalian enzymes differ in several respects from those of plant origin, i.a., with regard to substrate specificity. However, there are also interesting common features, especially for the lipoxygenase isolated from rabbit reticulocytes and for the related enzyme cyclooxygenase purified from sheep vesicular glands.

In this section, a summary of the metabolic pathways of arachidonic acid catalyzed by cyclooxygenase and mammalian lipoxygenases will be presented. A number of these enzymes will be described. Finally, the physiological role of mammalian lipoxygenases and cyclooxygenase is briefly summarized.

The Metabolism of Arachidonic Acid

The substrate of mammalian lipoxygenases and cyclooxygenase is often arachidonic acid (eicosatetraenoic acid, ETE) rather than linoleic acid, which is the best substrate for soybean lipoxygenase-1. In vivo, arachidonic acid is obtained i.a. from cell membrane phospholipids through the action of phospholipases. Due to the presence of 4 *cis* double bonds in an 1,4,7,10-undecatetraenoic system, lipoxygenase-catalyzed dioxygenation can occur at several positions (Scheme 6). So far, from the 12 possible hydroxy derivatives of arachidonic acid (HETE), the 5-S, 11-S/R, 12-S, and 15-S compounds have been isolated. In reaction with (partly) purified enzyme preparations, the presence of the corresponding precursor hydroperoxy arachidonic acid (HPETE) has been demonstrated (e.g., 12-S-HPETE in platelets). The peroxy radical of 11-R-HPETE is an intermediate in the reaction of cyclooxygenase. 5-S and 15-S-HPETE are the precursors of the leukotrienes. The leukotrienes derived from 5-S-HPETE (Scheme 7) have important physiological functions.

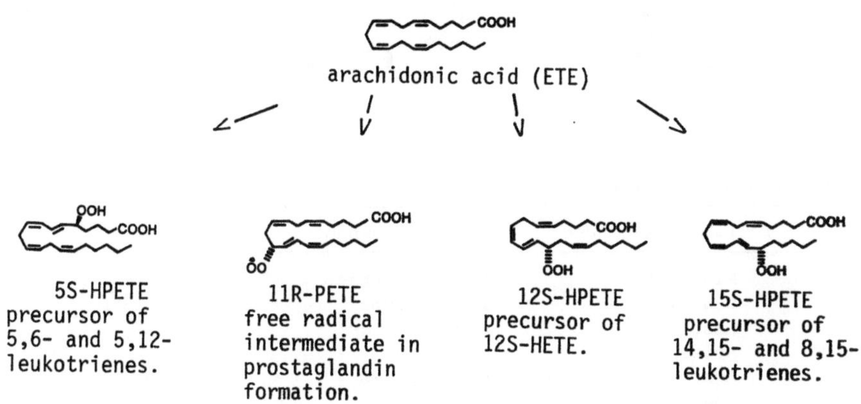

Scheme 6. The hydroperoxydation of arachidonic acid (ETE)

Lipoxygenases from Mammalian Origin

A listing of mammalian lipoxygenases arranged according to the position of dioxygen insertion is presented in Table 4. The lipoxygenases are indicated by the number of the carbon atom (counted from the carboxylic group) to which dioxygen is attached. A number of mammalian enzymes, which have been (partly) purified, will be described with emphasis on the metabolic aspects and compared with soybean lipoxygenase.

15-Lipoxygenase from Reticulocytes. A lipoxygenase has been isolated from rabbit reticulocytes by Rapoport et al. (1979). The enzyme has a molecular weight of 78 000 and an isoelectric point of 5.5. Furthermore,

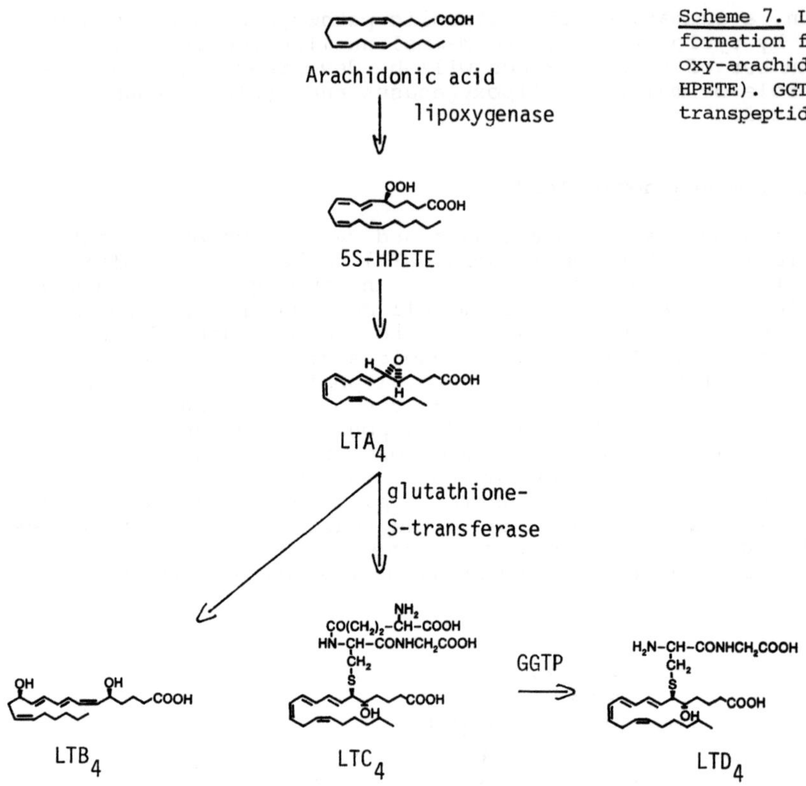

Scheme 7. Leukotriene (LT) formation from 5-hydroperoxy-arachidonic acid (5S-HPETE). GGTP = γ-glutamyl transpeptidase

Table 4. Sources of mammalian lipoxygenases

5-Lipoxygenase	12-Lipoxygenase	15-Lipoxygenase
Granulocytes	Platelets	Reticulocytes
Lymphocytes	Granulocytes	Lymphocytes
Mast cells	Mast cells	Granulocytes
Mastocytoma cells	Spleen	Monocytes
RBL-1 cells	Aorta	Macrophages
Monocytes	Monocytes	
Macrophages	Macrophages	
Lung tissue	Lung tissue	
Spleen		

it contains 5% neutral sugars and one mol iron per mol protein. The environment of iron is thought to be identical to that of soybean lipoxygenase, because it has similar spectroscopic properties and there is no evidence for either a porphyrin system or an iron-sulfur cluster. Reticulocyte lipoxygenase converts linoleic acid into 13-S-HPOD and arachidonic acid into 15-S-HPETE similar to soybean lipoxygenase-1. However, its substrate specificity is less pronounced compared to the soybean enzyme. All polyunsaturated fatty acids having a 1,4-*cis*,*cis* pentadiene system and all classes of phospholipids containing polyunsaturated fatty acids are attacked. The enzyme also acts on intact mitochondrial membranes and causes repiratory inhibition of submitochondrial particles of various origin. Besides the dioxygenation reaction reticulocyte lipoxygenase has a number of features in common with soybean lipoxygenase (Härtel et al. 1982). For the activation of the enzyme, 13-S-HPOD can be used. Addition of one molar equivalent of 13-S-HPOD leads to the conversion of iron from the Fe(II) into the Fe(III) state, while concomitantly the fluorescence and absorption properties of the enzyme change. The enzyme is also capable of converting 13-S-HPOD in the presence of linoleic acid under anaerobic conditions. The products of this reaction are similar to those obtained in the anaerobic reaction of soybean lipoxygenase (cf. Scheme 3). The catalytic activity of reticulocyte lipoxygenase has a suicidal nature, i.e., an inactivation occurs by the products formed during catalysis at temperatures above 20°C. Both conversion products of 13-S-HPOD as well as the products of the anaerobic reaction are responsible for the inactivation of the enzyme. Self-inactivation has also been reported for cyclooxygenase.

12-Lipoxygenase from Platelets. Hamberg and Samuelsson (1974) were the first to describe the conversion of arachidonic acid into 12-S-HETE catalyzed by a lipoxygenase in human platelet homogenates. Nugteren (1975) purified a platelet lipoxygenase from cows and a novel preparation of human platelet lipoxygenase has recently been described by Wallach and Brown (1981). This 12-lipoxygenase is not inhibited by aspirin and indomethacin like cyclooxygenase, but it is inhibited by tetraynoic acids and a series of phenylhydrazone inhibitors. These compounds also inhibit dioxygenation catalyzed by soybean lipoxygenase and cyclooxygenase. Cyclooxygenase is relatively less sensitive to these inhibitors than human platelet lipoxygenase. Inhibition of platelet lipoxygenase by toluene-3,4-dithiol and other Fe(III) chelators has been reported by Aharony et al. (1981). Fe(II) chelators have no influence on the platelet lipoxygenase catalysis, suggesting that the enzyme activity depends on iron in the Fe(III) state. This is in agreement with the observation that the platelet enzyme is activated by 12-S-HPETE as is seen with soybean lipoxygenase-1 by its product 15-S-HPETE. Hamberg and Hamberg (1980) have investigated the mechanism of the dioxygenation of arachidonic acid by human platelet lipoxygenase. Similar to soybean lipoxygenase-1, an antarafacial relation between the hydrogen abstraction from C-10 and the insertion of dioxygen at C-12 was found. Furthermore, the hydrogen abstraction is probably the initial step in the dioxygenation reaction. The main product of the catalytic reaction with arachidonic acid as substrate is 12-S-HPETE, which is converted by a hydroperoxidase into 12-S-HETE. In addition, minor quantities of epoxy-hydroxy and trihydroxy derivatives of arachidonic acid are formed, possibly via mechanisms similar to those observed with soybean lipoxygenase.

5-Lipoxygenase from Rat Basophilic Leukemia Cells. So far, only a few 5-lipoxygenases have been (partly) purified and little is known about the properties of these enzymes. Rat basophilic leukemia cells (RBL-1 cells) contain a lipoxygenase which catalyzes the conversion of arachidonic acid into 5-S-HPETE. This product is further converted into several products including leukotriene B_4 and D_4 and 5-S-HETE. The 5-lipoxygenase from RBL-1 cells has recently been purified by Parker and Aykent (1982). The enzyme has a molecular weight of 90 000. In the presence of Ca^{2+}, a dimer is formed which possesses the enzymatic activity. So far, most experiments have been done with homogenates and intact cells. As yet, information on the presence and possible involvement of iron in the dioxygenation catalyzed by this enzyme is not available.

Cyclooxygenase

Cyclooxygenase or prostaglandin endoperoxide synthase (EC 1.14.99.1) catalyzes the insertion of dioxygen into polyunsaturated fatty acids. When linoleic acid is the substrate, the main product is 9-R-hydroxy-octadecadienoic acid (9-R-HOD) (Hamberg and Samuelsson 1980), which is also the main product formed by soybean lipoxygenase-2 (Van Os et al. 1979). Arachidonic acid is further converted into 15-hydroperoxy-9α,11α-peroxidoprosta-5,13-dienoic acid (PGG_2) by cyclooxygenase. This hydroperoxy-endoperoxide is then reduced by the same enzyme to a hydroxy-endoperoxide (PGH_2), which is, in turn, the precursor of other prostaglandins, prostacyclin, and thromboxanes (Scheme 8).

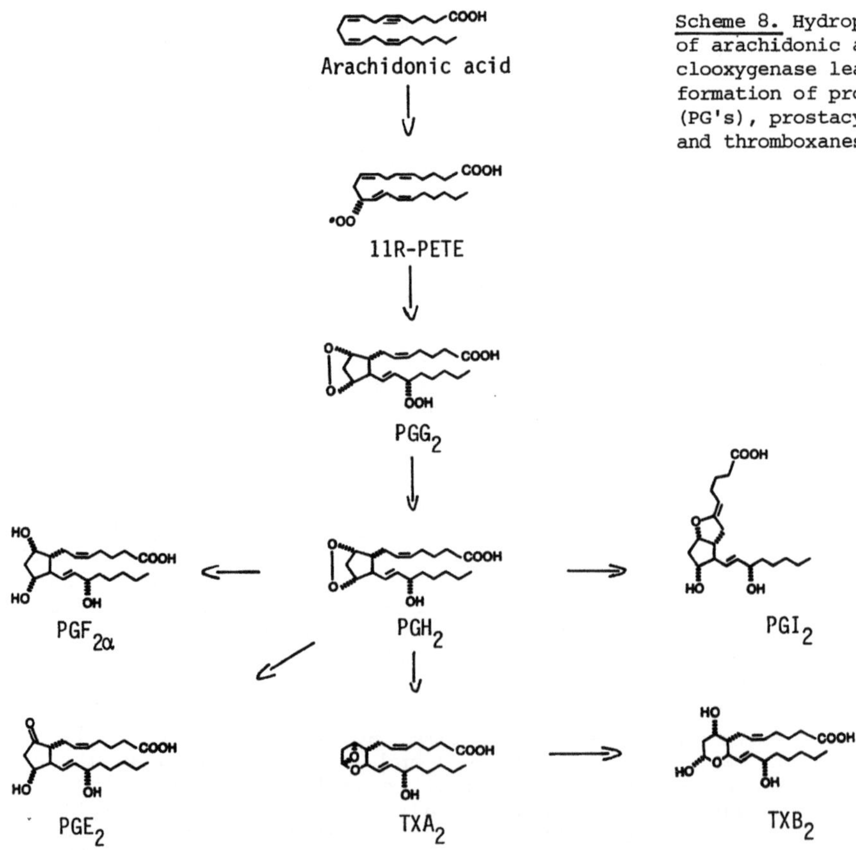

Scheme 8. Hydroperoxidation of arachidonic acid by cyclooxygenase leads to the formation of prostaglandins (PG's), prostacyclin (PGI), and thromboxanes (TX's)

Cyclooxygenase has been isolated from sheep vesicular glands and has a molecular weight of approx. 70 000 (Hemler et al. 1976; Van der Ouderaa et al. 1977). Hemler et al. (1976) have reported that the isolated enzyme contains nonheme iron similar to soybean lipoxygenase, but for full activity, hemin has to be bound to the enzyme in at least stoichiometric amounts. However, Van der Ouderaa et al. (1979) have excluded the presence of nonheme iron in the enzyme. Similar to soybean lipoxygenase, the enzyme is activated by hydroperoxide (Hemler et al. 1979). This suggests that iron in the Fe(III) state is a requirement for activity (cf. yellow Fe(III)-lipoxygenase in Scheme 2).

Conversely, Peterson et al. (1980) put forward the hypothesis that Fe(II) is the active form of iron in cyclooxygenase. EPR studies may contribute to settle this controversy. So far, EPR studies on the cyclooxygenase reaction have been confined to only a free radical formed during a reaction of PGG_2 with ram seminal vesicles. For a long time, the radical has been described as an oxygen-centered free radical, which is formed during the reduction of PGG_2 to PGH_2, but recently, Kalyanaraman et al. (1982) have attributed this radical to a hemoprotein free radical, possibly formed by the oxidation of an amino acid, which is a part of the heme environment. Radicals formed during the reduction of PGG_2 to PGH_2 are thought to be responsible for the self-inactivation of the enzyme similar to the suicidal action observed for lipoxygenase from reticulocytes.

Inhibition studies on cyclooxygenase are rather complicated from a mechanistic point of view (Kuehl et al. 1981). For example, reducing agents act both on the peroxides necessary for the activation of the enzyme (inhibition) and on the oxidants released during the reduction of PGG_2 (protection against self-inactivation). The antiinflammatory effect of most nonsteroidal antiinflammatory agents, like aspirin and indomethacin, is based on the inhibition of cyclooxygenase. Differences in the inhibition mechanism of these agents are observed. Aspirin irreversibly modifies the active site by transferring an acetyl group to a serine residue, while indomethacin reversibly blocks the uptake of dioxygen. Furthermore, often different effects of inhibitors on the two steps of the cyclooxygenase-catalyzed reaction are observed.

Physiological Role of Mammalian Lipoxygenases and Cyclooxygenase

Concerning the physiological role of mammalian lipoxygenases and cyclooxygenase, a lot of information has become available when compared to the enzymes from plant origin. Cyclooxygenase plays a key role in the synthesis of prostaglandins, prostacyclin, and thromboxanes. These compounds are involved in many physiological processes (Table 5). This has led to a growing number of medical applications of these compounds and their derivatives.

Table 5. Biological effects of arachidonic acid metabolites

Prostaglandins, prostacyclin and thromboxanes	Leukotrienes LTB_4	LTC_4 and LTD_4
- contraction and relaxation of smooth muscle - stimulation (TXA_2) and prevention (PGI_2) of platelet aggregation - regulation of gastric secretion	- chemotactic for leukocytes - promotes sticking of leucocytes to blood vessel wall (capillary vessels)	- contraction of smooth muscle (slow-reacting substance of anaphylaxis, bronchoconstriction, asthma) - increase of vascular permeability (edema) - induction of TXA_2 formation (guinea pig lung)

An important function of the mammalian lipoxygenases is the production of the precursors to leukotrienes. Known biological effects of the leukotrienes are given in Table 5. The physiologically active compounds probably play a role in asthma and in various allergic reactions. It is expected that research in the field of lipoxygenase-catalyzed reactions of polyunsaturated fatty acids, especially leukotriene formation, will lead to a better understanding of the biochemical events, physiological processes, and therapies related to different diseases. The fundamental knowledge of the reactions catalyzed by soybean lipoxygenases forms a sound basis for these research activities.

Acknowledgements. Most of the EPR studies described in this article have been performed in collaboration with Prof. Dr. BG Malmström and Dr. R Aasa (Chalmers Institute of Technology and University of Göteborg, Sweden). The magnetic susceptibility measurements have been carried out in collaboration with Prof. Dr. A Ehrenberg and Dr. L Petersson at the Department of Biophysics, University of Stockholm Sweden.

Discussion

Hamprecht: From one of your tables, I took that in macrophages lipoxygenases act on arachidonic acid at positions 5, 12, and 15. Do you know whether the three different enzymes occur in the same cell or in different subtypes of macrophages or whether they occur under different induction conditions?

Vliegenthart: The production of 5-hydroperoxy arachidonic acid and its derivatives (leukotrienes and 5-HETE) by macrophages from different sources is dependent on the presence of calcium, ionophore, and glutathione. In the absence of calcium, the formation of other hydroxy compounds of arachidonic acid also takes place. Unfortunately, the literature is rather confusing at this point. However, it could well be that some cell types are more outspoken in the production of specific compounds than others.

Stoffel: I have two questions: (1) What is the reason for hydroperoxide formation in different positions of linoleic acid? (2) What is the function of lipoxygenase in soybeans, rich in polyunsaturated fatty acids?

Vliegenthart:
1) The stereospecificity of the product formation from the incubation of linoleic acid is determined i.a. by the source of lipoxygenase and by the reaction conditions (pH, substrate concentration, temperature).
2) The function of lipoxygenase in plants is still obscure. Suggestions given in the literature are dealing with a possible role during the germination of seeds when normal respiration has not yet been started, with an involvement in wound healing through the production of the wound hormone traumatic acid and with the formation of volatile compounds.

Weller: With regard to the lectures this morning, it appears of interest where the iron is bound in the protein. Is there no idea about that, e.g., from Resonance Raman Spectroscopy of the ferric enzyme?

Vliegenthart: The iron coordination to the protein backbone is as yet unknown. Resonance Raman Spectroscopy was not applicable because of strong fluorescence from amino acid residues and instability of the ferric enzyme in the laser beam. Other spectroscopic studies including EXAFS are now in progress.

Walsh: How do you account for the antarafacial stereochemistry on the pentadiene system. Given C-H breakage as $-C^{\cdot} + H^{\cdot}$, where does $H^{\cdot}$ go? To Fe? If so, what coordinates the O_2?

Vliegenthart: For the study of the course of the stereochemical attack of the pentadiene system [$11L_S$-^3H], linoleic acid has been prepared by incubation of [$11L_S$-^3H] stearic acid with the green algae *Chlorella vulgaris*. Incubation of this specifically labelled linoleic acid with soybean lipoxygenase-1 leads to the product $13L_S$-hydroperoxy linoleic acid with a very low tritium content. This experimental evidence has led us to propose an antarafacial relationship between hydrogen abstraction and dioxygen insertion.

The formation of the linoleyl radical in the dioxygenation reaction is coupled with a reduction of iron in lipoxygenase. Concomitantly, the hydrogen atom is converted into a proton. Studies concerning the coordination of dioxygen are in progress. So far, no solid evidence could be obtained for binding O_2 to iron in native lipoxygenase.

References

Aharony J, Smith JB, Silver MJ (1981) Inhibition of platelet lipoxygenase by toluene-3,4-dithiol and other ferric iron chelators. Prostaglandins Med 6:237-242

Bild GS, Ramadoss CS, Axelrod B (1977) Multiple dioxygenation by lipoxygenase of lipids containing all-*cis*-1,4,7-octatriene moieties. Arch Biochem Biophys 184: 36-41

Chan HW-S (1973) Soya-bean lipoxygenase: an iron-containing dioxygenase. Biochim Biophys Acta 327:32-35

De Groot JJMC, Garssen GJ, Vliegenthart JFG, Boldingh J (1973) The detection of linoleic acid radicals in the anaerobic reaction of lipoxygenase. Biochim Biophys Acta 326:279-284

De Groot JJMC, Veldink GA, Vliegenthart JFG, Boldingh J, Wever R, Van Gelder BF (1975a) Demonstration by EPR spectroscopy of the functional role of iron in soybean lipoxygenase-1. Biochim Biophys Acta 377:71-79

De Groot JJMC, Garssen GJ, Veldink GA, Vliegenthart JFG, Boldingh J, Egmond MR (1975b) On the interaction of soybean lipoxygenase-1 and 13-L-hydroperoxylinoleic acid, involving yellow and purple coloured enzyme species. FEBS Lett 56:50-54

Egmond MR, Vliegenthart JFG, Boldingh J (1972) Stereospecificity of the hydrogen abstraction at carbon atom n-8 in the oxygenation of linoleic acid by lipoxygenases from corn germs and soya beans. Biochem Biophys Res Commun 48:1055-1060

Egmond MR, Veldink GA, Vliegenthart JFG, Boldingh J (1973) C-11 H-abstraction from linoleic acid, the rate-limiting step in lipoxygenase catalysis. Biochem Biophys Res Commun 54:1178-1184

Egmond MR, Fasella PM, Veldink GA, Vliegenthart JFG, Boldingh J (1977) On the mechanism of action of soybean lipoxygenase-1. A stopped-flow kinetic study of the formation and conversion of yellow and purple enzyme species. Eur J Biochem 76:469-479

Finazzi-Agrò A, Avigliano L, Egmond MR, Veldink GA, Vliegenthart JFG (1975) Fluorescence perturbation in soybean lipoxygenase-1. FEBS Lett 52:73-76

Garssen GJ, Vliegenthart JFG, Boldingh J (1971) An anaerobic reaction between lipoxygenase, linoleic acid and its hydroperoxides. Biochem J 122:327-332

Garssen GJ, Vliegenthart JFG, Boldingh J (1972) The origin and structures of dimeric fatty acids from the anaerobic reaction between soya-bean lipoxygenase, linoleic acid and its hydroperoxide. Biochem J 130:435-442

Hamberg M, Hamberg G (1980) On the mechanism of the oxygenation of arachidonic acid by human platelet lipoxygenase. Biochem Biophys Res Commun 95:1090-1097

Hamberg M, Samuelsson B (1967) On the specificity of the oxygenation of unsaturated fatty acids by soybean lipoxidase. J Biol Chem 242:5329-5335

Hamberg M, Samuelsson B (1974) Prostaglandin Endoperoxides. Novel transformations of arachidonic acid in human platelets. Proc Natl Acad Sci USA 71:3400-3404

Hamberg M, Samuelsson B (1980) Stereochemistry in the formation of 9-hydroxy-10, 12-octadecadienoic acid and 13-hydroxy-9,11-octadecadienoic acid from linoleic acid by fatty acid cyclooxygenase. Biochim Biophys Acta 617:545-547

Härtel B, Ludwig P, Schewe T, Rapoport SM (1982) Self-inactivation by 13-hydroperoxylinoleic acid and lipohydroperoxidase activity of the reticulocyte lipoxygenase. Eur J Biochem 126:353-357

Hemler M, Lands WEM, Smith WL (1976) Purification of the cyclooxygenase that forms prostaglandins. Demonstration of two forms of iron in the holoenzyme. J Biol Chem 251:5575-5579

Hemler M, Cook HW, Lands WEM (1979) Prostaglandin biosynthesis can be triggered by plant peroxides. Arch Biochem Biophys 193:340-345

Kalyanaraman B, Mason RP, Tainer B, Eling TE (1982) The free radical formed during the hydroperoxide-mediated deactivation of ram seminal vesicles is hemoprotein-derived. J Biol Chem 257:4764-4768

Kuehl FA, Egan RW, Humen JL (1981) Prostaglandin cyclooxygenase. Progress Lipid Res 20:97-102

Nugteren DH (1975) Arachidonate lipoxygenase in blood platelets. Biochim Biophys Acta 380:299-307

Parker CW, Aykent S (1982) Calcium stimulation of the 5-lipoxygenase from RBL-1 cells. Biochem Biophys Res Commun 109:1011-1016

Peterson DA, Gerrard JM, Rao GHR, White JG (1980) Reduction of ferric heme to ferrous by lipid peroxides: possible relevance tothe role of peroxide in the regulation of prostaglandin synthesis. Prostaglandins Med 4:73-78

Rapoport SM, Schewe T, Wiesner R, Halangk W, Ludwig P, Janicke-Höhne M, Tannert C, Hiebsch C, Klatt D (1979) The lipoxygenase of reticulocytes. Purification, characterization and biological dynamics of lipoxygenase; its identity with the respiratory inhibitors of the reticulocyte. Eur J Biochem 96:545-561

Roza M, Francke A (1973) Soybean lipoxygenase: an iron-containing enzyme. Biochim Biophys Acta 327:24-31

Schaich KM, Karel M (1976) Free radical reactions of peroxidizing lipids with amino acids and proteins: an ESR study. Lipids 11:392-400

Schewe T, Halangk W, Hiebsch Ch, Rapoport SM (1975) A lipoxygenase in rabbit reticulocytes which attacks phospholipids and intact mitochondria. FEBS Lett 60: 149-152

Slappendel S, Veldink GA, Vliegenthart JFG, Aasa R, Malmström BG (1980) EPR spectroscopy of soybean lipoxygenase-1. Determination of the zero-field splitting constants of high-spin Fe(III) signals from temperature and microwave frequency dependence. Biochim Biophys Acta 642:30-39

Slappendel S, Veldink GA, Vliegenthart JFG, Aasa R, Malmström BG (1981) EPR spectroscopy of soybean lipoxygenase-1. Description and quantification of the high-spin Fe(III) signals. Biochim Biophys Acta 667:77-86

Slappendel S, Aasa R, Malmström BG, Verhagen J, Veldink GA, Vliegenthart JFG (1982a) Factors affecting the line-shape of the EPR signal of high-spin Fe(III) in soybean lipoxygenase-1. Biochim Biophys Acta 708:259-265

Slappendel S, Aasa R, Falk K-E, Malmström BG, Vänngård T, Veldink GA, Vliegenthart JFG (1982b) ^1H-NMR spectroscopic study on the binding of alcohols to soybean lipoxygenase-1. Biochim Biophys Acta 708:266-271

Slappendel S, Aasa R, Malmström BG, Veldink GA, Vliegenthart JFG (1982c) An EPR study on the binding of alcohols to soybean lipoxygenase-1. Acta Chem Scand B36: 569-572

Slappendel S, Malmström BG, Petersson L, Ehrenberg A, Veldink GA, Vliegenthart JFG (1982d) On the spin and valence state of iron in native soybean lipoxygenase-1. Biochem Biophys Res Commun 108:673-677

Slappendel S, Veldink GA, Vliegenthart JFG, Aasa R, Malmström BG (1983) A quantitative optical and EPR study on the interaction between soybean lipoxygenase-1 and 13-L-hydroperoxylinoleic acid. Biochim Biophys Acta 747:32-36

Spaapen LJM, Veldink GA, Liefkens TJ, Vliegenthart JFG, Kay CM (1979) Circular dichroism of lipoxygenase-1 from soybeans. Biochim Biophys Acta 574:301-311

Spaapen LJM, Verhagen J, Veldink GA, Vliegenthart JFG (1980) The effect of modification of sulfhydryl groups in soybean lipoxygenase-1. Biochim Biophys Acta 618:153-162

Theorell H, Holman RT, Åkeson Å (1947) Crystalline lipoxidase. Acta Chem Scand 1: 571-576

Van der Ouderaa, Buytenhek M, Nugteren DH, Van Dorp DA (1977) Purification and characterisation of prostaglandin endoperoxide synthetase from sheep vesicular glands. Biochim Biophys Acta 487:315-331

Van der Ouderaa FJ, Buytenhek M, Slikkerveer FJ, Van Dorp DA (1979) On the haemoprotein character of prostaglandin endoperoxide synthetase. Biochim Biophys Acta 572:29-42

Van Os CPA, Rijke-Schilder GPM, Vliegenthart JFG (1979) 9-L*R*-linoleyl hydroperoxide, a novel product from the oxygenation of linoleic acid by type-2 lipoxygenases from soybeans and peas. Biochim Biophys Acta 575:479-484

Van Os CPA, Rijke-Schilder GPM, Van Halbeek H, Verhagen J, Vliegenthart JFG (1981) Double dioxygenation of arachidonic acid by soybean lipoxygenase-1. Kinetics and regio-stereo specificities of the reaction steps. Biochim Biophys Acta 663: 177-193

Verhagen J (1978) The conversion of 9-D- and 13-L-hydroperoxy-linoleic acid by soybean lipoxygenase-1. Thesis, University of Utrecht

Verhagen J, Bouman AA, Vliegenthart JFG, Boldingh J (1977) Conversion of 9-D- and 13-L-hydroperoxylinoleic acids by soybean lipoxygenase-1 under anaerobic conditions. Biochim Biophys Acta 486:114-120

Verhagen J, Veldink GA, Egmond MR, Vliegenthart JFG, Boldingh J, Van der Star J (1978) Steady-state kinetics of the anaerobic reaction of soybean lipoxygenase-1 with linoleic acid and 13-L-hydroperoxylinoleic acid. Biochim Biophys Acta 529: 369-379

Verhagen J, Veldink GA, Vliegenthart JFG, Boldingh J (1979) The conversion of 9-D- and 13-L-hydroperoxylinoleic acid by soybean lipoxygenase-1 under aerobic conditions. In: Appelquist L-Å and Liljenberg C (eds) Advances in the biochemistry and physiology of plant lipids. Elsevier North Holland, Amsterdam, pp 231-236

Vliegenthart JFG, Veldink GA (1982) Lipoxygenases. In: Pryor WA (ed) Free radicals in biology, vol V. Academic, New York, pp 29-64

Wallach DP, Brown VR (1981) A novel preparation of human platelet lipoxygenase. Characteristics and inhibition by a variety of phenyl hydrazones and comparisons witt other lipoxygenases. Biochim Biophys Acta 663:361-372

Yamamoto A, Yasumoto K, Mitsuda H (1970) Isolation of lipoxygenase isoenzymes and comparison of their properties. Agr Biol Chem 34:1169-1177

Physiological Reactions of Arachidonic Acid Oxygenation Products

R. J. Flower[1]

Introduction

I am very honoured to have been asked to speak at the Colloquium which commemorates the 100th birthday of Otto Warburg. Although Warburg had many scientific interests, he never worked directly in the field of prostaglandin biosynthesis or action. As the previous speaker Vliegenthart has pointed out, the cyclo-oxygenase enzyme, which biosynthesises the prostaglandin endoperoxides is a fascinating and unique dioxygenase and there can be few other enzymes in the body, which are capable of producing such a wide range of biologically active substances.

The prostaglandin literature is enormous and the catalogue of their biological activities very extensive. This makes the task of any lecturer a very difficult one. However, the organising committee have specifically asked me to talk about the *physiological* effects of the prostaglandins and I would like to develop a theme first introduced by Collier (1971) that prostaglandins are involved in "defense reactions" within the mammalian organism and that this represents their true physiological function.

How Prostaglandins are Formed

The previous speaker has explained the details of the cyclo-oxygenase reaction, which transforms arachidonic and certain other unsaturated fatty acids, into the prostaglandin endoperoxide intermediates G_2 and H_2 (Hamberg and Samuelsson 1973; Hamberg et al. 1974). These intermediates are labile substances and if allowed to decompose spontaneously give rise to a mixture of prostaglandins E, F and D as well as certain other breakdown products. In most cells however, further transformation of the endoperoxides occurs by enzymatic mechanisms and the type of prostaglandin product obtained depends to a large extent upon the cell or tissue type. This is important since nearly all cells in the mammalian organism can generate the endoperoxides and the only way in which any physiological specificity may be achieved is if these cells produce different end products. Two good examples of this are the blood platelets, which produce almost exclusively thromboxane from the endoperoxides (Hamberg et al. 1975) and blood vessel endothelial cells, which produce only prostacyclin (Moncada et al. 1976; Johnson et al. 1976).

It was earlier stated that prostaglandins are formed from unsaturated fatty acids. In most non-marine mammals arachidonic acid is the major substrate for the prostaglandin synthesising system and I will not

1 Department of Prostaglandin Research, Wellcome Research Laboratories, Langley Court, Beckenham, Kent BR3 3BS

34. Colloquium - Mosbach 1983
Biological Oxidations

discuss the products formed from other fatty acids. If you add arachidonic acid to cells, they will immediately generate prostaglandins and this implies that the availability of substrate is the rate limiting step for prostaglandin biosynthesis. All mammalian cells contain large amounts of arachidonic acid, but this is not "free" arachidonic acid. In fact, if you are careful with your analytical procedures, it is very difficult to detect any free acid at all within most cells. The majority of the arachidonic acid is esterified in some way such as neutral lipids, cholesterol esters or phospholipids. It therefore follows that to generate prostaglandins, we must first liberate free arachidonic acid from one of these ester pools.

In nearly all cells, the most likely source of the arachidonic acid is the phospholipid pool. In most phospholipid species, the arachidonic acid is esterified onto the β-position of the glyceryl backbone and to liberate the fatty acid, the enzyme phospholipase A_2 is required (although there may be other phospholipases, too). The plasma membrane of cells contains a great deal of phospholipid and phospholipase A_2 enzymes and it also contains many receptors for different biolgically active substances, such as histamine, bradykinin, thrombin and so on. It seems virtually certain that many of these substances, which are known to liberate prostaglandins from cells do so by stimulating the activity of phospholipase A_2, probably by some receptor coupled mechanism, thus liberating arachidonic acid from the membrane stores and leading to a generation and release of prostaglandins from the cell (see Flower and Blackwell 1976 for complete list of references). Incidentally, prostaglandins are not stored within cells; their release is immediately preceeded by de novo synthesis.

We can see from the foregoing discussion that the enzymatic equipment for the initiation of prostaglandin release is situated in the plasma membrane and this has a conceptual advantage, since the cell membrane is the part of the cell which comes into contact with the external environment. This is obviously important if as we think, prostaglandins are important in mediating defensive responses.

Prostaglandins and the Inflammatory Response

Inflammation may be regarded as the body's response to injury of any type. Many people regard inflammation and the related conditions of pain and fever as being pathological in nature, but this is a superficial view. It is a matter of fact that most injuries heal and the function of the inflammatory response is to remove the injurious agent and the dead or injured cells and thus clear the way for the healing processes to start. It is therefore an indispensible part of our physiology — although like many other of our bodily functions, it does occasionally fail to operate properly. This is particularly evident when we examine chronic inflammatory diseases, such as rheumatoid arthritis, in which (probably because of the persisting antigen, possibly a virus) there is a continual inflammation, which never resolves and which eventually causes substantial destruction of the joint. In this case we have an example of the body's response to a disease becoming the disease itself and we do our best to combat this using anti-inflammatory drugs.

At the cellular level, the characteristic signs and symptoms of inflammation seem to be caused by the release of chemical mediators. Substances such as histamine, 5-HT and bradykinin have been implicated, but we are chiefly interested in the prostaglandins. It has been known for a long time that any sort of damage to cells lead to the release of prostaglandins from cells. Such damage could comprise mechanical stimulation, antigen-antibody combination, the presence of bacteria, or stimulation by certain mediators such as bradykinin or histamine. With this in mind then it is not surprising to find that prostaglandins, particularly of the E-type are found in large amounts in inflammatory exudates of all types (Flower 1974). Not only that, but prostaglandins can themselves reproduce a great many of the signs and symptoms of the inflammatory response. For example, prostaglandin E_2 produces an increase in vascular permeability and wheal and flare responses (Crunkhorn and Willis 1971) and can produce inflammation of the joints in dogs (Rosenthale et al. 1972). Another interesting finding is that the prostaglandins can synergise with a great many other inflammogenic stimuli to greatly enhance their effect (Ferreira 1972).

One feature of the inflammatory response, which is not controlled by prostaglandins, is leucocyte migration. This very important phenomenon may be regulated by another product of the arachidonic acid cascade, leucotriene B_4, a dihydroxy acid formed by a lipoxygenase enzyme in leucocytes (Borgeat and Samuelsson 1979a,b).

Pain

Inflammatory conditions are almost always painful. *Overt pain* may not always be present, but there is frequently a condition known as "hyperalgesia." That is, when a normally painless stimulus becomes painful. There is evidence to suggest that this also depends upon local prostaglandin generation too. Large doses of prostaglandins cause pain (Karim 1971), but are unlikely to be encountered in vivo, but slow infusions of E-type prostaglandins given subdermally (to mimic the slow release during inflammation) produced a very marked and long-lasting hyperalgesia (Ferreira 1972). This was not obtained with any other substance tested. An interesting feature of this hyperalgesia was that there was a greatly increased sensitively to the algesic effects of other substances such as bradykinin.

Fever

Amongst the sequelae of the inflammatory condition or infection of any sort is fever. Although the exact reason for this response may not always be clear, there is no doubt that it is an important protective biological response, practised even by those animals which cannot regulate their own body temperature. Once again, we find prostaglandins are involved. Injection into animals, such as cats or rabbits, of bacterial endotoxins that cause the development of a fever liberate prostaglandins into the CSF (Feldberg and Gupta 1973; Feldberg et al. 1972). Prostaglandins of the E-type are the most potent pyrogens known, even nanogram doses injected into the CSF of cats cause a large increase in core temperature (cf. Milton and Wendlandt 1970, 1971). In this respect the prostaglandins are considerably more potent than other pyretic substances such as 5-HT (which is only active in certain species anyway).

Action of Anti-Inflammatory Drugs

Aspirin and Its Congeners

Since prostaglandin generation takes place at the site of inflammation and seems to be intimately linked to the development of the inflammatory response, it is not surprising to find that anti-inflammatory drugs interfere in some way with the prostaglandin system. In fact their effects are very striking. The fundamental discovery was made in 1971 by Vane and colleagues (Vane 1971; Ferreira, Moncada, and Vane 1971; Smith and Willis 1971); they found that aspirin itself and indeed several of the so-called aspirin-like drugs possess the unique ability to inhibit the cyclo-oxygenase enzyme, thus they prevent prostaglandin generation. This discovery has been confirmed in vivo and in vitro countless times and the original finding has been greatly extended. It now appears that almost all the aspirin-like drugs are inhibitors of the enzyme and that there is a good correlation between their anti-inflammatory and anti-enzyme effects, including some spectacular instances involving the activity of pairs of isomers. In addition, it has been observed that the anti-cyclo-oxygenase effect of these drugs can be easily achieved within the plasma levels that these drugs attain after therapeutic dosage (Flower 1974).

We are now in a position to understand the anti-inflammatory, analgesic and antipyretic properties of the aspirin-like drugs; by blocking the formation of these defensive chemicals, aspirin actually reduces the inflammation, fever and pain. Indeed, it can be directly demonstrated that the administration of such drugs leads to an immediate reduction in the prostaglandin content of inflammatory exudates, or of the CSF of cats undergoing a febrile response. In the latter case, the fall in prostaglandins was accompanied by a defervescence (Feldberg et al. 1972).

As you might expect, the inflammation fever and pain *caused* by prostaglandins is not inhibited by aspirin.

The Glucocorticoids

The aspirin-like drugs are not the only drugs which produce an anti-inflammatory effect. The glucocorticoids are also used to this end, usually with great success. Unlike the aspirin-like drugs, the glucocorticoids have no direct inhibitory effect on the cyclo-oxygenase, but they can reduce prostaglandin biosynthesis in a more subtle way, by suppressing the activity of the membrane phospholipase A_2. This enzyme, as we mentioned earlier, is responsible for liberating the substrate required for prostaglandin generation. In other words, the glucocorticoids can "starve" the dioxygenase enzyme of sufficient substrate for effective prostaglandin synthesis.

How do they accomplish this? The glucocorticoids do not have a *direct* action on the phospholipase, but act by stimulating the target cells to synthesise and release an inhibitory protein variously called "macrocortin" or "lipomodulin" (Flower and Blackwell 1979; Blackwell et al. 1980; Hirata et al. 1980, 1981). Recent research suggests they are derived from the same protein. The action of the glucocorticoids requires combination with a cytoplasmic receptor and de novo RNA and protein synthesis.

All the glucocorticoids which are used therapeutically are analogues of the naturally occurring compound hydrocortisone, which is secreted by the adrenal cortex in response to ACTH. The glucocorticoids regulate so many important physiological events, it is not surprising therefore to find that they can regulate the intensity of the inflammatory response by attenuating the generation of the defensive prostaglandins.

Prostaglandins, Platelets and Haemostasis

Another aspect of our physiological defences in which prostaglandins are intimately involved is the haemostatic system. When the vascular system is punctured there is a loss of blood and in order to contain this as rapidly as possible, the haemostatic system comes into operation. Part of this system consists of the "classical" clotting cascade and the other component comprises the platelets. It is upon the latter that prostaglandins have such marked effects.

Platelet Aggregation and Thromboxane

The blood platelets are dedicated end cells whose main function is to form an "aggregate" or plug with haemostatic properties. Platelets very rapidly undergo an aggregation reaction when exposed to various "aggregating agents" such as collagen, ADP or thrombin. Until recently, the mechanism by which this occurred was obscure, but it now transpires that the prostaglandin system is involved. When collagen comes into contact with platelets, it activates one or other of the phospholipase pathways, which liberate arachidonic acid from membrane phosphatides. This is transformed by the cyclo-oxygenase into the intermediate endoperoxides, but instead of being transformed into the "stable" prostaglandins E_2 of $F_{2\alpha}$, the platelets generate almost exclusively a very labile product known as thromboxane A_2 (Hamberg et al. 1975). This remarkable compound, which has a half life of only about 30 s at physiological pH and temperature, has two important properties. Firstly, it is an exceedingly potent platelet aggregating agent and secondly, it is a powerful vasoconstrictor agent. The sequence of events at a wound site then would seem to involve contact of the platelet with subendothelial collagen, liberation of arachidonic acid from membrane phospholipids, generation of thromboxane followed by formation of a platelet plug and local casoconstriction. When it spontaneously decomposes, it forms the biologically inactive compound, thromboxane B_2.

It is important to realise that there are other routes of platelet aggregation, which may not, however, depend upon thromboxane formation. Thrombin, which is formed during the activation of the clotting cascade also triggers platelet aggregation, but apparently does so by a thromboxane independent mechanism.

Control of Aggregation by Prostacyclin

Whilst the formation of a platelet plug at the site of a wound is an obvious instance of a defence mechanism, another not quite so obvious example is the system which prevents platelets from aggregating continuously within the vascular system.

Virtually any foreign body in contact with fresh blood rapidly becomes coated with platelets, which under normal conditions, will quickly accumulate, forming a thrombus. In what sense are "foreign bodies" foreign? Why do platelets not aggregate when in contact with normal blood vessel endothelial cells? Again we can turn to the prostaglandin system for a partial answer to this problem. In the same way that platelets make almost exclusively thromboxane A_2 from the endoperoxide intermediates, blood vessel endothelial cells make yet another derivative—prostacyclin (Moncada et al. 1976; Johnson et al. 1976; Gryglewski et al. 1976). Like the endoperoxides, prostacyclin is an unstable compound, although not so unstable as thromboxane A_2. It has a half life of about 8 min at physiological pHs and temperatures and decays spontaneously to an inactive compound known as 6-keto prostaglandin $F_{1\alpha}$. Prostacyclin has exactly the opposite profile of activity to thromboxane A_2, it is a potent vasodilator and a very powerful inhibitor of platelet aggregation, which is effective against the effects of nearly all known pro-aggregatory agents.

The picture which emerges then is of the normal endothelial layer protecting itself against inappropriate platelet aggregation by generating prostacyclin. Whether it does this by generating the prostacyclin from its own internal supplies of arachidonic acid in response to phospholipase stimulation or whether the blood vessel "prostacyclin synthetase" utilises platelet endoperoxides, which have been released from the platelets is not sure, but by stimulating platelet c-AMP, the prostacyclin so generated stabilises the platelet and prevents aggregation. Clearly, when endothelial cells are damaged, say during wounding, then this inhibitory influence of prostacyclin would be absent and thus the effect of thromboxane would be dominant.

It is possible that inhibition of endogenous prostacyclin production by disease may be responsible for the growth of thrombi on diseased arterial walls in conditions such as atherosclerosis. Lipid peroxides, of the type known to occur in atherosclerotic plaques are known to be able to inhibit the prostacyclin synthetase enzyme (Moncada et al. 1976b).

Prostaglandins and Parturition

We now move away from the haemostatic system to touch upon a completely different type of "defence" reaction on which prostaglandins are involved.

The correct timing of the moment of parturition is important for the well-being of the mother and the survival of the foetus. Prostaglandins were first discovered in human semen (von Euler 1936) and one of the earliest observations was that they were extremely potent uterotropic agents (indeed they are now used clinically to induce labour). It was subsequently found that the pregnant uterus was considerably (10-50 times) more sensitive to the stimulating effects of prostaglandins (Horton 1969). Further research indicated that the amount of prostaglandins generated by the pregnant mammal was greater than that of the non-pregnant animal and furthermore that this rose explosively as the normal time of parturition drew near (Flower 1977). This causal connection between prostaglandin generation and parturition was strengthened when it was found that aspirin and the aspirin-like drugs significantly delayed parturition in rats (Aiken et al. 1972, 1974; Chester 1972). It was also observed that there was a high incidence of delayed labour amongst human females who abused aspirin-like drugs (Lewis and Schulman 1973).

As if to underline the important role of prostaglandins in the pregnant mammal, we find that striking changes take place in the amount of the prostaglandin metabolising enzymes during pregnancy. Obviously, it would be disastrous if prostaglandins were able to act upon the pregnant uterus before parturition was due and to prevent this, the levels of the prostaglandin metabolising enzymes are dramatically increased by the hormones of pregnancy (especially progesterone) (Bedwani and Marley 1975; Sun and Armour 1974). As parturition approaches and the amounts of progesterone declines, the capacity of the lungs and other organs to metabolise prostaglandins also decreases so that these substances are present in sufficient concentration to act on the uterus at the time of parturition to expel the fetus (Flower 1977). It is a good example of a nicely adjusted biological control system. Of course, as with many vitally important physiological events, there are "back-up" systems, thus, the administration of the aspirin-like drugs only delays parturition and does not abolish it completely.

Prostaglandins and the Smooth Muscle of the Gut

The myometrium is not the only smooth muscle, which responds to the prostaglandins; another important example is that of the gut. Again it is likely that prostaglandins participate in a variety of gastrointestinal defence mechanisms. The administration of prostaglandins particularly those of the E-type to human volunteers or to animals, causes nausea, vomiting, accumulation of fluid in the gut, decreased intestinal transit time and diarrhoea. These are defence reactions, which characteristically occur after infection or poisoning, and indeed some bacterial endotoxins have been shown to release prostaglandins directly from the gut. As one might anticipate, aspirin-like drugs have been tested and found to be active as anti-diarrhoeal agents in some cases. *E. coli* endotoxin induced diarrhoea in dogs and the diarrhoea vomiting and colicky pain caused by clinical X-irradiation of the abdomen have been treated with aspirin itself and it has been found to be a useful palliative agent (Collier 1974).

Prostaglandins and Cytoprotection

It is not only on the smooth muscle of the gut that prostaglandins are active, they also possess an extraordinary property known as "cytoprotection." The name was first coined by Robert (Robert 1976). Originally, it was found that many of the naturally occurring prostaglandins inhibited gastric acid secretion induced either by food or by histamine (Robert et al. 1976). The duration of action of the naturally occurring compounds was short (because of metabolism), but synthetic analogues were considerably longer-lasting. Subsequently, these prostaglandins were found to possess striking anti-ulcer properties as well as anti-secretory properties. The prostaglandins seemed active against ulceration and gastrointestinal damage caused a very diverse range of agents including not only the aspirin-like drugs (which are well-known gastric irritants), but a great variety of other procedures or agents such as pyloric ligation, strong acids and bases, ethanol and even boiling water (Robert et al. 1977).

One possibility – as yet unconfirmed – is that there is a continual generation of prostaglandins by the mucosa of the gastrointestinal tract which somehow preserves the integrity of the organ. It is by removing the tonic influence of these agents that the aspirin-like drugs may cause the gastrointestinal damage for which they are well-known. The mechanism by which the prostaglandins bring about such extraordinary actions is far from clear.

Conclusion

In this small review, we have only touched upon some of the actions of the prostaglandins. It is clear that these agents have a very important physiological role, a large part of which is bound up in the body's defence systems, but obviously it will be some years before we can say with confidence whether or not this is the only function of these unique oxygenated lipids.

Discussion

Gibian: Would you mind to add a remark on aspirin and platelet aggregation?

Flower: Aspirin is a potent and irreversible inhibitor of the platelet cyclo-oxygenase. Since these cells are unable to synthesise new proteins, this inhibition persists for the lifetime of the platelet.

The generation of thromboxane by platelets is an important feature of the aggregation action of certain agents (such as ADP), so aspirin can in some circumstances effectively block platelet aggregation.

Schuster: The irreversibility of aspirin action might not be a desirable property of the compound if strong, but reversible inhibitors of individual steps of PG, thromboxane-synthesis can be used. Do there exist highly specific inhibitors for the reactions?

Flower: Yes. Several types of drug are known, including prostaglandin endoperoxide analogues and certain N-substituted imidazole derivatives.

Question from the floor: If transacetylation is the inhibitory mechanism of aspirin, what about salicylate or indomethacin?
Has salicylic acid to be acetylated first?

Flower: Indomethacin appears to inhibit the reaction by a different mechanism to aspirin. No one knows how salicyclic acid acts. Possibly it must be first acetylated.

Ullrich: In order to distinguish between a direct action of salicyclic acid or a mechanism via acetylation, one could use the "slow" and "fast" acetylators known in drug metabolism. Is this a possibility?

Flower: That is a good idea and has never been tested to my knowledge.

Staudinger: Could you comment on the role of Ca++ and the action of steroids on phospholipase A_2?

Flower: Most phospholipase enzymes of the A_2 type display an absolute requirement for calicum. The inhibitory action of steroids is not direct, but is mediated by a protein synthesized and released from cells following steroid stimulation.

References

Aiken JW (1972) Aspirin and indomethacin prolong parturition in rats: evidence that prostaglandins contribute to expulsion of foetus. Nature (Lond) 240:21

Aiken JW (1974) Prostaglandins and prostaglandin synthetase inhibitors: studies on uterine motility and function. In: Robinson HJ and Vane JR (eds) Prostaglandin Synthetase Inhibitors. Raven, New York, p 289

Bedwani JR, Marley PB (1975) Enhanced inactivation of prostaglandin E_2 by the rabbit lung during pregnancy or progesterone treatment. Br J Pharmacol 53:547

Blackwell GJ, Carnuccio R, Di Rosa M, Flower RJ, Parente L, Persico P (1980) Macrocortin: a polypeptide causing the anti-phospholipase effect of glucocorticoids. Nature 287:147

Borgeat P, Samuelsson B (1979a) Transformation of arachidonic acid by rabbit polymorphonuclear leukocytes. J Biol Chem 254:2643

Borgeat P, Samuelsson B (1979b) Metabolism of arachidonic acid in polymorphonuclear leukocytes. J Biol Chem 254:7865

Chester R, Dukes M, Slater SR, Walpole AL (1972) Delay of parturition in the rat by anti-inflammatory agents which inhibit the biosynthesis of prostaglandins. Nature (Lond) 240:37

Collier HOJ (1971) Prostaglandins and aspirin. Nature 232:17

Collier HOJ (1974) Prostaglandin synthetase inhibitors and the gut. In: Robinson HJ and Vane JR (eds) Prostaglandin Synthetase Inhibitors. Raven, New York, p 121

Crunkhorn P, Willis AL (1971) Cutaneous reactions to intradermal prostaglandins. Br J Pharmac 41:49

Feldberg W, Gupta KP (1973) Pyrogen fever and prostaglandin like activity in cerebrospinal fluid. J Physiol 228:41

Feldberg W, Gupta KP, Milton AS, Wendlandt S (1972) Effect of bacterial pyrogen and antipyretics on prostaglandin activity in cerebrospinal fluid of unanaesthetized cats. Proc Br Pharm Soc September 1972, p 38

Ferreira SH (1972) Prostaglandins, aspirin-like drugs and analgesia. Nature (New Biol) 240:200

Ferreira SH, Moncada S, Vane JR (1971) Indomethacin and aspirin abolish prostaglandin release from the spleen. Nature (New Biol) 231:237

Flower RJ (1974) Recent studies on the mode of action of aspirin and other nonsteroid anti-inflammatory drugs. In: Spencer B (ed) Industrial aspects of biochemistry. FEBS, p 669

Flower RJ (1974) Drugs which inhibit prostaglandin synthesis. Pharmacol Rev 26:33

Flower RJ (1977) The role of prostaglandins in parturition with special reference to the rat. The Fetus and Birth. Ciba Foundation 47 (New Series) p 297

Flower RJ, Blackwell GJ (1976) The importance of phospholipase A_2 in prostaglandin biosynthesis. Biochem Pharmacol 25:285

Flower RJ, Blackwell GJ (1979) Anti-inflammatory steroids induce biosynthesis of a phospholpase A_2 inhibitor which prevents prostaglandin generation. Nature 278:456

Gryglewski RJ, Bunting S, Moncada S, Flower RJ, Vane JR (1976) Arterial walls are protected against deposition of platelet thrombi by a substance (Prostaglandin X) which they make from prostaglandin endoperoxides. Prostaglandins 12:658

Hamberg M, Samuelsson B (1973) Detection and isolation of an endoperoxide intermediate in prostaglandin biosynthesis. Proc Natl Acad Sci USA 70:899

Hamberg M, Svensson J, Wakabayashi T, Samuelsson B (1974) Isolation and structure of two prostaglandin endoperoxides that cause platelet aggregation. Proc Natl Acad Sci USA 71:345

Hamberg M, Svensson J, Samuelsson B (1975) Thromboxanes: a new group of biologically active compounds derived from prostaglandin endoperoxides. Proc Natl Acad Sci USA 72:2994

Hirata F, Del Carmine R, Nelson CA, Axelrod J, Schiffmann E, Warabi A, De Blas AL, Nirenberg M, Manganiello V, Vaughan M, Kumagai S, Green I, Dekcer JL, Steinberg AD (1981) Presence of autoantibody for phospholipase inhibitor protein, lipomodulin in patients with rheumatic diseases. Proc Natl Acad Sci USA 78:3190

Hirata F, Schiffman E, Venkatasubramanian K, Salomon D, Axelrod J (1980) A phospholipase A_2 inhibitory protein in rabbit neutrophils induced by glucocorticoids. Proc Natl Acad Sci USA 77:2533

Horton EW (1969) Hypothesis on physiological roles of prostaglandins. Physiol Rev 49:122

Johnson RA, Morton DR, Kiner JH, Gorman RR, McGuire JR, Sun FF, Whittaker N, Bunting S, Salmon JA, Moncada S, Vane JR (1976) The chemical structure of prostaglandin X (prostacyclin). Prostaglandins 12:915

Karim SMM (1971) Action of prostaglandin in the pregnant woman. Ann NY Acad Sci 180:483

Lewis RB, Schulman JD (1973) Influence of acetylsalicylic acid, an inhibitor of prostaglandin synthesis, on the duration of human gestation and labour. Lancaet 2:1159

Milton AS, Wendlandt S (1970) A possible role for prostaglandin E_1 as a modulator for temperature regulation in the central nervous system of the cat. J Physiol 207:76

Milton AS, Wendlandt S (1971) Effect on body temperature of prostaglandins of the A, E and F series on injection into the third ventricle of unanaesthetized cats and rabbits. J Physiol 218:325

Moncada S, Gryglewski RJ, Bunting S, Vane JR (1976a) An enzyme isolated from arteries transforms prostaglandin endoperoxides to an unstable substance that inhibits platelet aggregation. Nature (Lond) 263:663

Moncada S, Gryglewski RJ, Bunting S, Vane JR (1976b) A lipid peroxide inhibits the enzyme in blood vessel microsomes that generates from prostaglandin endoperoxides the substance (Prostaglandin X) which prevents platelet aggregation. Prostaglandins 12:715

Robert A (1976) Anti-secretory, anti-ulcer, cytoprotective and diarrhoegenic properties of prostaglandins. Adv Prostaglandin Thromboxane Res 2:507

Robert A, Nezamis JE, Lancaster C (1977) Gastric cytoprotective property of prostaglandins. Gastroenterology 72:

Robert A, Schultz JR, Nezamis JE, Lancaster C (1976) Gastric anti-secretory and anti-ulcer properties of PGE_2, 15-methyl PGE_2 and 16,16 dimethyl PGE_2. Gastroenterology 70:359

Rosenthale ME, Dervinis A, Kassarich J, Singer S (1972) Prostaglandins and anti-inflammatory drugs in the dog knee joint. J Pharm Pharmac 24:149

Smith JB, Willis AL (1971) Aspirin selectively inhibits prostaglandin production in human platelets. Nature (New Biol) 231:235

Sun FF, Armour SB (1974) Prostaglandin 15-hydroxydehydrogenase and Δ^{13}-reductase levels in the lungs of maternal, fetal and neonatal rabbits. Prostaglandins 7:327

Vane JR (1971) Inhibition of prostaglandin synthesis as a mechanism of action for aspirin-like drugs. Nature (New Biol) 231:232

Von Euler US (1936) On the specific vasodilating and plain muscle stimulating substances from accessory genital glands in man and certain animals (prostaglandin and vesiglandin). J Physiol (Lond) 88:213

Bacterial Cytochrome P-450 Monooxygenases

G. C. Wagner[1]

The P450 cytochromes are a distinct class of heme thiolate proteins characterized by interactions with organic compounds and carbon monoxide that produce unique spectroscopic properties (1). Together with associated redox components, the P450 heme proteins function as monooxygenases, catalyzing the two-electron reduction of dioxygen to form a molecule of water and an activated oxygen atom (2). The subsequent transfer of this oxygen atom to hydrocarbon substrates constitutes a major pathway of oxygenase activity in biological systems (1-7). Eukaryotes in particular are characterized by a wide distribution of organ and organelle specific P450 monooxygenases, whose diversification is further amplified by multiple isozymes of overlapping substrate specificity, for example in mammalian hepatic tissue. The broad range of substrate selectivity exhibited by P450 cytochromes parallels their involvement in many different types of biological functions. In the endoplasmic reticulum of hepatic cells, P450 cytochromes participate in the detoxification of xenobiotic compounds, such as the hydroxylation of the pharmaceutical sedative phenobarbital. Other liver microsomal P450 systems will catalyze the activation of carcinogenic DNA-alkylating agents, as represented by the epoxidation of benzo[a]pyrene derivatives. In contrast, mitochondrial P450 monooxygenases of the adrenal cortex mediate a cascade of anabolic pathways in the biosynthesis of regulatory steroid hormones, such as hydrocortisone.

The organization of specific P450 monooxygenases into either microsomal or mitochondrial fractions is accompanied by a distinction in the composition of associated redox transfer proteins. In both systems, electron transfer proteins are required to mediate sequentially the two reduction equivalents from NAD(P)H to the b-type heme thiolate monooxygenase. In microsomal systems, this is achieved with a single reductase protein, which is composed of two flavin prosthetic groups, FMN and FAD. In mitochondrial systems, two redox transfer proteins are associated with the P450 monooxygenase. First, an FAD-containing reductase couples the two-electron oxidation of NAD(P)H to one-electron reductions of an $Fe_2S_2Cys_4$ type redoxin. The iron sulfur protein then catalyzes the two sequential one-electron transfers to the P450. Microbial P450 systems, such as the camphor 5-monooxygenase (8-12) and the linalool 8-monooxygenase (13,14) from *Pseudomonas* strains, are remarkably similar to the mitochondrial systems in their molecular compositions of the three protein components. More significant, and in spite of the differences in substrate specificities and modes of redox flow, these bacterial P450 cytochromes remain homologous to essentially all mammalian systems in structural and catalytic properties of the heme thiolate active site. The extensive physical-chemical investigations of cytochrome $P450_{CAM}$, including sequence and crystallographic studies, have provided many of the first direct molecular characterizations of the reaction states and intermediates in the P450 monooxygenase cycle. On a comparative basis, these bacterial monooxy-

1 University of Illinois, at Urbana-Champaign, Department of Biochemistry, Roger Adams Laboratory, Urbana, Illinois, 61801 USA

34. Colloquium - Mosbach 1983
Biological Oxidations

genases represent significant biological model systems that contribute to a fundamental understanding of dioxygen activation and hydrocarbon oxidation.

Microbial transformations of naturally occurring terpenes into obligatory carbon and energy sources are well-known for aerobic soil bacteria. In two strains of *Pseudomonas*, one specific for camphor and the other for linalool, the oxidative degradation of these substrates is initiated by P450 monooxygenase systems. The camphor system is comprised of three protein components, which are in turn homologous in both structure and function to their counterparts in the linalool monooxygenase system. Both systems catalyze the stereospecific oxygenation of their respective unactivated hydrocarbon substrates. Figure 1 illustrates the primary and secondary reactions mediated by these bacterial terpene monooxygenases. There are essentially three facets of the microbial systems which have promoted them to the forefront of molecular P450 studies. First, the protein components are produced in high specific content, which are readily isolated as soluble forms (15). This contrasts notably with the integral membrane association of mammalian P450 cytochromes and their lipid requirement for activity. Second, isotopic enrichments and atomic substitutions of proteins (15-17) for a broad range of spectroscopic investigations are both feasible and practical in the bacterial systems. Finally, the camphor system has yielded three forms of the only crystalline P450 cytochrome (18). This has ultimately led to the first three-dimensional characterization by X-ray diffraction studies (19) and has the potential to examine various reaction states in molecular detail (11,20).

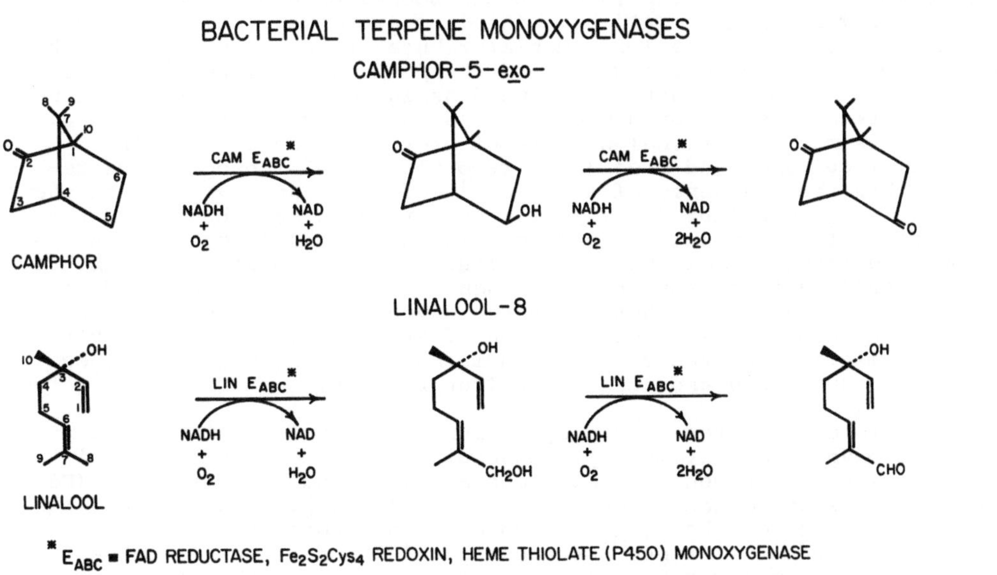

Fig. 1. Reactions catalyzed by the camphor 5-*exo*-monooxygenase and linalool 8-monooxygenase systems

The reaction states of the $P450_{CAM}$ monooxygenase cycle have been extensively characterized in terms of the structural, energetic, and dynamic properties of the heme thiolate active site (8-12). Dioxygen activation proceeds through a sequence of states starting with the native Fe^{3+} monooxygenase (m^{O}) binding the camphor substrate (m^{OS}), which ellicits the characteristic blue-shift of the Soret absorption maximum from 417 nm to 391 nm. Next, a one electron reduction of m^{os} to the reduced Fe^{2+} state (m^{rs}) enables binding of dioxygen to form a ternary complex ($m^{rs}_{O_2}$). The catalytically inactive carbon monoxide complex of the reduced state (m^{rs}_{CO}) gives rise to a characteristic optical absorption maximum at 450 nm (Soret band), and hence the basis for the trivial name of these heme proteins. In the overall rate limiting step of catalysis (21), the transfer of a second reduction equivalent from the iron sulfur protein, putidaredoxin, to $m^{rs}_{O_2}$ leads directly to product formation (5-*exo*-hydroxycamphor) and regeneration of m^{o}. A more definitive characterization of the steps and intermediates leading to product formation has not been fully resolved (21-24). Thus, elucidation of the basic P450 mechanism of dioxygen cleavage and hydrocarbon oxygenation remains under active investigation. Since the iron protoporphyrin IX prosthetic group of P450 cytochromes is common to all b-type heme proteins, various aspects of P450 monooxygenase function have been compared with other systems: electron transfer steps to respiratory redox cytochromes; the dioxygen adduct to oxygen binding globins; and active oxygen intermediates to peroxidase-type reactions. This latter analog is based partly on mechanistic studies of the P450 cytochrome reaction with peroxides or peracids. Oxidized Fe^{3+} $P450_{CAM}$ for example will catalyze in vitro the ROOH dependent hydroxylation of camphor without requirement of O_2, NADH, or the redox transfer proteins (25). This P450 peroxygenase type shunt has been utilized to search for interdiates that may be relevant to the monooxygenase mechanism (26-30).

Molecular characterizations of the bacterial cytochrome P450 reactions states have focused primarily on spectroscopic probes of the active site (8-12). Inherent to all heme thiolate proteins, the properties of P450 cytochromes have long been attributed to an iron sulfur linkage between the prosthetic group and the protein. Only recently, however, has a direct and straightforward assignment of an Fe-S structure been determined in the oxidized enzyme substrate complex of $P450_{CAM}$. Resonance Raman vibrational spectra of protein enriched in sulfur (bacterial culture) and iron (heme reconstitution) isotopes have revealed that the frequency of a 351 cm^{-1} mode is sensitive to both the mass of iron and sulfur (31). Thus, these results have confirmed the Fe-S axial ligand structure in a soluble P450 protein solution. Although the integrity of this iron sulfur linkage appears essential to enzymatically active protein, at present there is no direct evidence which implicates this structure in the oxygenation mechanism. In addition, any comprehensive description of P450 catalysis will require characterizations of the molecular domains, which modulate the specificity for substrate and the iron sulfur redoxin.

Protein-protein interactions play a significant role in microbial P450 monooxygenase systems. Spectroscopic measurements have identified equilibria between the FAD reductase (fp) and putidaredoxin (Pd) (32) and between Pd and cytochrome $P450_{CAM}$ (33-36). Figure 2 describes the binding isotherm between the reductase and redoxin that was deduced by optical difference spectroscopy. The formation of high-affinity dienzyme complexes in these systems underscores the importance of the interactions in catalysis. Kinetic studies of the redox transfer reactions between the metal centers of Pd and $P450_{CAM}$ have revealed that both of the one-electron transfers occur within the dienzyme complex (21). Modification of putidaredoxin by limited proteolytic hydrolysis of two residues at the C-terminus retains the native spectroscopic properties of the iron sulfur redox site (33). However, this modified protein produces a large decrease in both the affinity and catalytic constants for the P450 monooxygenase (25,33). These results emphasize the inherent specificity in the interaction and implicate

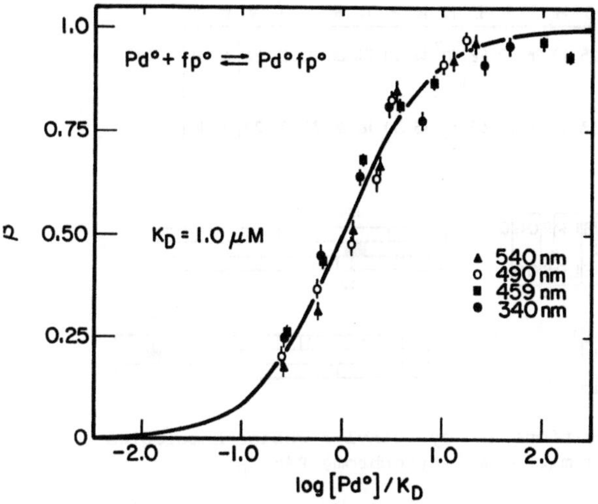

Fig. 2. Binding isotherm for oxidized putidaredoxin reductase (fp^o) titrated with oxidized putidaredoxin (*Pd*^{o}). Absorption difference spectroscopic measurements at 20° in 50 m*M* Tris-Cl, pH 7.4; 5% (v/v) glycerol. Reductase concentration = 0.026 m*M*

the C-terminal region of the iron sulfur protein in a specific effector mode in the oxygenation process (22,33). Additional evidence for the high degree of specific protein interactions in these complexes has developed from cross-reactivities of the three homologous components from the linalool and camphor systems. Kinetic studies have shown that the interchange of a single protein from either system will induce nearly a complete inhibition of reconstituted monooxygenase activity (13). This regulation of monooxygenase activity through specific protein interaction has emphasized the importance of the ongoing characterizations into the full three-dimensional structures of the P450 monooxygenase components.

The molecular structure of the $P450_{CAM}$ monooxygenase has progressed to the point where reactions are now discussed in terms of specific residues of the amino acid sequence. The first complete primary structure of a P450 protein deduced from amino acid sequence techniques was recently reported for $P450_{CAM}$ (37-39). Concurrently, there have appeared partial sequence studies of various microsomal P450 proteins (40,42), including a complete structure deduced from DNA sequence methodology (40). Considerable interest has evolved from these results in elucidating a peptide sequence homology, which would indicate the essential cysteinyl residue of the active site. Figure 3 shows the two regions of possible homology which center around $P450_{CAM}$ cysteinyl residues 134 and 355. Chemical modification studies of $P450_{CAM}$ with sulfhydryl reagents have assigned the 355 residue to an apparently nonessential and highly reactive cysteine (43). Thus, if active site homology persists in P450 heme proteins, then Cys-134 is implicated as the essential active site region. This conclusion has now been verified directly from the continuing X-ray diffraction study of the $P450_{CAM}$ tertiary structure.

Crystallographic measurements of $P450_{CAM}$ have been obtained to 2.5 Å resolution and electron densities have been assigned to nearly 80% of the 412 amino acid residues (19). Although further analysis is necessary, the following conclusions can be made. The overall shape of $P450_{CAM}$ is ellipsoidal with approximate dimensions of 55 × 55 × 39 Å. The heme prosthetic group is buried within α-helical domains and is not directly accessible to the surface. Fitting of the primary sequence to the electron density map unambiguously identifies Cys-134 in one of the α-helical structures to the axial ligand in the oxidized camphor

	A	C	D	E	F	G	H	I	K	L	M	N	P	Q	R	S	T	V	W	Y	
CYTOCHROME P450$_{CAM}$ Haniu *et al.*, 1982	31	8	21	32	18	25	12	26	12	42	10	14	30	24	26	25	19	24	4	9	412
CYTOCHROME P450$_{RAT-PB}$ Fujii-Kuriyama *et al.*, 1982	23	6	23	33	39	33	17	30	24	63	11	13	30	18	30	33	27	23	1	14	491

Fig. 3. Comparisons of amino acid compositions and cysteinyl sequence homologies for cytochromes P450$_{CAM}$ and rabbit liver microsomal cytochrome P450$_{PB}$

bound complex of P450$_{CAM}$. Finally, the orientation of the heme relative to this proximal ligand is identical to that found in hemoglobin (44) and opposite to cytochrome c peroxidase. The current state of refinement focuses on the specific orientation and interactions of the camphor molecule within the enzyme substrate complex. At present, the molecular domains which modulate the enzyme substrate interaction in P450 monooxygenases are in general not well-defined (45). In anticipation of isolating P450$_{LIN}$ crystallographic forms, a structural comparison of these bacterial proteins could ultimately provide a molecular basis for specificity in substrate recognition and P450 monooxygenation function.

References

1. Sato R, Omura T (1978) Cytochrome P450. Academic, New York
2. Hayaishi O (1974) Molecular Mechanisms of Oxygen Activation. Academic, New York
3. White RE, Coon MJ (1980) Ann Rev Biochem 49:315-356
4. Coon MJ, White RE (1980) In: Spiro TG (ed) Dioxygen Binding and Activation by Metal Centers. John Wiley, New York, pp 73-123
5. Ullrich V, Roots I, Hildebrant A, Estabrook RW, Conney AH (1977) Microsomes and Drug Oxidations. Pergamon, Oxford
6. Coon MJ, Conney AH, Estabrook RW, Gelboin HV, Gillette JR, O'Brien PJ (1980) Microsomes, Drug Oxidations, and Chemical Carcinogenesis. Academic, New York
7. Sato R, Kato R (1982) Microsomes, Drug Oxidations, and Drug Toxicity. Wiley-Interscience, New York
8. Gunsalus IC, Meeks JR, Lipscomb JD, Debrunner PG, Munck E (1974) In: Hayaishi O (ed) Molecular Mechanisms of Oxygen Activation. Academic, New York, pp 559-613
9. Gunsalus IC, Sligar SG (1978) Adv Enzymol 47:1-44
10. Debrunner PG, Gunsalus IC, Sligar SG, Wagner GC (1978) In: Sigel H (ed) Metal Ions in Biological Systems, Vol 7. Marcel Dekker, New York, pp 241-275
11. Gunsalus IC, Wagner GC, Debrunner PG (1980) In: Coon MJ, Conney AH, Estabrook RW, Gelboin HV, Gillette JR, O'Brien PJ (eds) Microsomes, Drug Oxidations, and Chemical Carcinogenesis. Academic, New York, pp 233-242
12. Wagner GC, Gunsalus IC (1982) In: Dunford HB, Dolphin D, Raymond KN, Sieker L (eds) The Biological Chemistry of Iron. D Reidel, Boston, pp 405-412
13. Ullah AJH, Bhattacharyya PK, Bakthavachalam J, Wagner GC, Gunsalus IC (1983) Fed Proc 42:819
14. Wagner GC, Jung C, Shyamsunder E, Bowne S (1983) Fed Proc 42:820

15. Gunsalus IC, Wagner GC (1978) Methods Enzymol 52:166-188
16. Wagner GC, Toscano WA, Perez M, Gunsalus IC (1981) J Biol Chem 256:6262-6265
17. Wagner GC, Gunsalus IC, Wang R, Hoffman BM (1981) J Biol Chem 256:6266-6273
18. Poulos TL, Perez M, Wagner GC (1982) J Biol Chem 257:10427-10429
19. Poulos TL, Alden RA, Wagner GC, Gunsalus IC (1983) J Biol Chem (submitted)
20. Devaney P, Wagner GC, Debrunner PG, Gunsalus IC (1980) Fed Proc 39:1139
21. Pederson TC, Austin RH, Gunsalus IC (1977) In: Ullrich V et al (eds) Microsomes and Drug Oxidations. Pergamon, Oxford, pp 275-283
22. Sligar S, Kennedy K, Pearson D (1980) Proc Natl Acad Sci USA 77:1240
23. Heimbrook D, Sligar SG (1981) Biochem Biophys Res Commun 99:530
24. Gelb M, Heimbrook D, Malkonen P, Sligar SG (1982) Biochemistry 21:370-377
25. Sligar SG, Shastry BS, Gunsalus IC (1977) In: Ullrich V et al (eds) Microsomes and Drug Oxidations. Pergamon, Oxford, pp 202-209
26. White RE, Sligar SG, Coon MJ (1980) J Biol Chem 255:11108-11111
27. Blake RC, Coon MJ (1980) J Biol Chem 255:4100-4111
28. Blake RC, Coon MJ (1981) J Biol Chem 256:5755-5763
29. Wagner GC (1981) Fed Proc 40:710
30. Wagner GC (1982) Fed Proc 41:1406
31. Champion PM, Stallard BR, Wagner GC, Gunsalus IC (1982) J Am Chem Soc 104: 5469-5472
32. Wheatley R (1981) BS Thesis. University of Illinois, Urbana
33. Sligar S, Debrunner P, Lipscomb J, Namtvedt M, Gunsalus IC (1974) Proc Natl Acad Sci USA 71:3906
34. Lipscomb J, Sligar SG, Namtvedt MJ, Gunsalus IC (1976) J Biol Chem 251:1116
35. Sligar SG, Gunsalus IC (1976) Proc Natl Acad Sci USA 73:1078
36. Sligar SG, Gunsalus IC (1979) Biochemistry 18:2290
37. Haniu M, Armes LG, Tanaka M, Yasunobu KT, Shastry BS, Wagner GC, Gunsalus IC (1982) Biochem Biophys Res Commun 105:889-894
38. Haniu M, Tanaka M, Yasunobu KT, Gunsalus IC (1982) J Biol Chem 257:12657-12663
39. Haniu M, Armes LG, Yasunobu Kt, Shastry BS, Gunsalus IC (1982) J Biol Chem 257:12664-12667
40. Fujii-Kuriyama Y, Mizukami Y, Kawajiri K, Sugawa K, Muramatsu M (1982) Proc Natl Acad Sci USA 79:2793-2797
41. Heinemann FS, Ozols J (1982) J Biol Chem 257:14988-14999
42. Black SD, Tarr GE, Coon MJ (1982) J Biol Chem 257:14616-14619
43. Haniu M, Yasunobu KT, Gunsalus IC (1982) Biochem Biophys Res Commun 107:1075-1081
44. Ortiz de Montellano PR, Kunze KL, Beilan HS (1983) J Biol Chem 258:45-47
45. Murray R, Gunsalus IC, Dus KM (1982) J Biol Chem 257:12517-12525

Reactive Intermediates Derived from Cytochrome P-450 Monooxygenases

D. Mansuy[1]

Introduction

Cytochrome P-450-dependent monooxygenases are widely distributed in living organisms like mammals, fish, birds, yeasts and plants. In man, they exist in several organs and tissues with a maximum concentration in the liver. Some of them are in charge of the elimination of foreign compounds (xenobiotics) from the body and have to hydroxylate a broad spectrum of lipophilic compounds making them more hydrophilic and more easily excreted (Ullrich 1979). These cytochromes P-450 play a key role not only in pharmacology since they control the elimination rate of drugs, but also in toxicology since they seem to be the privileged sites of formation of reactive intermediates inside the cell.

RH (drug) ⟶ [Monooxygenases] ⟶ ROH ⟶ conjugation and elimination from the body

Monooxygenases ↓ $[R^+]$ (reactive metabolite) ⟶ toxic effects for the cell

Their unique ability to form reactive intermediates from substrate oxidation is due to their two following main characteristics: 1. they exhibit an active site very accessible to an extremely wide range of organic compounds, 2. they involve in their catalytic cycle several intermediate complexes with different oxidation states of the iron, some of them being very reactive and able to activate a great variety of substrates including inert alkanes.

Many reactive intermediates are presently known to be involved in chemical reactions. When a transition metal complex is present in such reactions, the reactive intermediate can exist either free or bound to the transition metal. Among the carbon-centered reactive intermediates, free radicals, carbanions, carbocations and carbenes are well-known. Intermediate metal complexes involving free radicals or carbanions bound to the metal are α-alkyl complexes containing a simple metal-carbon bond, whereas complexes formed upon binding of a carbene moiety to the metal are carbene complexes with a multiple metal-carbon bond.

$R_3C^{\bullet}$	R_3C^-	R_3C-Met.	σ-alkyl complex
R_2C		R_2C=Met.	carbene complex

1 Laboratoire de Chimie de l'Ecole Normale Supérieure, 24 rue Lhomond, F-75231 Paris Cedex 05, France

34. Colloquium-Mosbach 1983
Biological Oxidations

Nitrenium ions (R_2N^+) and nitrenes (RN) are nitrogen-centered reactive intermediates and several nitrene complexes involving a multiple metal-nitrogen bond are presently known. Finally, the involvement of an oxene, an oxygen atom with six peripheral electrons, has been postulated in oxidation reactions, and transition metal complexes containing a multiple metal-oxygen bond, that are called oxo or oxene complexes, seem to play a key role in metal-catalyzed oxidations (Ullrich 1979).

The purpose of the present paper is to show that most of these reactive intermediates are formed during cytochrome P-450-dependent activation of substrates.

The Reactive Iron Complexes Involved in the Catalytic Cycle of Cytochrome P-450

The catalytic cycle of dioxygen activation and substrates oxidation by cytochrome P-450 has been already discussed in the preceding paper. The first reactive intermediate of this catalytic cycle is cytochrome P-450-Fe(II) which is considerably electron-enriched not only by the porphyrin ring but also by its cysteinate axial ligand. It is important to realize that, by this complex, cytochrome P-450 is first of all an electron-transferring agent. In most cases, its electron is transferred to dioxygen and this is the most relevant reaction of cytochrome P-450-Fe(II) from the point of view of physiology. However, it can also transfer its electron toward several reducible substrates such as halogenated compounds, nitroarenes or arene-oxides (Mansuy 1981). These reductions are a first source of reactive metabolites formation.

The second reactive iron complex of cytochrome P-450 catalytic cycle is the so-called active oxygen complex. Although the precise structure of this complex is not known, several data are in favour of a high-valent oxo, formally Fe(V)=O, structure for this complex (Ullrich 1979). Whatever its structure may be, this active oxygen complex is one of the most powerful oxidizing agents known so far; it is able to insert its oxygen atom into many substrates including alkanes, performing alkanes hydroxylation, alkenes and arenes epoxidation and N- or S-oxidation.

Formation of Reactive Intermediates from Reduction of Substrates

Formation of Free Radicals and σ-Alkyl-Iron Complexes Upon Reduction of Benzylhalides and Halothane

In anaerobic conditions, microsomal cytochromes P-450 are able to catalyze the reduction of benzyl halides such as pNO_2-$C_6H_4CH_2Cl$ and $C_6H_5CH_2Br$ by NADPH, leading to the corresponding toluenes pNO_2-$C_6H_4CH_3$ and $C_6H_5CH_3$. During these reductions, new cytochrome P-450-metabolite complexes, characterized by unusually redshifted Soret peaks around 478 nm, are formed in steady-state concentrations (Mansuy et al. 1983). A study of the reduction of these benzyl halides by a heme model system demonstrates the intermediate formation of a ferric σ-benzyl complex (Mansuy et al. 1982). These results support the following mechanism for microsomal cytochrome P-450-dependent reduction of benzyl halides:

$$\text{P-450Fe(II)} + ArCH_2X \xrightarrow{-X^-} \text{P-450Fe(III)} + {}^{\bullet}CH_2Ar \longrightarrow ArCH_3$$

$$\text{P-450Fe(III)} \xrightarrow{+e^-} \text{P-450Fe(II)} \qquad {}^{\bullet}CH_2Ar + \text{P-450Fe(III)} \xrightarrow{+e^-} [\text{P-450Fe(III)}-CH_2Ar]$$

$$\text{P-450Fe(III)} + ArCH_3 \xleftarrow{H^+} [\text{P-450Fe(III)}-CH_2Ar]$$

During these reductions, intermediate free radicals, $ArCH_2^{\bullet}$, and σ-benzyl ferric complexes exhibiting a Soret peak around 478 nm are formed.

Similar intermediates seem to occur during microsomal reduction of the volatile anaesthetic halothane, $CF_3CHClBr$. In that case, the intermediate σ-alkyl complex P-450Fe(III)-$CHClCF_3$ exhibiting a Soret peak at 470 nm has another possible evolution, the β-elimination of a fluoride ion leading to $CHCl{=}CF_2$ (Ahr et al. 1982):

$$\text{P-450Fe(II)} + CF_3CHClBr \xrightarrow{-Br^-} CF_3\dot{C}HCl + \text{P-450Fe(III)} \xrightarrow{e^-}$$

$$[\text{P-450Fe(III)}-CHClCF_3] \xrightarrow{-F^-} \text{P-450Fe(III)} + CHCl = CF_2$$

Here also, heme model studies of halothane reduction allowed the isolation of a σ-alkyl complex, porphyrin-Fe^{III}-$CHClCF_3$ (Mansuy et al. 1982).

<u>Formation of Iron-Carbene Complexes upon Reduction of Polyhalogenomethanes Such as CCl_4</u>

Actually, this Fe(III)-$CHClCF_3$ complex is the only so far detected iron-σ-alkyl complex of cytochrome P-450 or of an iron-porphyrin containing a halogen substituent in α position relative to iron. The microsomal reduction of polyhalogenomethanes such as CCl_4, CBr_4 or CCl_3F is believed to involve intermediate ${}^{\bullet}CX_3$ radicals, and one could expect, by analogy to the aforementioned results, that σ-alkyl Fe(III)-CX_3 complexes are also involved as intermediates. However, the only species that have been so far detected in these reductions are characterized by Soret peaks between 450 and 460 nm (Wolf et al. 1977) and are not paramagnetic ferric complexes. Model reductions of these polyhalogenomethanes by iron(II)-porphyrins, in the presence of a reducing agent in excess, led to a general preparation method of porphyrin-iron-carbene complexes (Mansuy 1980):

$$\text{(P)Fe(II)} + RR'CX_2 \xrightarrow[-2X^-]{+2e^-} [\text{(P)Fe(II)} \leftarrow CRR'] \leftrightarrow [\text{(P)Fe(IV)}{=}CRR']$$

It thus seems highly probable that the cytochrome P-450 complexes formed upon reduction of CCl_4, CBr_4, $CFCl_3$ are ferrous-carbene complexes.

All these results are in favor of the following general mechanism for the reduction of halogenated compounds by cytochrome P-450 and iron-porphyrins:

$$Fe(II) + RR'CX_2 \xrightarrow{-X^-} Fe(III) + {}^{\bullet}CRR'X \xrightarrow{+e^-} Fe(II) + {}^{\bullet}CRR'X$$

$$\rightarrow [Fe(III)\text{—}CRR'X] \xrightarrow{+e^-} [Fe^-(II)\text{-}\underset{\displaystyle X}{\underset{|}{C}}RR'] \xrightarrow{-X^-} [Fe(II) \leftarrow CRR']$$

The free radicals that are formed inside the active site of cytochrome P-450 may either react with the iron leading to the σ-alkyl- or carbene-iron complexes, or escape from the active site and react with cell macromolecules leading to irreversible modifications of these macromolecules and to cytotoxic effects (Mason 1979).

Formation of Reactive Intermediates from Oxidation of Substrates

When reacting with certain substrates, the cytochrome P-450 active oxygen complex, which will be written as an iron(V)=o species in the following though this structure is not yet really established, is also able to form reactive intermediates derived from the substrate such as free radicals or σ-alkyl (or aryl)-iron complexes.

Formation of Free Radicals and σ-Alkyl-Iron Complexes upon Oxidation of Monosubstituted Hydrazines

Very recent results show that this is the case for monosubstituted hydrazines $RNHNH_2$. These compounds are known to be oxidatively metabolized by hemoglobin or hepatic cytochrome P-450 leading to irreversible modifications of these hemoproteins. The dioxygen-dependent microsomal oxidation of phenylhydrazine has been shown to lead to a cytochrome P-450-iron-metabolite complex characterized by a Soret peak at 480 nm (Jonen et al. 1982). In order to understand this reaction, we have studied the oxidation of several hydrazines $RNHNH_2$ by iron-porphyrins, and found that iron(II)-diazene and σ-alkyl-iron(III) complexes were formed during these oxidations, according to the following scheme (Battioni et al. 1983):

$$(P)Fe(III) + RNHNH_2 \xrightarrow{+O_2} [(P)Fe(II) \leftarrow NH = NR] \xrightarrow{+O_2} [(P)Fe(III)\text{—}R]$$

With R = C_6H_5 or CH_3, the final σFe(III)-R complexes have been isolated and completely characterized, the σ-phenyl complex being much more stable towards dioxygen than the σ-methyl complex.

Similar studies on reactions between microsomal cytochrome P-450 and several monosubstituted hydrazines $RNHNH_2$ (R = CH_3, C_2H_5, $CH_2C_6H_5$ or C_6H_5) or the corresponding diazenes RN = NH showed that iron(II)-diazene complexes (λ_m = 446 nm) are formed as well as other complexes characterized by a Soret peak around 480 nm (Battioni et al. 1983). The great similarity between the routes of formation and properties of these cytochrome P-450 complexes and the aforementioned iron-porphyrin model complexes strongly suggest that the 480 nm-absorbing complexes are cytochrome P-450 Fe(III)-R complexes.

Moreover, since free radicals $R^{\bullet}$ have been detected during microsomal oxidation of hydrazines $RNHNH_2$ (Augusto et al. 1981), the following mechanism is proposed for cytochrome P-450-dependent oxidation of these hydrazines:

$$RNHNH_2 + P\text{-}450Fe(III) \xrightarrow{ox.} [P\text{-}450Fe(II) \leftarrow NH = NR]$$

$$\xrightarrow[-H^+]{+ox.} [P\text{-}450Fe(II) + RN = N^{\bullet}] \xrightarrow{-N_2} [P\text{-}450Fe(II) + R^{\bullet}]$$

$$\downarrow R^{\bullet}$$

$$\longrightarrow [P\text{-}450Fe(III)—R] \quad (\lambda_m = 480 \text{ nm})$$

Formation of Iron-Carbene Complexes upon Oxidation of 1,3-Benzodioxole Derivatives

Some derivatives of 1,3-benzodioxole are oxidatively metabolized by cytochrome P-450-dependent monooxygenases with the formation of very stable iron-metabolite complexes which are characterized in the ferric and ferrous state respectively by Soret peaks around 439 and 455 nm (Franklin 1977). A comparison of studies performed both on these cytochrome P-450 complexes and heme models (Mansuy et al. 1979) led to the conclusion that they are iron-1,3 benzodioxole-2-carbene complexes. If one admits an iron-oxo structure for the active oxygen cytochrome P-450 complex, the formation of the iron-carbene bond upon 1,3-benzodioxole oxidation would correspond formally to the replacement of the oxo ligand by its carbon analogue, the carbene ligand:

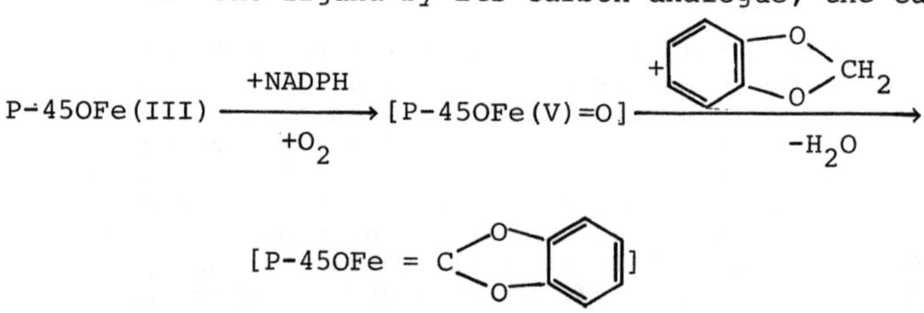

Formation of Iron-Nitrene Complexes upon Oxidation of 1,1-Dialkylhydrazines

Certain 1,1-dialkylhydrazines have been found to form, upon metabolic oxidation by hepatic microsomes, cytochrome P-450-metabolite complexes with Soret peaks at 438 and 455 nm for the ferric and ferrous state respectively (Hines et al. 1980). Experiments performed recently between heme models and 1,1-dialkylhydrazines led to the isolation of the first nitrene complex of a metalloporphyrin either from the dioxygen-dependent oxidation of 2,2,6,6-tetramethyl-piperidyl hydrazine in the presence of an iron(III)-porphyrin, or by direct interaction of the corresponding dialkylaminonitrene with the iron(II)-porphyrin at -70°C (Mansuy et al. 1982):

$$R_2N\text{-}NH_2 \xrightarrow[+O_2]{+(P)Fe(III)}$$

$$R_2\overset{+}{N}{=}\overset{-}{N} \longleftrightarrow R_2N\text{-}N \xrightarrow[(-70^{\circ}C)]{+(P)Fe(II)}$$

$$[(P)Fe(II) \leftarrow N\text{-}NR_2] \longleftrightarrow [(P)Fe(IV) = N\text{-}NR_2]$$

An X-ray analysis of the 2,2,6,6-tetramethyl-piperidyl-N-nitrene complex of tetra (parachlorophenyl)porphyrin-iron(II) definitely established the existence of the iron-nitrene bond, the Fe-N-N moiety being linear (Mahy et al. 1983):

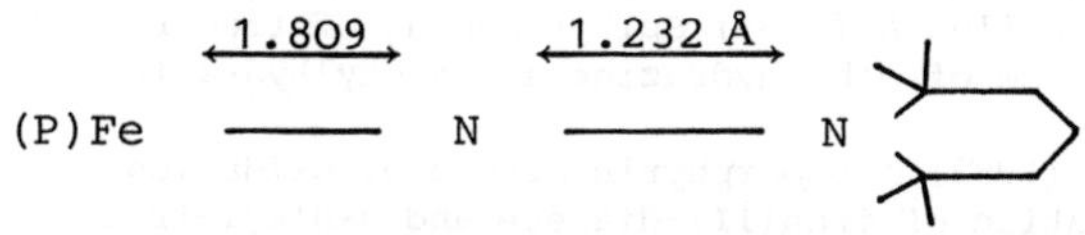

The formation of these nitrene complexes upon dioxygen-dependent oxidation of a 1,1-dialkylhydrazine in the presence of an iron-porphyrin is a strong argument in favor of the proposed nitrenic structure for the cytochrome P-450 complexes formed upon dioxygen-dependent oxidation of 1,1-dialkylhydrazines in the presence of microsomes. If one admits the Fe(V) = O structure for the active oxygen cytochrome P-450 complex, the formation of these nitrene complexes would formally correspond to the replacement of the oxo ligand by its nitrogen analogue, the nitrene ligand.

$$\text{P-450Fe(III)} \xrightarrow[+\ O_2]{+\text{NADPH}} [\text{P-450Fe(V)=O}] \xrightarrow[-H_2O]{+R_2NNH_2} [\text{P-450Fe=N-NR}_2]$$

Conclusion

It is remarkable that all the reactive intermediates or their metal complexes mentioned in the introduction have been detected upon oxidation or reduction of substrates by cytochrome P-450. Thus free radicals and carbanions as well as σ-alkyl (or aryl)-iron complexes and iron-carbene complexes have been detected upon the oxidation of certain substrates or the reduction of other ones. Moreover, the nitrogen analogues of these carbene complexes, the iron-nitrene complexes, seem to occur upon 1,1-dialkyl-hydrazines oxidation by cytochrome P-450, while their oxygen analogue, the iron-oxo complex is believed to be the hydroxylating complex of the catalytic cycle of cytochrome P-450:

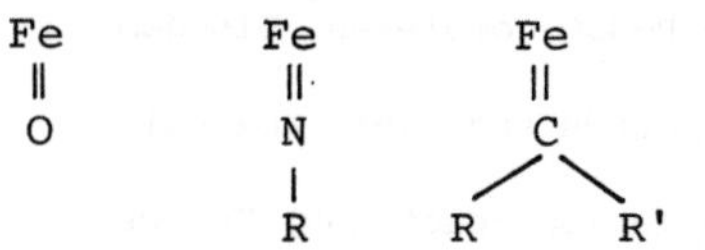

These results underline the unique richness of the chemistry and coordination chemistry of cytochromes P-450 and show the existence of an important organometallic chemistry of this hemoprotein during substrates activation. The variety of reactive intermediates formed during such substrates oxidative or reductive activation allows to explain that cytochromes P-450 are the privileged sites for reactive metabolites formation inside the cell.

References

Ahr HJ, King LJ, Nastainczyk W, Ullrich V (1982) The mechanism of reductive dehalogenation of halothane by liver cytochrome P-450. Biochem Pharmacol 13:383

Augusto O, Ortiz de Montellano PR, Quintanilha A (1981) Spin-trapping of free radicals formed during microsomal metabolism of ethylhydrazine and acetylhydrazine. Biochem Biophys Res Comm 101:1324

Battioni P, Mahy JP, Gillet G, Mansuy D (1983) Ironporphyrin dependent oxidation of methyl- and phenylhydrazine: isolation of iron(II)-diazene and σ-alkyl-iron (III) complexes. Relevance to the reactions of hemoproteins with hydrazines. J Amer Chem Soc 105:1399

Battioni P, Mahy JP, Delaforge M, Mansuy D (1983) Reaction of monosubstituted hydrazines with rat liver cytochrome P-450. Eur J Biochem 134:241

Franklin MR (1977) Inhibition of mixed-function oxidations by substrates forming reduced cytochrome P-450 metabolic intermediate complexes. Pharmacol Ther A2: 227

Hines RN, Prough RA (1980) The characterization of an inhibitory complex formed with cytochrome P-450 and a metabolite of 1,1-disubstituted hydrazines. J Pharm Exp Therap 214:80

Jonen HG, Werringloer J, Prough RA, Estabrook RW (1982) The reaction of phenylhydrazine with microsomal cytochrome P-450. J Biol Chem 257:4404

Mahy JP, Battioni P, Mansuy D, Fischer M, Weiss R, Mispelter J, Morgernstern I, Gans P (in press) Iron-porphyrin-nitrene complexes. J Am Chem Soc

Mansuy D, Battioni JP, Chottard JC, Ullrich V (1979) Preparation of a porphyrin-iron-carbene model for the cytochrome P-450 complexes obtained upon metabolic oxidation of 1,3-benzodioxoles. J Am Chem Soc 101:3971

Mansuy D (1980) New iron-porphyrin complexes with metalcarbon bonds. Biological implications, Pure and Applied Chem 52:681

Mansuy D (1981) Use of model systems in biochemical toxicology: heme models. Rev Biochem Toxicol 3:283

Mansuy D, Fontecave M, Battioni JP (1982) Intermediate formation of a σ-alkyl-iron (III) complex in the reduction of 4-nitro-benzylchloride by iron(II)-porphyrins. JCS Chem Comm 317

Mansuy D, Battioni JP (1982) Isolation of σ-alkyl-iron(III) or carbene-iron(II) complexes from reduction of polyhalogenated compounds by iron(II)-porphyrins: the particular case of halothane, $CF_3CHClBr$. JCS Chem Comm 638

Mansuy D, Battioni P, Mahy JP (1982) Isolation of an iron-nitrene complex from the dioxygen and iron-porphyrin-dependent oxidation of a hydrazine. J Amer Chem Soc 104:4487

Mansuy D, Fontecave M (1983) Reduction of benzyl halides by liver microsomes: formation of 478 nm absorbing σ-alkyl ferric cytochrome P-450 complexes. Biochem Pharmacol 32:1871

Mason RP (1979) Free radical metabolites of foreign compounds and their toxicological significance. Rev Biochem Toxicol 1:151

Ullrich V (1979) Cytochrome P-450 and biological hydroxylation reactions. Topics Curr Chem 83:68

Wolff CR, Mansuy D, Nastainczyk W, Deutschmann G, Ullrich V (1977) The reduction of polyhalogenated methanes by liver microsomal cytochrome P-450. Molec Pharmacol 13:698

Integrated Systems

Regulation of Glycolysis[1]

B. Hess, A. Boiteux, and D. Kuschmitz[2]

Introduction

A study of regulation of glycolysis is directly related to the mechanisms coupling glycolytic reactions to other cellular processes to which glycolysis donates phosphoryl groups, reducing equivalents and carbon fragments for further oxidation, reduction, and biosynthesis. Any activation of the latter processes will lead to a corresponding and well-balanced activation of glycolysis. The problem of regulation is therefore related to the mechanisms of coupling: what are the coupling components and how do they interact, what are their stoichiometric relationships, what are the enzymic mechanisms involved, and to what extent do they have controlling functions? Over the years of Warburg's century, the investigations of metabolic balance and stoichiometries made good use of the various effects which had been observed whenever metabolic states of glycolysis were influenced by oxygen, light, or multiple activators or inhibitors of cellular metabolism. These studies led to our current knowledge of the balance between processes of energy generation and energy utilization.

Following the classical observation of Pasteur on yeast in the states of aerobiosis and anaerobiosis, Harden and Young (1905) discovered the requirement of inorganic phosphate for fermentation [1]. When Warburg and Christian (1939) found the enzymic basis for the phosphate requirement [2], Lynen [3] and Johnson [4] readily suggested that the phosphate cycle is the reason for Pasteur's observation. However, this rather simple picture was soon challenged by multiple observations on regulatory interactions with glycolysis which were reported whenever a more detailed analysis of the phenomenon was carried out [1].

As soon as the glycolytic enzymes were identified, isolated, and ready for kinetic studies, regulating interactions of multiple metabolic

Abbreviations. ALD aldolase. ENO enolase. FBPase I and II fructosebisphosphatase. GAPDH glyceraldehyde phosphate dehydrogenase. G6PDH glucose-6-phosphate dehydrogenase. PFK I and II phosphofructokinase. PGK phosphoglycerate kinase. PGM phosphoglycerate mutase. PK pyruvate kinase. PPC phosphopyruvate carboxylase. TIM triose phosphate isomerase. DAP dihydroxyacetone phosphate. FBP D-fructose-1,6-bisphosphate. F6P D-fructose-6-phosphate. GAP D-glyceraldehyde-3-phosphate. PEP phosphoenolpyruvate. EthOH ethanol.

1 In Memoriam Eraldo Antonini

2 Max-Planck-Institut für Ernährungsphysiologie, Rheinlanddamm 201, 4600 Dortumund, FRG

34. Colloquium-Mosbach 1983
Biological Oxidations

intermediates were discovered with these enzymes and later with the help of the allosteric concept, points of control within the enzymic pathway could be defined [5,6]. Simultaneously, balance studies as well as enzymology were complemented with the analysis of the dynamic behavior of the glycolytic pathway [7], which revealed transient reactions in the form of overshoots, undershoots, and finally oscillations in glycolysis. The latter experiments directly demonstrated the localization of steps of inhibition and activation, respectively, at control points in glycolysis resulting in a whole network of control interactions [8]. A classical experiment is shown in Fig. 1, which demonstrates Helmut Holzer's indication of glycolytic control phenomena in an analysis of the transition following the addition of glucose to yeast cells, which evokes the overshoot phenomenon of fructose-1,6-bisphosphate [9,10].

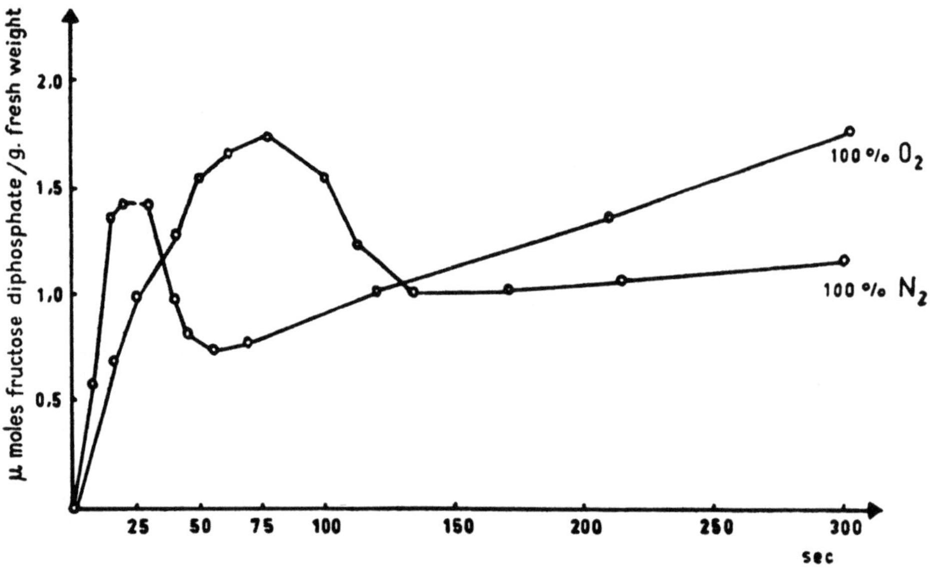

Fig. 1. Kinetic of fructose-1,6-bisphosphate following addition of glucose to starved yeast under anaerobic (in presence of nitrogen) and aerobic (in presence of oxygen) conditions. (From Holzer and Freytag-Hilf 1959)

Today, the phosphokinases in glycolysis are considered to be the most important regulatory steps and ATP, ADP, phosphoenolpyruvate, β-fructose-1,6-bisphosphate, and the recently discovered β-fructose-2,6-bisphosphate are the most prominent control ligands [11,12].

The present discussion will be limited to a presentation of a general regulatory circuit of glycolysis in yeast, a short review of important regulatory interactions, and a discussion of dynamic states of glycolysis resulting from interaction with respiration and the plasma membrane H^+-transport reaction. In addition, the result of recent studies on the efficiency problem will be reported. A discussion of the problems of enzyme interconversion, biosynthesis, and proteolysis is out of the scope of this paper.

Design of the Regulatory Network

The network of Fig. 2 illustrates the major part of the glycolytic pathway in an abbreviated form and identifies the well-known glycolytic intermediates with respect to their stereochemical configuration, indicating the anomeric specificity of the enzymes involved. The arrows indicate the direction of glycolytic or glucogenetic fluxes, respectively. Dashed lines show the allosteric control sites exerted by prominent regulatory ligands, serving as activators (+) and inhibitors (-).

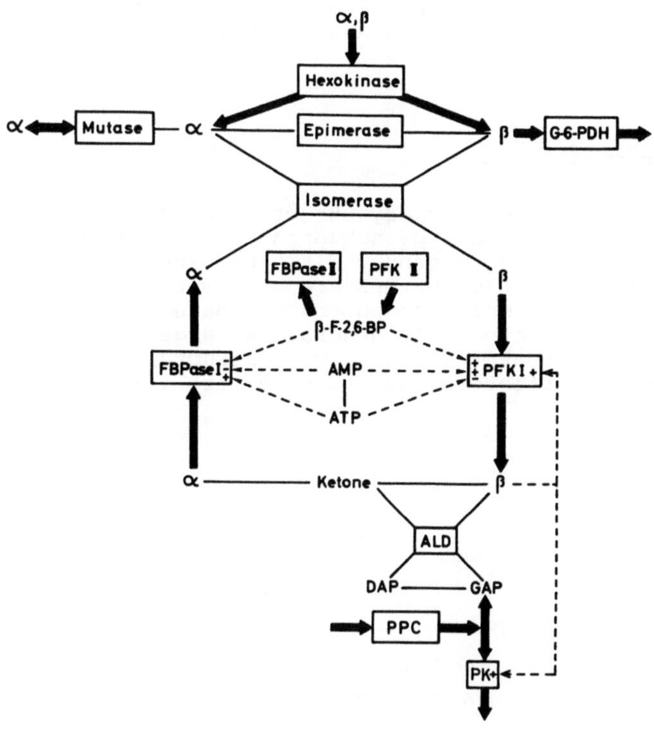

Fig. 2.

The distribution of anomeric specificity between enzymes of glycolytic and glucogenetic functions suggests that the key enzymes of the anabolic glucogenetic pathway are stereospecific for the α-anomers of their carbohydrate substrates, whereas those of the katabolic glycolytic process are specific for the β-anomers. This conclusion, drawn from a kinetic analysis of the stereospecificity of glycolytic enzymes is strongly supported by studies of the distribution of tautomeric and anomeric configurations in intact cells as well as organs with the use of N.M.R. techniques under glucogenetic and glycolytic conditions (see [11]).

In addition to the well-known controlling action of AMP, ATP, and β-fructose-1,6-phosphate serving as activators and inhibitors of the phosphofructokinase I, pyruvate kinase, and fructose-bisphosphatase I, the interaction of β-fructose-2,6-bisphosphate must be considered. In general, as indicated in the figure, this compound is specifically produced from fructose-6-phosphate and ATP by a second phosphofructo-kinase II and degradated by a second fructose-bisphosphatase II to yield as hydrolysis product fructose-6-phosphate and inorganic phosphate. Thus, this controlling ligand of phosphofructokinase I and fructose-bisphosphatase I is generated and degradated through bypass reactions of glycolysis and does not form an intermediate of either the glycolytic nor the glucogenetic pathway. This is of great interest especially because in this respect it resembles the properties of cyclic AMP to which it serves in many respects as direct antagonist [12].

The figure illustrates the reciprocal activation and inhibition mechanisms of three different controlling ligands on the key enzymes controlling glycolysis and glucogenesis, respectively. Analysis of the dynamics of reciprocal control of glycolysis and its reversed process will be given below.

Design of Glycolytic Enzymes in Yeast

Table 1 summarizes the data on the quaternary structure of the glycolytic enzymes with respect to the number of subunits, their Hill coefficients, and prominent controlling ligands. With the exception of phosphoglycerate kinase all glycolytic enzymes are oligomers composed of two to four identical subunits. Only phosphofructokinase with eight subunits has an α-, β-peptide structure. In general, a larger number of enzyme subunits indicates a greater complexity of structural changes upon binding of substrates and effectors, and thus, a more intricate kinetic control [11].

Table 1. Properties of Glycolytic Enzymes in Yeast

Enzymes	n	n_H	Ligands
hexokinase	2	1	ATP, H^+
phosphoglucoisomerase	2	1	6-phosphogluconate
phosphofructokinase	8	2-6	F-2,6-BP,F-6-P,ATP,AMP,ADP,K^+,H^+
aldolase	2	1	
triosephosphate isomerase	2	1	
glyceraldehydephosphate dehydrogenase	4	1-2	NAD
phosphoglycerate kinase	1	1	
phosphoglycerate mutase	4	1	
enolase	2	1	
pyruvate kinase	4	1-3	F-1,6-BP,PEP,ATP,K^+,H^+
pyruvate decarboxylase	2	2	pyruvate
alkohol dehydrogenase	4	1	

n is the number of subunits, n_H is the Hill coefficient, and the last column gives the prominent controlling ligands

Unfortunately, X-ray studies in the case of phosphofructokinase and pyruvate kinase are not resolved to the extent to allow a detailed discussion of allosteric control mechanisms (for other enzymes see contribution by Rossmann of this symposium). In case of phosphofructokinase from Bacillus stearothermophilus, the crystalline structure has been resolved at 2.4 Å to allow a tentative location of active as well as effector sites within each subunit. The analysis reveals that each subunit consists of two domains. The active site lies in a cleft between the two domains of the subunit and it is interesting to note that fructose-6-phosphate can be located between two subunits, which suggests a rearrangement of the subunits upon binding as an explanation of the cooperative function of this substrate. The effector site lies in a cleft between the other two subunits. Thus, here again the binding of ADP involves both subunits. It has been suggested that the allosteric effectors ADP and phosphoenolpyruvate crosslink the subunits yielding the active and inactive conformations of the enzyme respectively [13].

The mechanism of allosteric control of phosphofructokinase from yeast is more complex. The higher state of oligomerization is reflected in the large variation of the Hill coefficient as a function of the controlling ligands within their physiological concentration range. The Hill plot of Fig. 3 illustrates the range of activities as a function of the substrates and controlling ligands obtained in a crude preparation of the enzyme [14]. In addition, the substrate affinity of this enzyme is affected by the recently discovered fructose-2,6-bisphosphate as shown in Fig. 4, which demonstrates in a semilogarithmic plot that increasing amounts of fructose-2,6-bisphosphate increase over more than one order of magnitude the affinity of the enzyme obtained from

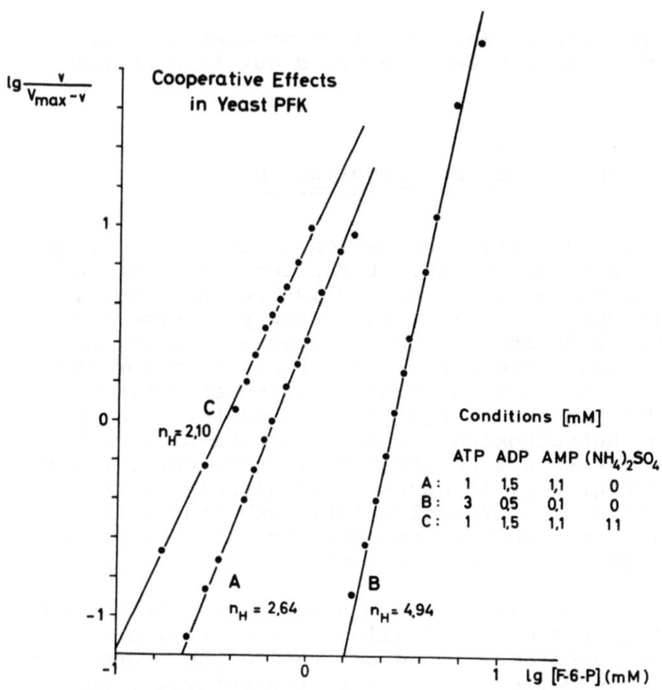

Fig. 3. (From Hess et al. 1975)

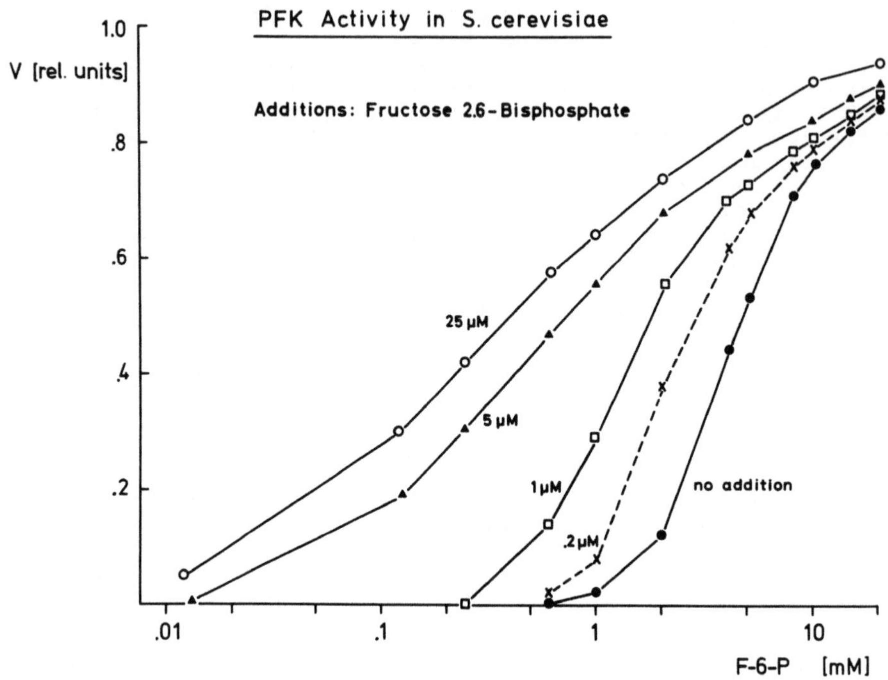

Fig. 4

yeast towards fructose-6-phosphate [15]. Detailed studies of this new controlling ligand are presently being carried out in a number of laboratories.

Coupling of Glycolysis, Glucogenesis, and Respiration

The most complex problem of glycolytic regulation is the switch from glycolysis to glucogenesis and vice versa. It implies a reversal of flux direction all over the pathway. The Michaelis-Menten enzymes, operating near their thermodynamic equilibrium, simply reverse the direction of catalysis. The regulatory enzymes, however, operating far from equilibrium in a quasi-irreversible reaction, need for reversed flux the proper functioning of a counter enzyme operating in the opposite direction. The important regulatory sites for flux reverse are the enzymes phosphofructokinase I and fructose-bisphosphatase I as well as pyruvate kinase and phosphoenolpyruvate carboxylase, as shown schematically in Fig. 5 [11]. Whereas the glycolytic state is maintained by a suitable supply of a glycolytic substrate, the glucogenetic state is maintained only in the presence of ethanol as a substrate and of oxygen for active respiration.

Given these outer conditions, the intrinsic control properties of the regulatory enzymes result in a reciprocal allosteric control of each of the two enzymes couples, suppressing and activating the proper direction of glycolytic and glucogenetic turnover, respectively. The schematic representation of Fig. 6 illustrates the concerted regula-

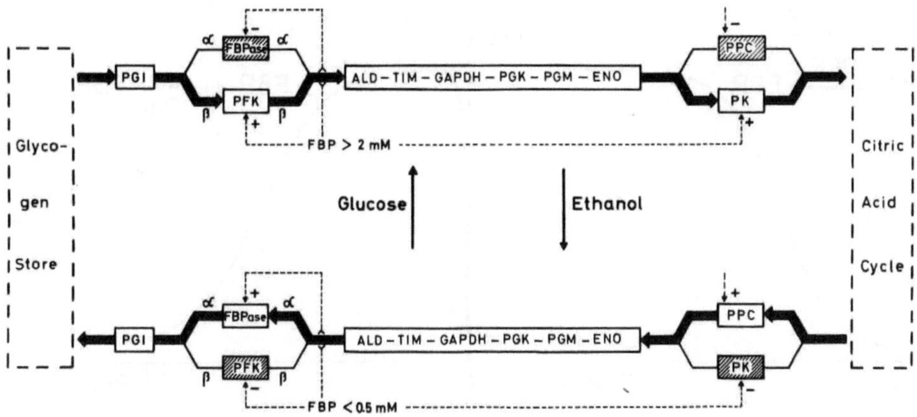

Fig. 5. (From Boiteux and Hess 1981)

RECIPROCAL ALLOSTERIC CONTROL

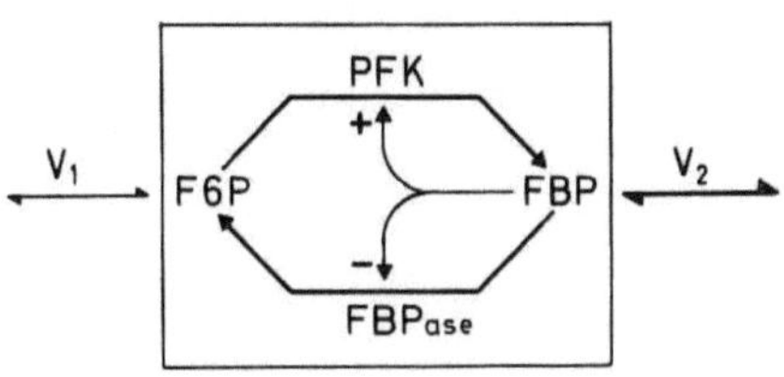

$$\frac{d\,[FBP]}{dt} = \vec{v}_{PFK} - \overleftarrow{v}_{FBPase} = V_{1,2}$$

dynamic states	
I. Glycolysis	$\vec{v}_{PFK} \gg \overleftarrow{v}_{FBPase}$
II. Glucogenesis	$\vec{v}_{PFK} \ll \overleftarrow{v}_{FBPase}$

Fig. 6. (From Hess 1983)

tion of the antagonistic enzyme couples phosphofructokinase I - fructose-bisphosphatase I. The scheme presents an open system with v_1 and v_2, the velocities of source and sink reactions and v_{PFK} and v_{BPFase} the velocities of the enzymic fluxes through phosphofructokinase I and fructose-bisphosphatase I, respectively. Fructose-1,6-bisphosphate exerts cooperative product activation and substrate inhibition on the enzyme couple. This scheme - although not including the additional action of fructose-2,6-bisphosphate - indicates the principle of regulation: the activity balance of the two antagonistic enzymes determines the direction of the general flux [16,17].

A simple analysis of the allosteric properties of both enzymes illustrates the stability domains for glycolysis and glucogenesis, respectively, Fig. 7 shows the kinetics of both enzymes as a function of the concentration of fructose-1,6-bisphosphate. In case of phosphofructokinase I, cooperative product activation is observed; in case

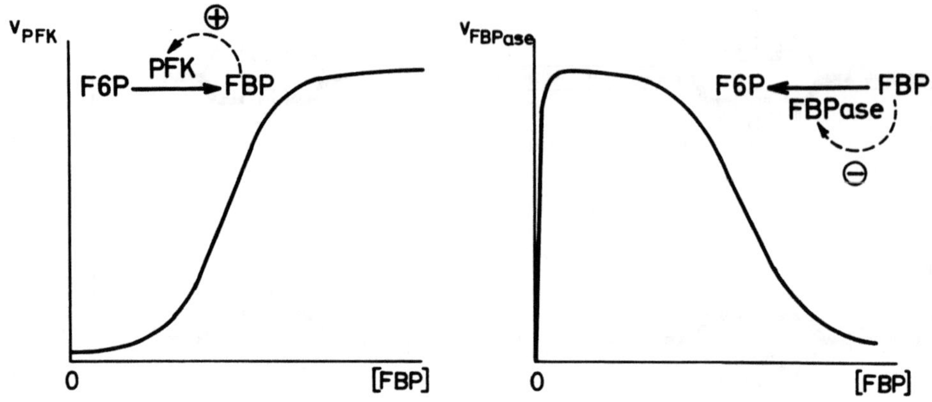

Fig. 7. (From Boiteux et al. 1980)

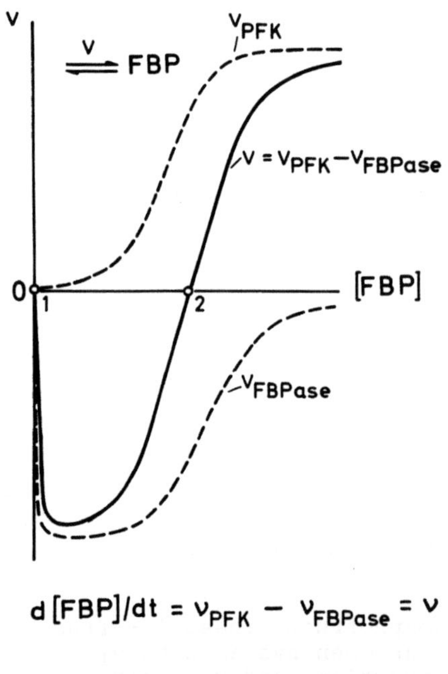

Fig. 8. (From Boiteux et al. 1980)

of fructose-bisphosphatase I, cooperative substrate inhibition. A graphical difference diagram shown in Fig. 8 yields a characteristic curve with positive and negative signs for the fluxes describing glycolysis (positive sign, upper trace) and glucogenesis (negative sign, lower trace). The two intersecting points mark zero net flux and maximum futility. Adding to this characteristic relationship the sink reaction (v_2), which describes the reactions of the aldolase coupled glycolytic reactions (see Fig. 9), we obtain a representation of the stable quasi-stationary states, namely, state S_1 at negative values of the flux (glucogenetic state) and S_3 at positive values (glycolytic state). The points of intersections are obtained by drawing the characteristic curve for the kinetic sink reactions. The quasi-stationary state S_2 is unstable. Any change of the concentration of fructose-1,6-bisphosphate causes an imbalance of the conditions and shifts the flux towards S_1 or towards S_3 [11,16,17].

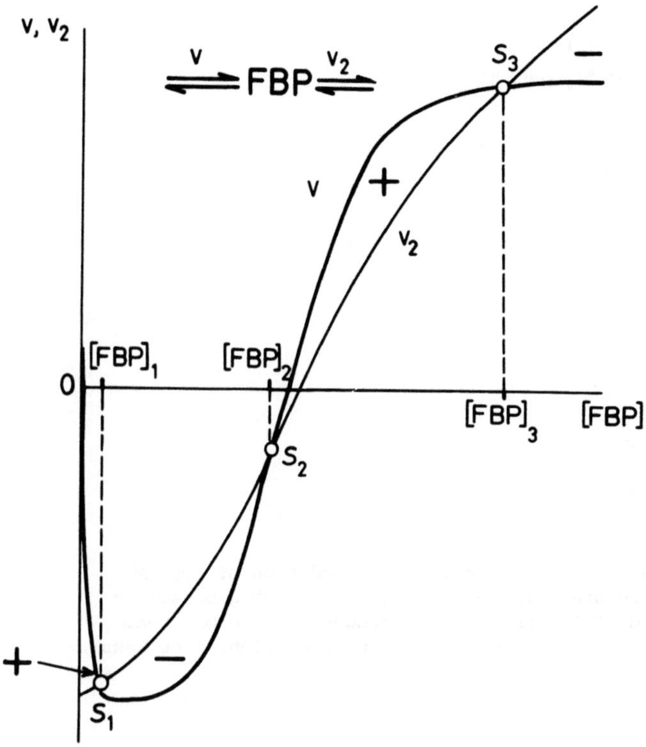

Fig. 9. (From Boiteux et al. 1980)

This type of antagonistic regulation is widely found in nature, often supplemented with additional reciprocal effectors like AMP and fructose-2,6-bisphosphate. The same stability analysis holds for the second regulatory enzyme couple, namely, pyruvate kinase and phosphoenolpyruvate carboxylase.

Experimentally, the transition in yeast from the glucogenetic state towards the glycolytic state can be followed by double-beam spectroscopy recording the redox state of the pyridine nucleotides as indicators of glycolysis. Addition of ethanol to a suspension of starved yeast cells in the presence of oxygen yields a monotonic transition towards the glucogenetic state indicated by a strong reduction of pyridine nucleotides. Upon addition of glucose, the glucogenetic state is switched to the glycolytic state in a complex transition towards a more oxidized state of the pyridine nucleotides [5] (see Fig. 10).

The analysis of the metabolite levels during the glucogenetic as well as glycolytic states are summarized in Table 2. As can be seen, the concentrations of the three adenosine phosphates vary little on switching from one state to the other, whereas there is a great change in the level of fructose-6-bisphosphate and an extreme change in the level of fructose-1,6-bisphosphate. The saturations of phosphofructokinase I and pyruvate kinase given in the last column were calculated from the known parameters of the enzymes. The fluxes through both enzymes are switched on for the glycolytic and switched off for the glucogenetic states. A comparable analysis is given for *E. coli*. It is interesting to see that the change of the level of fructose-1,6-bisphosphate

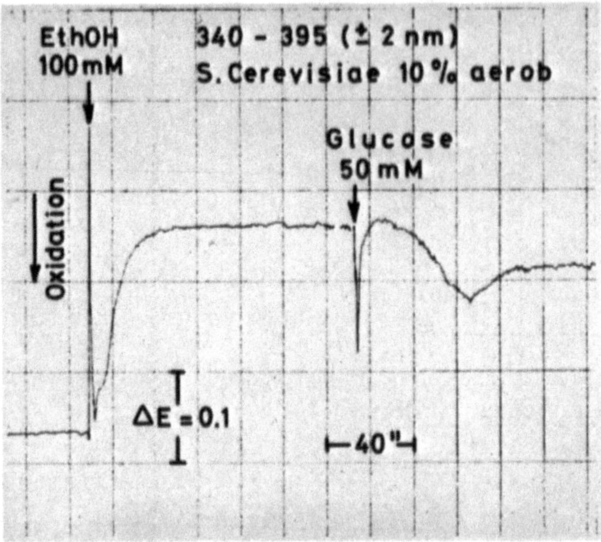

Fig. 10. Transition from endogenous to glucogenic (after addition of 100 m*M* ethanol) and finally to glycolytic state (after addition of 50 m*M* glucose) in yeast. Representation of a record of the absorbancy changes in a double beam spectrophotometer at 340-395 (± 2 nm) in a 10% anaerobic suspension. (see Hess 1973)

Table 2. Analysis of the metabolic levles during the glucogenetic as well as glycolytic states

Metabolic State	Concentrations in mM						v/V_{max}	
	ATP	ADP	AMP	PEP	F6P	FBP	PFK	PK
Yeast								
Glycolytic	3.12	2.00	0.37	0.26	0.88	4.31	.24	.29
Glycogenetic	3.36	2.08	0.37	0.45	0.37	0.04	<.001	<.001
E. coli								
Glycolytic	2.46	1.40	0.36	0.29	0.71	3.64	.15	.08[a]
Glycogenetic	2.24	1.76	0.40	0.96	0.23	0.52	<.001	<.01[a]

[a] calc. for the FBP activated pyruvate Kinase from E. coli

though not as large as observed in yeast, is quite extensive here also. The relatively incomplete switch off of the activity of pyruvate kinase is probably due to an unknown effector not being implied in the calculation [11].

The experiments show as indicated schematically in Fig. 5 that the main effector of the two states is the concentration of fructose-1,6-bisphosphate, which exerts a threshold function: below a certain concentration, which differs somewhat from organism to organism, it inactivates in a combined feedback and feedforward action the katabolic enzymes phosphofructokinase I and pyruvate kinase and releases antagonistically the inhibition from the anabolic fructose-bisphosphatase I. Above the threshold level, fructose-1,6-bisphosphate activates the glycolytic enzymes and inhibits the reverse pathway. It can be concluded that the physiological transition inside a cell from glycolysis to glucogenesis and vice versa is an all- or non-effect, quite different from the subtile and integrate regulation of fluxes in either direction.

These considerations leave the mechanism of coupling of respiration to glycolysis and glucogenesis open. Since Pasteur's discovery of the influence of oxygen on fermentation, the synergetics of glycolysis and respiration are today still not understood. Although the simple problem of energy balance is properly clarified, the regulatory contributions of chemi-osmotic parameters for a coupling of scalar processes with vectorial processes needs further analysis (see Klingenberg, this volume). In this context, it may suffice to state that the decreased rate of uptake of carbohydrates in the presence of oxygen is an open problem. A simplified interaction between glycolysis and a plasma membrane-bound function is given in the next chapter.

Coupling of Glycolysis with the Plasma Membrane Potential

One of the most delicate problems in cell metabolism is the mechanism of coupling of scalar metabolism with vectorial processes. Indeed, because all scalar processes are bounded by membranes, controlled interaction between both functions is decisive for the maintainance of all cellular functions and thus for survival.

For coupling of glycolysis and the cellular plasma membrane potential, the situation is demonstrated in Fig. 11. In case of suppressed respiration in the absence of oxygen and in the presence of glucose, glycolysis is the only ATP-generating system and drives the proton translocating ATPase of the plasma membrane, which builds up a proton gradient and, with or without companion movement of other ions, sets up a plasma membrane potential. It is interesting to note that the membrane potential generated might well affect the glucose uptake process and the glycolytic generation of the ATP potential will feedback on the glucose uptake in a closed control loop, combining scalar and vectorial elements.

In order to study the coupling process, steady state analysis of glycolysis and cellular ionic distribution will lead to a stoichiometric relationship, however, it is not very informative with respect to possible mechanisms of regulation. Therefore, we selected the oscillatory state of glycolysis for experimentation [17]. This dynamic performance of glycolysis represents a steady series of repetitive transitions in which the coupled systems run from maximum to minimum and reverse all

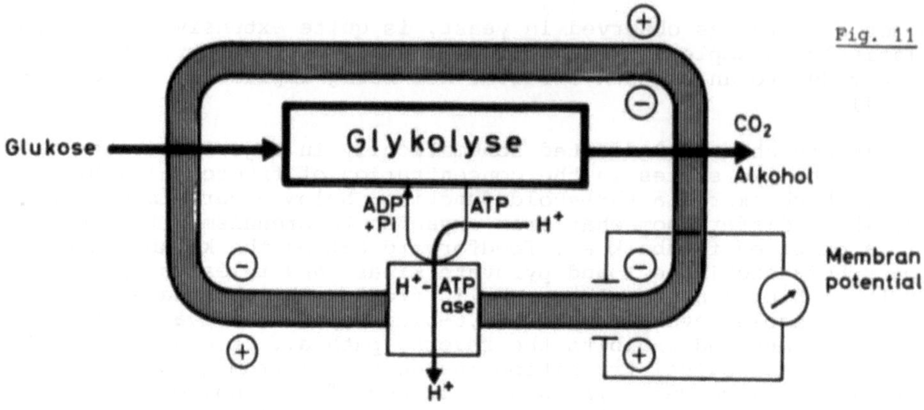

Fig. 11

over again and deliver not only the difference of concentration (given in the amplitudes), but also the time scale of the coupling process (in the frequency) as well as the time relationship between the processes given in the form of the phase angle.

The simultaneous time analysis of the change of the activity of glycolysis and the plasma membrane potential is based on a record of NADH fluorescence as indicator of glycolysis and Rhodamin 6G fluorescence as indicator of plasma membrane potential. From independent experiments, it is known that fluorescence quenching indicates increase in the plasma membrane potential [18]. In order to exclude interference of respiration, the experiments described here are carried out with respiratory deficient mutants of yeast. For recording, the double fluorimetry technique was used.

A typical record is given in Fig. 12, obtained after addition of a single dose of glucose, which activates glycolysis as well as the plasma membrane potential in a single sweep until the glucose is fully utilized [19]. This allows the observation of the transition from the state of the yeast in the absence of glucose to the maximum of glycolytic activity and return to the glucose exhausted state. One clearly sees that the fluorescence traces run through a transient until an oscillatory and finally the resting state is reached. The same holds for the NADH record as well as the Rhodamin 6G record, although it is obvious that glycolysis reaches the oscillatory state somewhat earlier and the original lag times of both parameters are not synchronized at all up to about 2 min. Long time oscillations of glycolysis and the plasma membrane potential (Fig. 13) demonstrate the relationship of the two processes. Here, the membrane potential as indicated by the Rhodamin 6G fluorescence oscillates between 50% and 100% of total and the NADH fluorescence between 80% and 100% with a period of approx. 60 s for both. Comparing the maxima of the two oscillations, it is obvious that both systems are running with the same frequency.

The relationship of NADH and Rhodamin fluorescence over one period is given in the phase plane plot of Fig. 14. Here, we can distinguish three different phases (I, II, III), each defining different dynamic states during a limit cycle. As already evident from Fig. 13, phase II is controlled by a relatively slow rate of reduction of NADH during the initial part of the transition from NADH minimum to NADH maximum. However, this is not reflected in the plasma membrane potential trace, which obviously runs during this phase with its autonomous pace. On the other hand, in phase III (at the end of the transition towards potential minimum) glycolysis seems to be running on its own, whereas

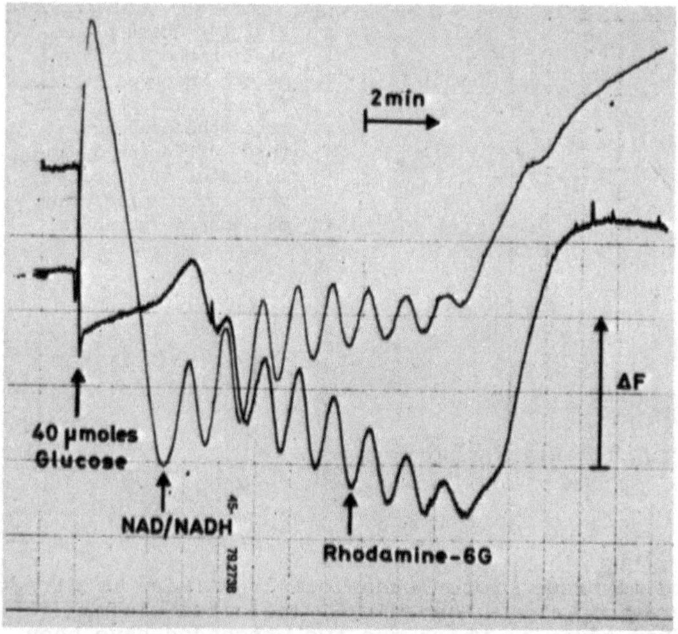

<u>Fig. 12.</u> Transitions following the addition of a single dose of glucose to a suspension of yeast, represented in a record of the fluorescence changes of cellular NADH as well as membrane-bound Rhodamin-6G displayed by a double fluorimeter selecting fluorescence emission modes of both indicators

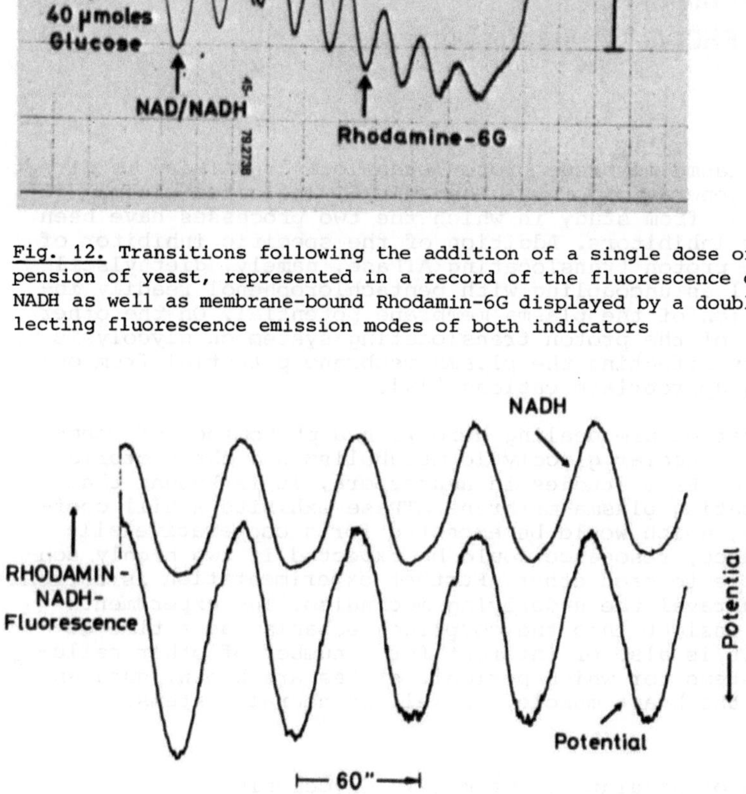

<u>Fig. 13.</u> Steady oscillations of glycolysis as well as plasma membrane potential induced by addition of glucose, recorded as given in Fig. 12

the rates of both processes are perfectly synchronized during phase I. At the present time, a complete designation of the rate controlling steps in both coupled processes is not at hand.

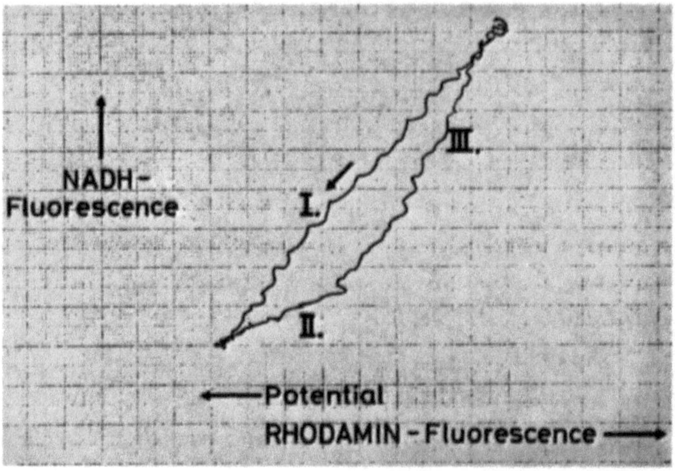

Fig. 14. Phase plane plot of the two parameters recorded in Fig. 13. The plot indicates the presence of three different dynamic states of glycolysis and coupled plasma membrane potential

Evidence that the plasma membrane proton translocating ATPase is directly involved in the generation of the Rhodamin 6G indicated plasma membrane potential comes from study in which the two processes have been uncoupled by proper inhibitors. Addition of the specific inhibitor of the plasma membrane proton translocating ATPase, namely, diethylstilbestrol [20] as well as uncoupling with pentachlorophenol readily inhibits the oscillation of the plasma membrane potential. On the other hand, the influence of the proton translocating system on glycolysis can be documented by affecting the plasma membrane potential from outside the cell using appropriate cations [19].

It could well be that we are dealing here with a phenomenon of chemical resonance of the scalar glycolytic metabolism and the vectorial ion transport system. From studies in neurospora, it is known, that the proton translocating plasma membrane ATPase exhibits a Hill coefficient of two [20], which would be expected for a cooperative allosteric enzyme. Indeed, resonance would be expected if two highly nonlinear systems couple to each other. Further experimentation is currently carried out to unravel the underlying mechanism. The experiments illustrate a novel insight into the coupling mechanism as a time-dependent function. It is also of interest for a number of other cellular glycolyzing systems for which periodic states are known, such as the smooth muscle, the heart muscle, as well as neural systems.

Extremal Properties of Regulatory Enzymes in Glycolysis

Some time ago, it has been observed experimentally that the reactions in glycolysis catalyzed by regulatory enzymes run far off their thermodynamic equilibrium, whereas all other enzymes run rather near equilibrium [21,22]. Indeed, the near equilibrium enzymes of glycolysis are displaced from equilibrium by an amount, which is directly proportional to the resulting flux.

The question of the larger deviation of the reactions catalyzed by regulatory enzymes is open. In a recent theoretical investigation [23], the dependencies of the size of the deviation of the pyruvate kinase reaction from equilibrium with respect to a large variety of stationary metabolite levels at constant fluxes were analyzed. Since the stationary flux for pyruvate kinase is determined by the concentrations of phosphoenolpyruvate, ADP, ATP, fructose-1,6-bisphosphate, and magnesium ions on the basis of the well-known rate law of this enzyme, a large range of concentrations of these ligands might yield identical fluxes. All these concentration patterns define for a given flux a plane in a three-dimensional concentration space of the metabolites keeping the concentration of fructose-1,6-bisphosphate and free magnesium constant. The form of the plane is determined by the rate law.

Thus, the steady state mass action ratio of the pyruvate kinase is given by

$$K_{ss} = \frac{[ATP] \cdot [PYR]}{[ADP] \cdot [PEP]} \quad .$$

This ratio can acquire on a plane of steady state flux a rather large range of values. The mathematical analysis shows a maximum in K_{ss} (Fig. 15) for concentrations which are relevant for physiological conditions. Indeed, the concentrations of the metabolites ATP, ADP, and phosphoenolpyruvate given by $(K_{ss})_{max}$ are those which are found in *E. coli* and in yeast cell extracts under glycolyzing conditions. This interesting observation could well point to an optimal principle in glycolysis. Since K_{ss} is a measure for the deviation of the reaction from thermodynamic equilibrium, it can be used to compute the

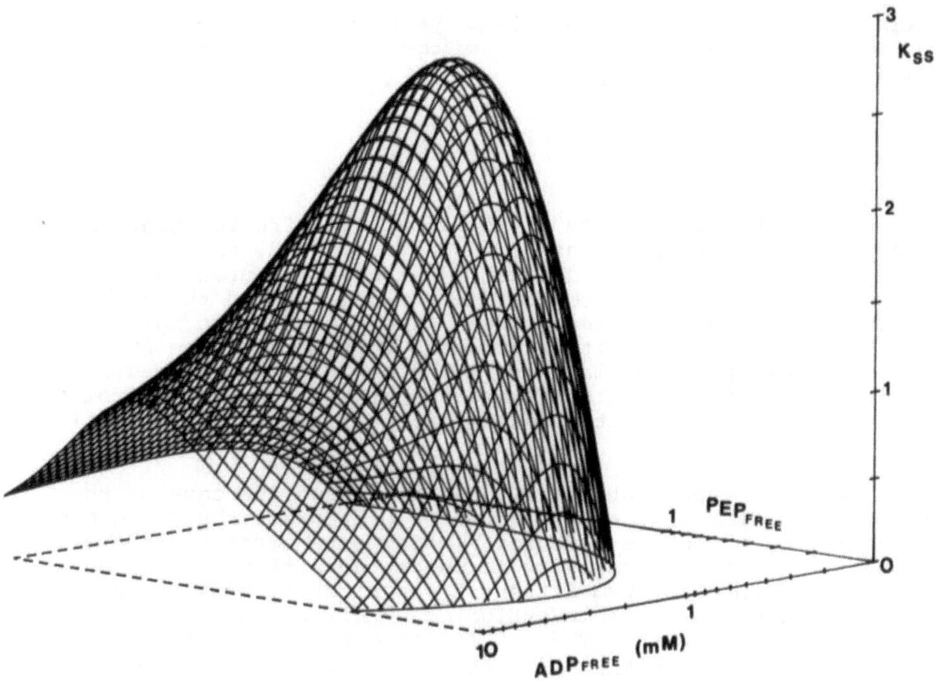

Fig. 15. Relationship between K_{ss} and the concentrations of free phosphoenolpyruvate and free ADP at a steady state flux of 0.4 V_{max} and corresponding concentrations of ATP

dissipation function Φ_{ss}, which defines the dissipation of free energy per unit time at a given glycolytic flux:

$$\Phi_{ss} = V_{ss} \cdot V_{max} \cdot RT(\ln K - \ln K_{ss}) \geq 0 \quad .$$

This equation illustrates that a maximum in K_{ss} corresponds to a minimum in the dissipation function. This minimum is not related to the principle of minimum entropy production, since it is a minimum within a continuum of stationary states and is not a minimum with respect to the deviation from a stationary state.

Upon coupling the reaction of phosphofructokinase and pyruvate kinase, the following relationship for the glycolytic dissipation function is obtained:

$$\Phi_{ss} \approx v_{ss} \left(\frac{1}{2} A_{PFK} + A_{PK}\right) \quad .$$

A_{PFK} is the affinity of the PFK reaction, A_{PK} the affinity of the pyruvate kinase reaction. The stationary flux is the one of pyruvate kinase. Since both affinities are positive, it can well be assumed that both enzymes are optimized with respect to their contribution to the dissipation of free energy. It is suggested that the maximization of the mass action ratios (K_{ss}) under stationary conditions is significant with respect to the maintainance of an efficient cell metabolism. Since the metabolite concentrations under the conditions of the maximum of K_{ss} are determined by the corresponding binding constants and turnover numbers of the enzymes, it is assumed that these conditions are a result of the evolution of glycolytic enzymes in phylogenesis.

Acknowledgements. With pleasure we acknowledge the generous support of our work by the Fritz-Thyssen-Stiftung, the Fonds der Chemischen Industrie, as well as the IBM Deutschland GmbH.

Discussion

Bühner: You pointed out a case of an allosteric effector binding in the interface region between enzyme subunits. Indeed, there are more examples to that: our (with R. Hensel, Munich) structural investigation of the allosteric LDH from *Lactobacillus casei* showed the activator FDP to bind in the contact area between subunits and also rather close to the other two subunits of the tetramer. Furthermore, the AMP binding site of muscle phosphorylase is in the subunit contact region. Thus, allosteric effectors linking and coordinating subunits from a position within their contact areas appears to be a rather general way of action.

Hess: In my short representation I only discussed the case of the allosteric phosphofructokinase I of yeast. Of course, there are many other examples in which the effector binding site stretches between two subunits, just in the same way in which the active site of an enzyme might be composed of two different subunits; a classical case is the phosphoglucose isomerase investigated by Hilary Muirhead quite some time ago (for ref. see: The Enzymes of Glycolysis: Structure, Activity and Evolution, A Royal Society Discussion, D Philipps, CCF Blace and HC Watson, Philosophical Transaction of the Royal Society of London, series B, Vol 293 (No. 1063), pp 1-218, 1981).

Sies: Evidence is accumulating that several enzymes of glycolysis may be modified by mixed disulfide formation. Is it known whether there are changes in the titrable -SH groups during oscillation?

Hess: As far as I know no titration data on the state of -SH groups during glycolytic oscillations has been published.

Brand: You have shown a coupling between glycolysis and the membrane potential via ATPase in oscillating yeast cells. Changes in the membrane potential should affect ion transport across the plasma membrane. My question is: did you observe in your experiments changes in ion fluxes in an oscillatory manner, e.g., for potassium or sodium ions or did you measure changes in intracellular ion concentrations especially with respect to calcium, which is known to be an important modular of a variety of metabolic processes in mammalian cells.

Hess: For the sake of simplicity, our experiments were designed to control fluxes of other ions and to demonstrate that only the proton translocating ATPase is coupled to glycolysis. Intracellular ion concentrations will be analyzed at a later stage of our experimentation.

Schäfer: You suggested that the proton is the coupling species for the chemical resonance of glycolysis and cell membrane potential, with other words: the intracellular pH. I wonder what the experimental evidence for this mechanism is and whether the intracellular buffering capacity would allow propagation of ΔpH-waves.

Hess: Since protons are the substrate for the proton translocating plasma membrane ATPase, they certainly must be one of the coupling components, the other one being ATP. As I mentioned, we do not know at the present time, whether there is an intracellular oscillation of pH. It might well be that the intracellular oscillation of ATP is a sufficient condition for coupling as long as the intracellular pH-level is maintained between an upper and lower bound. I should repeat that the coupling between the two processes in terms of chemical resonance is only a suggestion for which experimental evidence still has to be presented.

Hamprecht: If I compare the dimensions of your Petri dish and of a yeast cell, I wonder whether a spatial distribution of the oscillation of glycolysis is possible in something as small as a yeast cell.

Hess: The conditions, which determine the dimensions of dynamic chemical patterns, are not too well-known and can certainly not be extrapolated to the conditions prevailing in a yeast cell. Of course, there are examples of spatial pattern formation and propagation in cell biology. Perhaps it would be reasonable to ask for the conditions which are necessary and obligatory to reduce and prevent spatial pattern formation, respectively. Since spatial pattern formation depends on the hydrodynamics, the chemical reaction, as well as the diffusion process, suitable conditions can well be as such to generate dynamic spatial conditions in one single yeast cell. Evidence, however, has not been presented as yet for this type of cell.

Holzer: What is known about effect of AMP on PFG II and fructose-2,6-bisphosphatase?

Hess: So far we have not yet analyzed this effect (see ref. 12).

Schirmer: The African trypanosomes, e.g., *T. Brucei*, contain a specialized organelle, the glycosome, inside which all glycolytic enzymes are assembled. Is this system of structured glycolysis suitable for basic research and/or as a target of drug design?

Hess: I thank you very much for this interesting reference, which I don't know. Sorry, I cannot answer your question.

References

1. Racker E (1975) In: Energy transducing mechanisms 3, MTP international review of science, University Park Press, Baltimore pp 163-183
2. Warburg O, Christian W (1939) Biochem Z 303:40-68
3. Lynen F (1941) Justus Liebigs Ann Chem 546:120-141
4. Johnson MJ (1941) Science 94:200-202
5. Hess B (1973) Organization of glycolysis: Oscillatory and stationary control. In: Rate control of biological processes, Cambridge pp 105-131
6. Sols A, Gancedo C, DelaFuente G (1971) Energy-Yielding Metabolism in Yeasts. In: Rose AH, Harrison JS (eds) The Yeast, Academic, New York
7. Hess B, Change B (1959) Naturwissenschaften 46:238-257
8. Hess B, Boiteux A (1971) Annu Rev Biochem 40:237-258
9. Holzer H, Freytag-Hilf R (1959) Hoppe-Seyler's Z Physiol Chem 316:7-30
10. Hess B, Chance B (1961) J Biol Chem 236:239-246
11. Boiteux B, Hess B (1981) Phil Trans R Soc Lond 293:5-22
12. Hers H-G, van Schaftingen E (1982) Biochem J 206:1-12
13. Evans PR, Farrants GW, Hudson PJ (1981) Phil Trans R Soc Lond 293:53-62
14. Hess B, Boiteux A, Busse HG, Gerisch G (1975) Spatiotemporal Organization in Chemical and Cellular Systems. In: Nicolis G, Lefever R (eds) Advances in Chemical Physics. John Wiley 29:137-168
15. Boiteux A, Hess B, unpublished experiments
16. Boiteux A, Hess B, Sel'kov EE (1980) Curr Top Cell Regul 17:171-203
17. Hess B (1983) Hoppe-Seyler's Z Physiol Chem 364:1-20
18. Aiuchi T, Daimatsu T, Nakaya K, Nakamura Y (1982) Biochim Biophys Act 685: 289-296
19. Kuschmitz D, Hess B, unpublished experiments
20. Goffeau A, Slayman CW (1981) Biochim Biophys Act 639:197-223
21. Hess B (1963) In: Karlson P (ed) Funktionelle und morphologische Organisation der Zelle. Springer, Berlin Göttingen Heidelberg
22. Bücher Th, Rüssmann W (1964) Angew Chem internat Edit 3:426-439
23. Plesser Th, Markus M (1982) Hoppe-Seyler's Z Physiol Chem 363:546

Regulation of ATP Synthesis in Mitochondria

M. Klingenberg[1]

Introduction

Regulation of ATP synthesis in mitochondria has been an early example for recognizing regulatory problems in a sequence of metabolic reactions. Still 20 years later it attracts intense attention and is presently again in the focus of new formalism on systems control. Early, respiratory control was mainly regarded as dependent on the concentration of ADP [1,2]. However, a few years later in context with the reversibility of electron transport, we observed that the control is also dependent on the ATP concentration, such that more the ratio of ATP/ADP seemed to be the controlling entity [3,4]. At that state, it was not yet known that oxidative phosphorylation in mitochondria is masked by a transport system which exports the ATP synthesized inside the mitochondria against the uptake of ADP. The discovery of this exchange system [5] changed drastically the scene for analyzing the control of ATP synthesis in mitochondria. In fact, control of ATP production by mitochondria is fairly well-understood on the level of transport, but not yet on the level of synthesis per se.

At first, intramitochondrial ADP serves as substrate for the ATP synthase, which then releases the ATP back into the inner mitochondrial pool. At this state, the system resembles that of aerobic bacteria. For providing the ATP to the cytosol, the membrane is equipped with a highly specific counter-exchange system in which one molecule ATP is exchanged for the uptake of one molecule ADP. Phosphate is taken up by a separate phosphate transporter. These transport systems are not linked to the ATPase, although this has been suggested repeatedly [6,7]. The ADP/ATP translocator acts first of all between the intra- and extramitochondrial pools of nucleotides and thus represents the most characteristic entity which differentiates mitochondria from bacteria. In other words, mitochondria have incorporated the ADP/ATP translocator in addition to the other components of the oxidative phosphorylation already present in prokaryotes. The exchange system superimposes on the mitochondrial oxidative phosphorylation a number of characteristics; these are the high specificity for ADP and ATP, strong temperature dependence, and last but not least, the control characteristics which will be the major concern in the following.

Respiratory Control by ADP or ATP/ADP

Still today a publication entitled *"Evidence against regulation of respiration by extramitochondrial ATP/ADP ratios"* [8] illustrates the unsettlement

1 Institut für Physikalische Biochemie, Universität München, Goethestraße 33, 8000 München 2, FRG

34. Colloquium-Mosbach 1983
Biological Oxidations

in this field, although it is now 22 years since evidence for the dependence of respiration, not only on extramitochondrial ADP, but also on ATP was presented [3]. Nevertheless, in the long intervening time, progress has been achieved in the understanding of the interaction of nucleotides with the oxidative phosphorylation system. However, it seems that the achievements in the time up to 1972 are occasionally not fully recognized. Therefore, a brief excursion into these developments will be given.

In context with the studies on the reversibility of electron transport, Klingenberg and Schollmeyer observed in 1960 that ATP added to mitochondria on top of ADP partially inhibits respiration [3,4]. Concomitantly, the redox state of respiratory carriers is changed, as shown in Fig. 1 from this early paper. A crossover is observed between cytochrome b and c, reflecting the transition to a more controlled state of electron transport under the influence of ATP. On activation with dinitrophenol or ADP and phosphate, the reversed crossover at the same site was obtained.

Together with Bode, a quantitative relation to the ATP/ADP ratio and the control of electron transport was obtained by following the cytochrome c redox level in respiring mitochondria (Fig. 2; 9). For enhancing the response of cytochrome c, the respiration was partially inhibited by small amounts of azide. The main result was a linear relation between ATP and the redox ratio of cytochrome c, the slope of which shifts with the ADP and phosphate concentrations. In fact, the evaluation given in Table 1 shows that the redox ratio correlates to the ATP/ADP-phosphate ratio, resulting in a constant of about $K_1 = 12$ mM^{-1}. This value should roughly approximate the free energy term in ΔG_{ATP} =

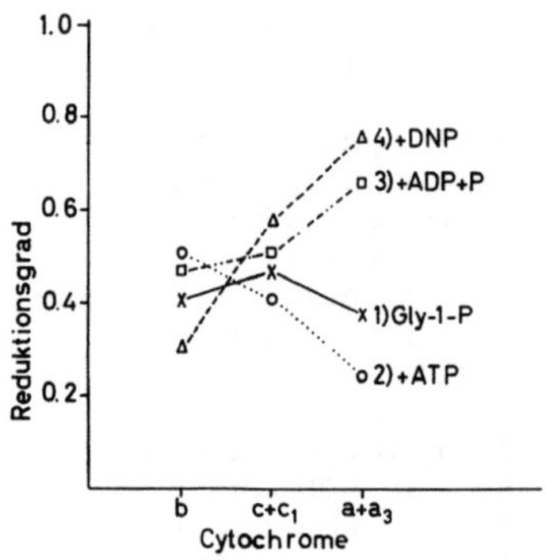

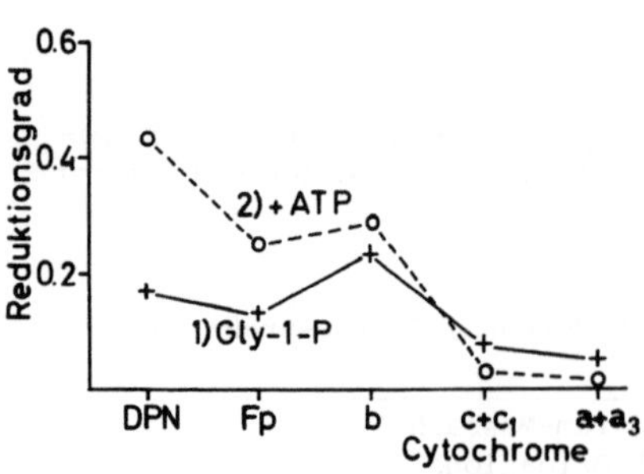

Fig. 1. "Redox patterns" of the respiratory chain in various states of electron transport activity. Rat skeletal muscle mitochondria incubated with various additions as indicated *A* 4m*M* glycerolphosphate, 1 m*M* ATP; *B* partially inhibited respiration by 60 μm azide in order to enhance the redox state response of the cytochromes. Further additions: 4 m*M* glycerol-phosphate, 0.1 m*M* dinitrophenol, 0.2 m*M* ADP, 2 m*M* Pi, and 1 m*M* ATP. (From Klingenberg 1969)

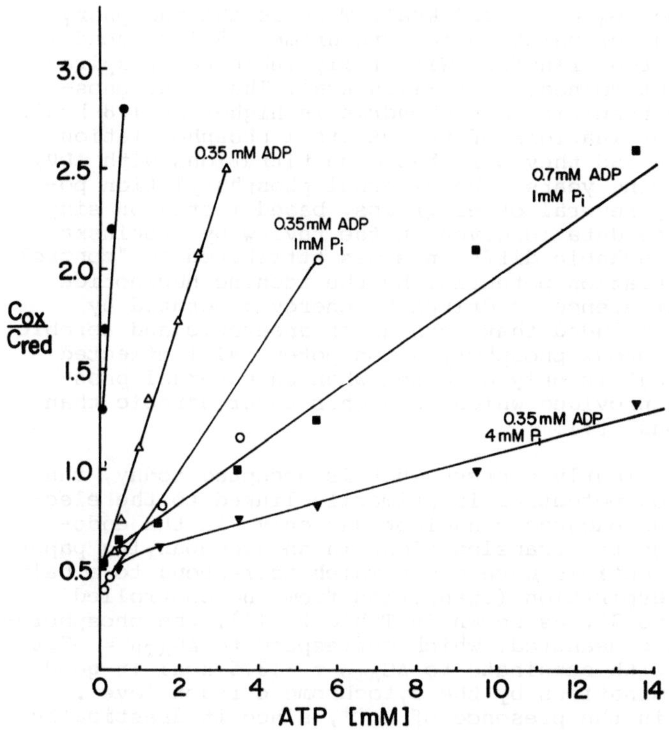

Fig. 2. The relation between the redox ratio of cytochrome c and the ratio [ATP]/[ADP] · [Pi] was followed on the basis of a titration with ATP in the presence of various concentrations of ADP and Pi. Evaluation of the C_{ox}/C_{red} ratio from spectrophotometric recordings of a suspension of rat liver mitochondria in O_2-saturated sucrose-EDTA at 20°C with the addition of 0.3 m*M* azide and 2 m*M* β-hydroxybutyrate. (From Klingenberg 1969)

Table 1. The ratio $\frac{ATP}{ADP \cdot P}$ in equilibrium with C_{red}/C_{ox} under aerobic and anaerobic conditions (From Klingenberg 1969)

Respiration	P_i	ADP	Slope (S)	$\frac{S^{-1}}{ADP \cdot P_i} = K_1$
	(m*M*)	(m*M*)	(mM^{-1})	(mM^{-1})
Partially inhibited	1	0.35	0.23	13
(0.3 m*M* azide)	1	0.37	0.12	11.5
	4	0.35	0.06	12
Fully inhibited	1	0.1	0.055	0.45
(1 m*M* KCN)	1	0.2	0.12	0.6

$-7.8 - 1.39 \log K_1$, giving $\Delta G_{ATP} = -13.4$ kcal. This is the phosphorylation potential required for maintaining cytochrome c 50% reduced in partially inhibited electron transfer. With fully inhibited respiration, a lower value is determined: $\Delta G = -11.6$ kcal. Thus, the phosphorylation potential in respiring mitochondria is higher by 1.8 kcal. These were the first determinations of the external phosphorylation potential in mitochondria and they were based on titrations with ADP, ATP and P_i. In the following years, the external phosphorylation potential was determined by several other groups, based mostly on single states. A summary of these data is given in the review by Erecinska and Wilson [10]. This remarkable difference was attributed to "control of the external phosphorylation potential by the adenine nucleotide translocator.... This difference is driven by energy generated by respiration. It may be concluded that both under anaerobic and aerobic conditions the same endogenous phosphorylation potential (reflected in the same ratio C_{ox}/C_{red}) is only obtained when an external phosphorylation potential is provided which is higher under aerobic than under anaerobic conditions" [9].

These excerpts from 1969 clearly express what is accepted today, the endogenous phosphorylation potential is primarily linked to the electron transfer, whereas the exogenous nucleotides only via the endogenous nucleotides through the translocation. In an accompanying paper [11], phosphorylation potentials were given which correspond to a half maximum stimulation of respiration (transition from the controlled style active state or 4 to 3). As shown in Table 2 [11], the phosphorylation ratios ATP/ADP are measured, which correspond to $\Delta G_{ATP} = -7.8 - -5.9 = -13.7$ kcal and with succinate to $\Delta G_{ATP} = -13.5$ kcal in good agreement with the determinations by the cytochrome c redox level. The ATP effect is lower in the presence of Mg^{++}, since it drastically decreases the concentration of free ATP. For these reasons, other laboratories at first failed to see an ATP effect and insisted that respiration is dependent only on ADP [12]. It was noted by us that in the transition period, when ADP is largely consumed, the sharp break in the respiration is only observed in the presence of Mg^{++}, which largely traps the synthesized ATP, whereas in the absence of Mg^{++}, a more smooth transition is seen, corresponding to the increasing inhibition by ATP.

Table 2. Evaluation of the phosphorylation potential for half-maximal stimulation of respiration (From Klingenberg 1969)

Substrate, β-hydroxybutyrate

ATP (m*M*)	P_i (m*M*)	K_m for ADP (m*M*): Pyridine nucleotide oxidation	K_m for ADP (m*M*): Respiration	$\left(\frac{[ATP]}{[ADP][P_i]}\right)$ 1/2 (mM^{-1})	
5	0.5	0.6	0.6	17	17
2.5	1	0.15	0.2	17	13
5	1	0.25	0.3	20	16

The Role of the Endogenous Nucleotide Pool

Since 8 years, interest in the interaction of the ATP/ADP pool with the electron transfer has greatly increased. There were widely different conclusions on whether the external or internal ATP/ADP ratio was linked to respiratory activity (see discussion in ref. 13). Despite the earlier evidence, some of these suggestions seemed to be caused by the conclusions from two laboratories that the endogenous nucleotide pool is bypassed and that the exogenous nucleotides interact with the ATPase directly, either via the translocator or via different pathways [6,7]. Abundant evidence, however, is available to rebuke these postulates based on earlier and subsequent observations of the endogenous pool.

An early quantitative description of the phosphorylation kinetics of the endogenous and exogenous nucleotides by Heldt and Pfaff [14] was already a proof for the interaction of the endogenous pool with the ATPase and as an intermediary for the phosphate transfer to the exogenous nucleotides. From the phosphorylation and desphosphorylation kinetics of the endogenous ATP, first-order rate constants were determined. A rate equation was derived, incorporating these rate constants and those for the translocation, as well as the internal ADP/ATP ratios. It described the kinetics of the endogenous pool and the exogenous added ADP. Very good agreement of the computed curves with those actually measured for the appearance of endogenous and afterwards exogenous ATP was obtained (Fig. 3), demonstrating the full competence of the endogenous nucleotide pool as primary acceptor of phosphate for the subsequent translocation forming the exogenous ATP.

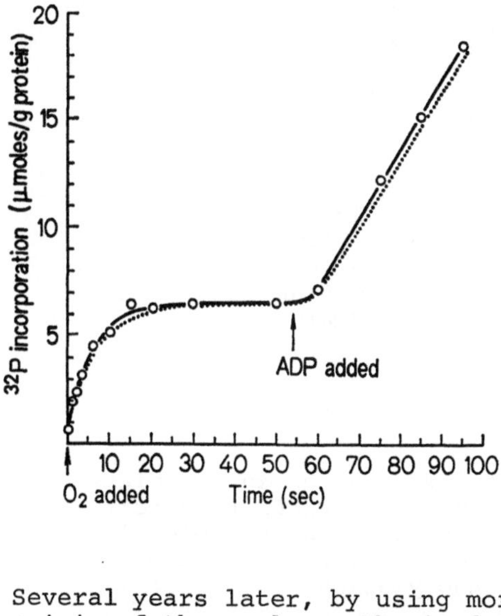

Fig. 3. Phosphorylation of endogenous and exogenous ADP in rat liver mitochondria at 10°C. Comparison between the measured and computed data. The *dotted line* represents the computed time course of phosphorylation. (From Heldt and Pfaff 1969)

Several years later, by using more advanced sampling techniques, the origin of the newly synthesized ATP for the endogenous pool was directly demonstrated in double-labelling experiments [15]. The appearance of endogenous ATP prior to that of exogenous ATP is clearly seen and the exogenous ATP is a result of an export from the endogenous pool. This direct demonstration of the movements in phosphorylation of both pools should have removed the last doubts about the endogenous ADP and ATP in phosphorylation.

A clear rate limitation by the ADP/ATP exchange was originally observed by following the phosphorylation rate of the endogenous and exogenous pools of nucleotides in dependence on temperature. At low temperature, the phosphorylation rate of external ATP was strongly inhibited, whereas that of endogenous nucleotides remained fairly active [16]. This could then be traced to the very high temperature dependence of the nucleotide translocator below 14°C, corresponding to a Q-10 = 4.5 [17,18].

In another unconventional approach, the problem of rate limitation was addressed by following the kinetic response of the respiratory chain to a sudden complete blockage of the translocator (19; Fig. 4). In this case, a lag period was observed, obviously until the endogenous pool reaches an ATP/ADP ratio, which would control electron transport. Here the rate control by the translocator was weak since respiration was partially inhibited by azide. There was no lag period when respiration was fully active, indicating a stronger rate control by the translocation.

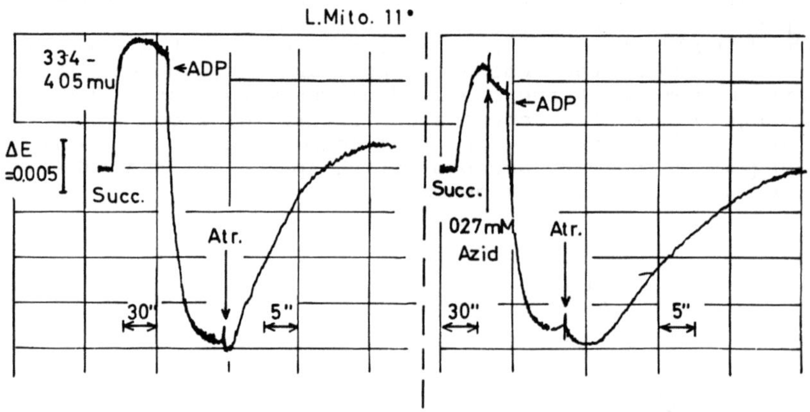

Fig. 4. The response of respiratory chain to blockage of the ADP/ATP carrier. Kinetics of the NADH + NADPH response to addition of atractylate (Atr). Rapid mixing by the moving mixing chamber. Rat liver mitochondria incubated at 11°C. In the second experiment, partial inhibition of respiration by azide in order to enlarge the delay after the Atr addition. (From Klingenberg 1967)

The Two Parameters Controlling ATP Export

Once the nucleotide pool was firmly established as an intermediary of oxidative phosphorylation, the rate data of the ATP/ADP transport, which are determined by an exchange with the endogenous nucleotide pool, retain also their significance for evaluating the rate control by these steps in overall oxidative phosphorylation.

The degree of participation in external ATP synthesis of the exchange is essentially controlled by two major parameters; the ATP/ADP ratios inside and outside and the membrane potential [20,21]. The nucleotide exchange is basically neutral with respect to the selection of ADP and ATP from either side of the membrane. Thus, the carrier will select ADP and ATP in about the ratio of the occurrence inside or outside, unless it is further modulated by other parameters (Fig. 5). The total

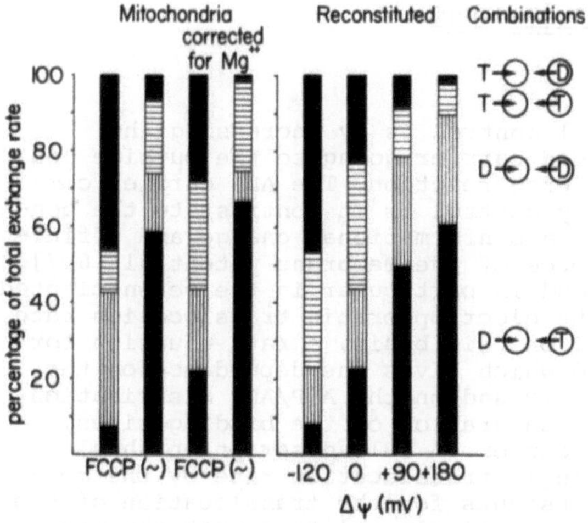

Fig. 5. Evaluation of the four combinations of exchange in competitive experiments. For mitochondria, the four combinations are also given after correcting for the binding of endogenous nucleotides to excess Mg^{++}. For the reconstituted system, purified translocator preparations are incorporated into egg yolk lecithin liposomes, which are loaded with both ADP and ATP. Membrane potential is generated by K^+ gradient + valinomycin. The four combinations are evaluated from appropriate combinations of ^{14}C- and ^{3}H-labelled ADP and ATP. (~) = energized state = high $\Delta\psi$; +FCCP = deenergized state = "low" $\Delta\psi$ (From Klingenberg 1980)

translocator activity is thought to be divided up into four exchange combinations, two homo-exchanges (ATP/ATP, ADP/ADP), which are unproductive, and two modes of hetero-exchange, one of which is counterproductive and one productive in oxidative phosphorylation (ATP/ADP) [21]. Unless the ATP/ADP ratio is higher inside than outside, no net transfer would occur. Contrary to expectations, the internal ATP/ADP was found to be even considerably lower than the external one. However, our early studies showed that the exchange is modulated in dependence on the energizing of the membrane, such that ADP is taken up even in large excess of external ATP [22]. With more elaborate techniques, we could show also that from the inside, ATP is preferentially excreted, even in the presence of internal ADP [15]. What is the regulatory parameter? It was proposed early by us that the charge difference between ADP and ATP makes the transport responsive to the membrane potential of the inner mitochondrial membrane [23]. In the productive exchange mode, movement of one negative charge to the outside would occur, promoted by the membrane potential as an overall electrophoretic effect. The same membrane potential would repulse the counterproductive exchange. The fact that the exchange is not electroneutral was directly demonstrated by several types of evidence [24-26]. As a result of the membrane potential, modulation of the exchange, a now well-established large increase of the ATP/ADP ratio from inside to outside is maintained. In fact, a linear correlation of the double ratio to $\Delta\psi$ was determined [27].

The Control Characteristics of the Exchange in Oxidative Phosphorylation

The mechanism of membrane potential control is by increasing the translocation rate of the ATP loaded carrier going to the outside and correspondingly inhibiting the reverse reaction. The ADP carrier complex is not affected. This activity control is in contrast to the conformational control, which assumes a conformational change and affinity change for ATP under the influence of the membrane potential [6,7]. The results in mitochondria [22] and in particular in the reconstituted system [28] clearly established the electrophoretic translocation rate control by the membrane potential. On this basis, a rate equation for the translocation could be derived which gives the dependence on the single rate constants for ADP and ATP and on the ATP/ADP distribution across the membrane. Even assuming saturation of the binding sites, this equation becomes relatively complex. A full description should include the modification of the single translocation rate by the membrane potential. Thus, the rate constants for the translocation of the ATP carrier complex should follow in a simple relation with the membrane potential:

$$\vec{v} = \vec{k} e^{-\Delta\psi\ F/RT} , \quad \overleftarrow{v} = \overleftarrow{k} e^{+\Delta\psi\ F/RT} .$$

Since, however, the membrane potential influences the carrier distribution, i.e., the carrier sites available for ATP export, more complex nonlinear dependencies on the membrane potential should be expected. Thus, the rate for ATP export can be described as a function of the ADP/ATP ratios inside and outside and of the membrane $\Delta\psi$ potential. The dependence on these parameters should be nonlinear:

$$\overleftarrow{v}_{ATP} = f(v_{max}, (ATP/ADP), (ATP/ATP)_e, e^{\Delta\psi\ F/RT}) .$$

To which degree this rate determines the overall rate of oxidative phosphorylation is of course dependent on the total system, i.e., the rate of ATP synthesis as well as of consumption in the cytosol. The rate of ATP consumption can be controlled by variable load with the hexokinase glucose system [22,29]. With this setup, the share of the energy consumed in translocation as compared to ATP consumption was compared. In addition, added creatine kinase [30] or ATPase [31] were used as outside loads. Very useful was the introduction of a load on the intramitochondrial pool by the intramitochondrial citrulline synthesis [32]. With these studies, different conclusions were drawn on the question of rate limitation by the nucleotide carrier. Davis and co-workers [31] as well as Kunz and his group [29], although using quite different systems, found a correlation of the respiration with the extramitochondrial ATP/ADP ratio, where the intramitochondrial remained fairly constant. Consequently, they concluded that this reflects rate limitation by the nucleotide exchange. Erecinska and Wilson [10], using a different experimental approach, viewed the respiratory chain as being in equilibrium with the extramitochondrial ADP/ATP system, similar as originally Klingenberg and his co-workers [3,4]. As a result, they regard the ADP/ATP translocator as not rate-limiting.

However, as already pointed out, the questions asked in these experiments were too simplified, as one cannot expect a rate limitation localized at one step in the reaction sequence of ATP synthesis. In fact, in further experiments by the group of Kunz [33] and also by the group of Tager [34,13], quite different conclusions were drawn. The intramitochondrial ADP/ATP ratio did change by raising the respiration with an increasing extra- or intramitochondrial load. The change of the intramitochondrial ratio was more linearly related to the respiration, whereas the admittedly much larger change of the extramitochondrial ADP/ATP ratio was more sigmoid [35]. This is exactly what should be expected in view of our early standpoint that the intramitochondrial nucleotide pool is more directly controlling respiration and that the translocation rate is a complex nonlinear function of the parameters involved. By switching the load from an intra- to an extramitochondrial sink, under conditions where respiratory activity remains constant, the intramitochondrial ATP/ADP ratio was also found to be unchanged, whereas the extramitochondrial ATP/ADP ratio decreases considerably [34,35]. This is to be expected for a high external load where the translocation capacity becomes increasingly limiting.

The description [34] of this situation in terms of an increasing "disequilibrium" of the carrier, e.g., being calculated to 8.2 kJ, is somewhat awkwardly expressing that the flux through the translocator is increased when switching from the internal to the external load. In this context, the authors stress that $\Delta\psi$ did not change significantly between both states. It seems, however, well possible that the local $\Delta\psi$ at the nucleotide translocator, which really modulates the translocation rate, is quite different from the overall $\Delta\psi$, because of the strong influence of surface potentials on the highly charged substrates (ATP^{4-}, ADP^{3-}). We understand that the localized variation of the membrane potential in context with the respiratory control is also considered by other groups [36].

It was clear from the beginning that in a sequential reaction system, such as ATP synthesis by isolated mitochondria using an ATP-trapping system, we are dealing with a reaction sequence each step of which will contribute to the overall rate (see also 37). For the control in this type of reaction sequence, Kacser and Burns [38] or Heinrich and Rapoport [39] introduced a new nomenclature and a quantitative though phenomenological description which gives a refined treatment of rate control. One procedure applied to this system is to pose incremental inhibitions at single steps and to observe the overall response [34]. Thus, the distribution of "control strength" of the various steps has been evaluated and it was shown that the influence of the translocator varies strongly according to the strenght of the source and sinks on both sides of the translocation steps in the ATP flux. Thus, these authors find that the translocator is more limiting with creatine kinase than with hexokinase [37].

However, this method does not give any explanation on the mechanism of the control parameters. According to our analysis, the findings can be understood in terms of activity distribution between productive and unproductive exchange modes of translocator capacity. Thus, the higher $(ATP/ADP)_e$ in the presence of creatine kinase idles away some capacity by un- or counter-productive exchange. The accumulation of control strength data may be very useful to obtain a quantitative value on the contribution of the exchange in the overall rate, but it is not more revealing on the mechanism and factors involved than the underlying simple experiments can furnish. In other words, by introducing new formalisms, not more can be extracted from the experiments than is actually invested in the design, the nature, and the sophistication of the experiments. The participation of the ADP/ATP exchange in the overall respiratory control remains a very complex function of $\Delta\psi$ and the ATP/ADP transmembrane distribution, which eventually will have to be resolved.

Discussion

Kuthan: There is an increasing number of applications of the phenomenological equations of near-equilibrium thermodynamics to the mitochondrial energy conversion and reactions related to this. In the former presentation by Benno Hess, we learned that glycolytic reactions are characterized by oscillations, which are not within the framework of near-equilibrium thermodynamics. Is it really so that on the one hand, reactions in mitochondria prefer to proceed in the "linear" range, whereas glycolytic reaction occur in the "nonlinear" range?

Klingenberg: Stucki has stressed first-order behavior in complex enzymatic reactions even far from equilibrium. In this sense, there should be no difference between glycolytic and mitochondrial reactions.

References

1. Chance B, Williams GR (1955) J Biol Chem 217:409-427
2. Lardy HA, Wellman H (1952) J Biol Chem 195:215-224
3. Klingenberg M, Schollmeyer P (1961) Biochem Z 335:231-242
4. Klingenberg M, Schollmeyer P (1961) Symp Nr V, Pergamon Oxford, pp 46-65 (V. Intern Congr of Biochemistry, Moscow)
5. Klingenberg M, Pfaff E (1966) In: Tager JM et al. (eds) Regulation of Metabolic Processes in Mitochondria, Elsevier, Amsterdam, pp 180-201
6. Out TA, Valeton E, Kemp A Jr (1976) Biochim Biophys Acta 440:697-710
7. Viganis PV (1976) Biochim Biophys Acta 456:1-12
8. Jacobus WE, Moreadith, RW, Vandegaer, KM (1982) J Biol Chem 25:2397-2402
9. Klingenberg M (1969) In: Papa S et al. (eds) The Energy Level and Metabolic Control in Mitochondria. Adriatica Editrice, Bari, pp 185-188
10. Erecinska M, Wilson DF (1983) J Membrane Biol 70:1-14
11. Klingenberg M (1969) In: Papa S et al. (eds) The Energy Level and Metabolic Control in Mitochondria. Adriatica Editrice, Bari, pp 189-193
12. Slater EC, Rosing J, Mol A (1973) Biochim Biophys Acta 292:543-553
13. Tager JM, Wanders RJA, Groen AK, Kunz W, Bohnensack R, Küster U, Letko G, Böhme G, Duszynski J, Wojtczak L (1983) FEBS Lett 151, 1-9
14. Heldt HW, Pfaff E (1969) Eur J Biochem 10:494-500
15. Klingenberg M (1977) In: van Dam K, van Gelder BF (eds) Structure and Function of Energy-Transducing Membranes, Elsevier, Amsterdam, pp 275-282
16. Heldt HW, Klingenberg M (1965) Biochem Z 343:433-451
17. Pfaff E, Heldt HW, Klingenberg M (1969) Eur J Biochem 10:484-493
18. Klingenberg M, Grebe K, Appel M (1982) Eur J Biochem 126:263-269
19. Klingenberg M (1967) In: Quagliariello E et al. (eds) Mitochondrial Structure and Compartmentation. Adriatica Editrice, Bari, pp 320-324
20. Klingenberg M (1980) J Membran Biol 56:97-105
21. Klingenberg M, Heldt HW, Pfaff E (1969) In: Papa S et al. (eds) The Energy Level and Metabolic Control in Mitochondria. Adriatica Editrice, Bari, pp 237-253
22. Klingenberg M (1972) In: Mitochondria:Biomembranes. Elsevier, Amsterdam, pp 147-162 (Proc 8th FEBS Meeting)
23. Pfaff E, Klingenberg M (1968) Eur J Biochem 6:66-79
24. Klingenberg M, Wulf R, Heldt, HW, Pfaff E (1969) In: Ernster L, Drahota Z (eds) Mitochondria: Structure and Function. Academic, London, pp 59-77
25. LaNoue K, Mizani SM, Klingenberg M (1978) J Biol Chem 253:191-198
26. Wulf R, Kaltstein A, Klingenberg M (1977) Eur J Biochem 82:585-592
27. Klingenberg M, Rottenberg H (1977) Eur J Biochem 73:125-130
28. Krämer R, Klingenberg M (1982) Biochemistry 21:1082-1089
29. Küster U, Bohnensack R, Kunz W (1976) Biochim Biophys Acta 440:391-402
30. Stucki JW, Brawand F, Walter P (1972) Eur J Biochem 27:181-191

31. Davis EJ, Lumeng L (1975) J Biol Chem 250:2275-2282
32. Küster U, Letko G, Kunz W, Duszynski J, Bogucka K, Wojtczak L (1981) Biochim Biophys Acta 636:32-38
33. Kunz W, Bohnensack R, Böhme G, Küster U, Letko G, Schönfeld P (1981) Arch Biochem Biophys 209:219-229
34. Wanders RJA, Groen AK, Meijer AJ, Tager JM (1981) FEBS Lett 132:201-206
35. Tager JM, Groen AK, Wanders RJA, Duszynski J, Westerhof HV, Verdoorn RC (1983) Biochem Soc Trans 11:40-43
36. Westerhoff HV, Colen AM, van Dam K (1983) Biochem Soc Trans 11:81-85
37. Higgins J (1965) In: Chance B et al. (eds) Control of Energy Metabolism. Academic, New York, pp 13-46
38. Kacser H, Burns JA (1979) Biochem Soc Trans 7:1149-1160
39. Heinrich R, Rapoport TA (1974) Eur J Biochem 42:97-105

Oxygenation Pathways in Bacteria

R. Müller and F. Lingens[1]

Introduction

If we look into biochemical textbooks, the role of oxygen in biological systems seems to be limited to its function as electron acceptor. Electrons abstracted from organic molecules are transferred to electron acceptors like NAD^+, $NADP^+$, or FAD. From there they are transported via a redox chain to oxygen and water is formed. The energy released in this process is in some cases conserved in the form of ATP (e.g., respiratory chain). In contrast to this, oxygenation reactions, where the oxygen is inserted into organic molecules, seem to be rather rare. One well-known example is the introduction of a hydroxy-group into various compounds by the microsomal cytochrome P-450 system of the liver, which is required for the glucuronation and excretion of toxic compounds in mammals. Another example for oxygenation reactions, which will be discussed in detail, is the microbial degradation of aromatic compounds.

Benzene, toluene, ethylbenzene, or napthalene, for example, are important industrial products, which contribute to the pollution of our environment. Many of the herbicides and pesticides belong to the aromatic compounds and the polymer lignin, of which annually about 10^{+9} tons are produced by plants, is composed mainly of aromatic monomers like coumarin alcohol, coniferous alcohol, and sinapyl alcohol. Since the resonance structure of the benzene nucleus is chemically rather inert, the evolution of bacterial systems that are able to attack such compounds was necessary to maintain the steady state of carbon turnover in nature.

The Degradation of Benzene

The basic principles of the degradation of aromatic compounds by bacteria will be illustrated by the example of the unsubstituted benzene. The first step in the degradation of benzene by bacteria is the simultaneous insertion of two atoms of oxygen. Gibson et al. (1970a) selected a mutant from *Ps. putida*, which accumulated the dihydrodiol from benzene. The same mutant oxidized toluene to the corresponding dihydrodiol. The absolute stereochemistry of these diols was determined. The comparison with other dihydrodiols showed that all dihydrodiols formed in bacteria from aromatic compounds by the insertion of oxygen are invariably in the *cis* configuration. This favors the explanation that oxygen reacts with benzene to form a cyclic peroxide and that NADH is required to reduce the peroxide to a dihydrodiol (Fig. 1a).

1 Institut für Mikrobiologie der Universität Hohenheim, Garbenstraße 30, D-7000 Stuttgart 70, FRG

34. Colloquium-Mosbach 1983
Biological Oxidations

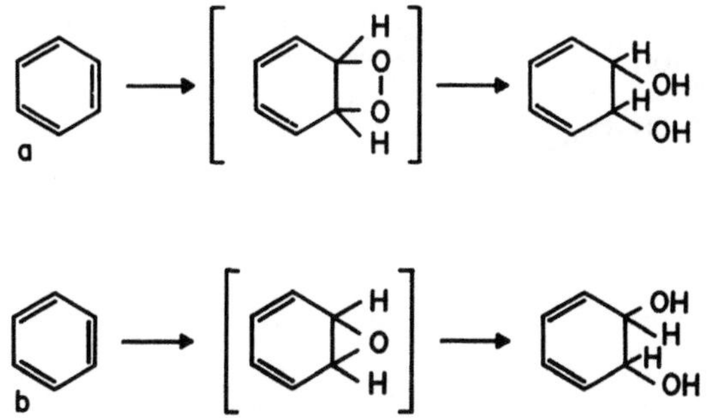

Fig. 1a,b. The initial step in the degradation of benzene. (a) by bacteria (b) by eukaryotic cells

This is in contrast to the eukaryotic cell, where an epoxide is proposed as intermediate (Jerina et al. 1968), which is then hydrolyzed to the *trans* diol (Fig. 1b). In this case one of the oxygen atoms is derived from water.

In the next step, the *cis*-dihydrodiol is dehydrogenated to the catechol by a dehydrogenase, which requires NAD^+. Thus, in the second step of the reaction sequence, the NADH required for the first step is recovered. At this stage of the degradation, a second oxygen molecule is introduced and the benzene nucleus is cleaved. This cleavage proceeds in two different ways. Either the ring is cleaved between the two hydroxyl groups (*ortho*-fission) and *cis, cis*-muconic acid is formed or it is cleaved adjacent to the hydroxyl groups forming 2-hydroxymuconic acid semialdehyde (*meta*-fission). The two modes of ring fission are the initial steps of two separate pathways which lead to the formation of products that can be fed into the citrate cycle. In the *ortho* pathway reviewed by Stanier and Ornston (1973), the *cis, cis* muconic acid is lactonized to muconolactone. This is isomerized to β-ketoadipate-enol-lactone, which is hydrolyzed to β-ketoadipate. The β-ketoadipate is transformed into its CoA-derivative and finally cleaved by an aldolase to form acetyl-CoA and succinate, both products of the citrate cycle. In the *meta* pathway, reviewed by Dagley (1971), the 2-hydroxymuconic acid semialdehyde can undergo two different reactions. Either it is oxidized by a NAD^+-requiring dehydrogenase into 2-hydroxymuconic acid or it is cleaved by a hydrolase into formic acid and 2-hydroxypenta-2,4-dienoic acid. The 2-hydroxymuconic acid can also be decarboxylated to form 2-hydroxypenta-2,4-dienoic acid. To the double bond of this compound, water is added to form 2-keto-4-hydroxyvaleric acid, which is finally cleaved by an aldolase into pyruvate and acetaldehyde. So after ring cleavage, no more oxygen is required for the formation of the products which can be fed into the citrate cycle.

The Degradation of Substituted Benzenes

For substituted aromatic compounds, in principle, two ways for the degradation are possible. Oxygen can attack directly the benzene moiety. This leads to substituted catechols, which can be metabolized via modified *ortho* or *meta* pathways. The other possibility is the modification of the substituent, which leads to simpler substituted compounds, which are then degraded via attack of the benzene nucleus. In some cases, both ways are described. This principle shall be illustrated by different examples.

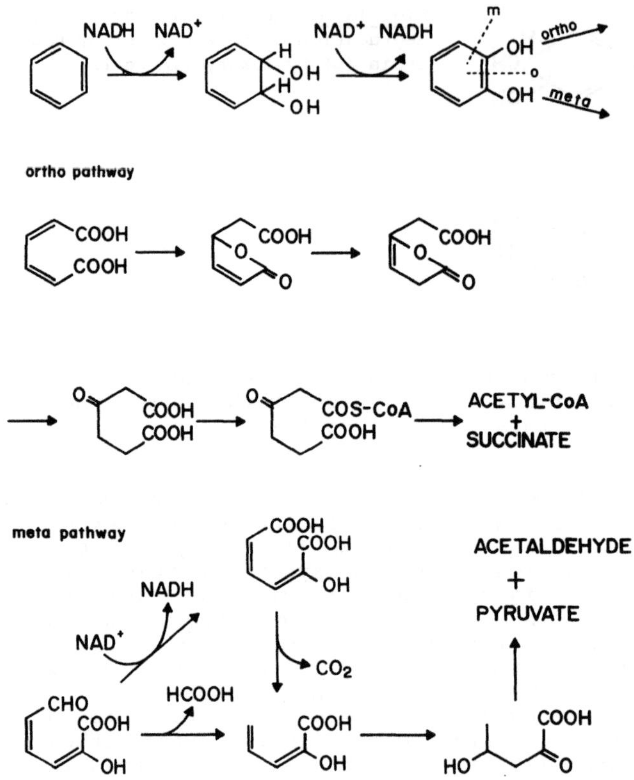

Fig. 2. The degradation of benzene

Ephedrine

The degradation of the alkaloid ephedrine by *Ps. putida* (Klamann and Lingens 1980) and *Arthrobacter globiformis* (Klamann 1976) is an example for a degradation pathway in which first the side chain is removed and then the benzene nucleus is attacked. The initial degradation step is the removal of methylamine from the side chain of ephedrine and methylbenzoylcarbinol is formed. This undergoes cleavage into acetaldehyde and benzaldehyde. The latter compound is subsequently oxidized to benzoic acid. (Fig. 3).

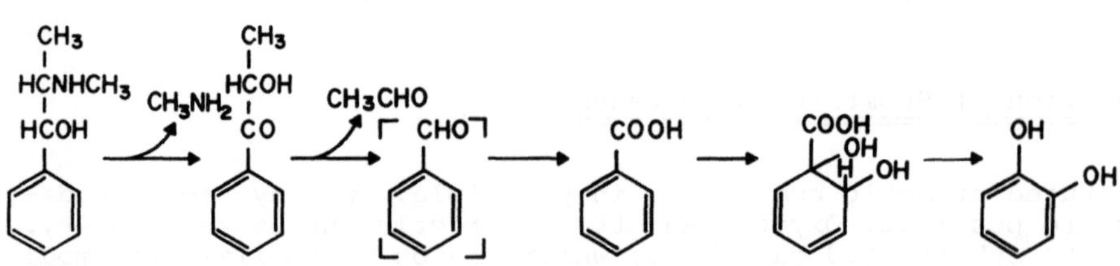

Fig. 3. The degradation of ephedrine

Benzoic Acid

The benzoic acid, formed in the degradation of several substituted aromatic compounds, can be oxidized to the 1,2-dihydrodiol (Reiner and

Hegeman 1971). This diol is then dehydrogenated with simultaneous decarboxylation and unsubstituted catechol is formed (Fig. 3).

Bidisin

The degradation of the herbicide bidisin (2-chloro-3-(4-chlorophenyl)-propionic acid methylester) is another example where first the side chain is removed and then the aromatic ring is degraded. Two types of bacteria belonging to the genera *Flavobacterium* and *Brevibacterium* were found (Köcher et al. 1976) that are able to degrade the side chain of bidisin. The initial step is the hydrolysis of the ester, followed by the removal of the chlorine of the side chain. The resulting 4-chlorocinnamate is then oxidized in analogy to the fatty acid oxidation to yield 4-chlorobenzoate. The degradation of 4-chlorobenzoate by a *Pseudomonas sp.* (Klages and Lingens 1980), a *Nocardia sp.* (Klages and Lingens 1979) and an *Arthrobacter sp.* (Ruisinger et al. 1976) were reported. First, the chlorine is eliminated and 4-hydroxybenzoate is formed. This is hydroxylated in 3-position. The resulting protocatechuate is cleaved in *ortho*-position by the *Pseudomonas sp.* and in *meta*-distal position by the *Nocardia sp.* and the *Arthrobacter sp.* (Fig. 4). So in this case, the bacteria that degrade the side chain of the substrate are different from the bacteria that degrade the aromatic ring.

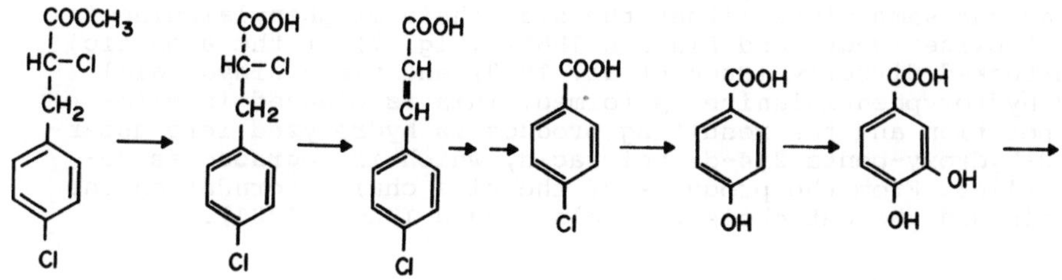

Fig. 4. The degradation of the herbicide bidisin

Cinnamic Acid

The degradation of cinnamic acid by *Phenylobacteria* is an example where only the direct attack at the benzene ring is found. Tittmann et al. (1980) reported the formation of the 2,3-dihydrodiol, which is dehydrogenated to the 2,3-catechol-derivative and then cleaved in *meta*-proximal position. The resulting product is hydrolyzed to fumarate and 2-hydroxypenta-2,4-dienoic acid (Fig. 5).

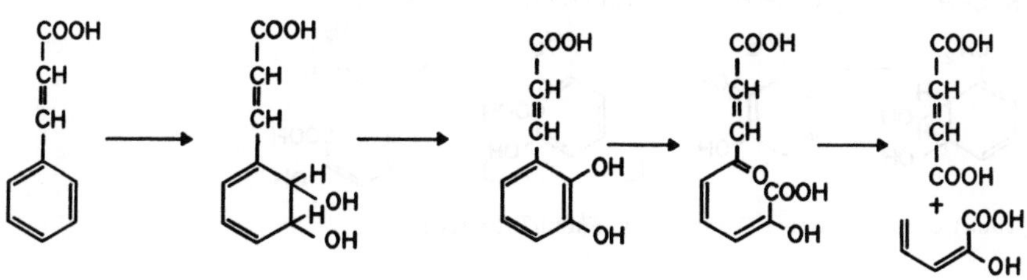

Fig. 5. The degradation of cinnamic acid by *Phenylobacteria*

Biphenyl

In the degradation of biphenyl, two rings are available for oxidation. But only one ring is attacked in the initial step and a dihydrodiol is formed (Catelani et al. 1971). This is dehydrogenated to the catechol and finally cleaved in the *meta*-proximal position (Catelani et al. 1973). The resulting product is hydrolyzed to benzoic acid and 2-hydroxy penta-2,4-dienoic acid, whose degradations have been described above (Fig. 6; see also Figs. 2 and 3).

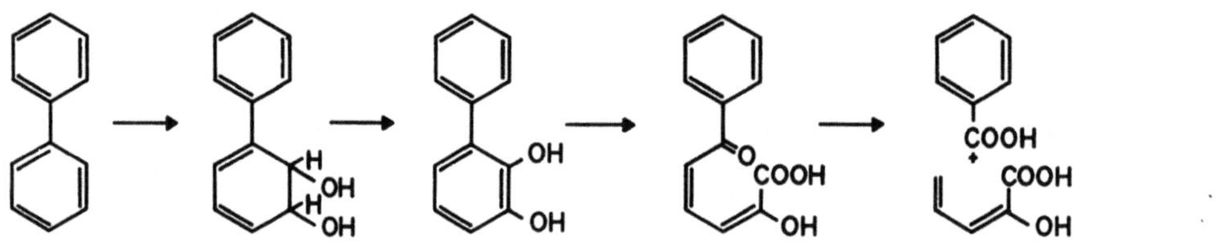

Fig. 6. The degradation of biphenyl by *Ps. putida*

Phenylalanine

In *Phenylobacteria*, two pathways for the degradation of phenylalanine are possible at the same time. Either the side chain of phenylalanine is degraded stepwise (Wegst and Lingens 1981) (Fig. 7) or the aromatic ring is attacked directly (Buck et al. 1979) and via a dihydrodiol the 2,3-dihydroxyphenylalanine is formed. This is cleaved in *meta*-proximal position and the resulting product is hydrolyzed into aspartate and 2-hydroxy-penta-2,4-dienoic acid, which is degraded as described earlier. From the products of the side chain degradation the dihydrodiols and the catechols were also found (Wegst 1981).

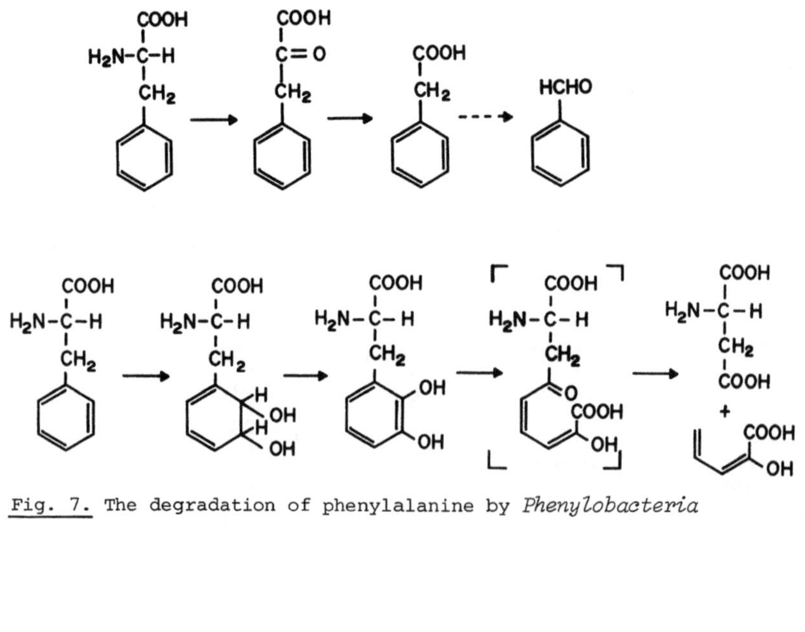

Fig. 7. The degradation of phenylalanine by *Phenylobacteria*

Toluene

In the case of toluene, two degradation pathways are also described. Kitagawa (1956) reported the conversion of toluene in *Pseudomonas aeruginosa* via benzylalcohol, benzaldehyde to benzoate, whereas Gibson et al. (1970b) showed in *Pseudomonas putida* the conversion to the dihydrodiol and 3-methyl-catechol (Fig. 8).

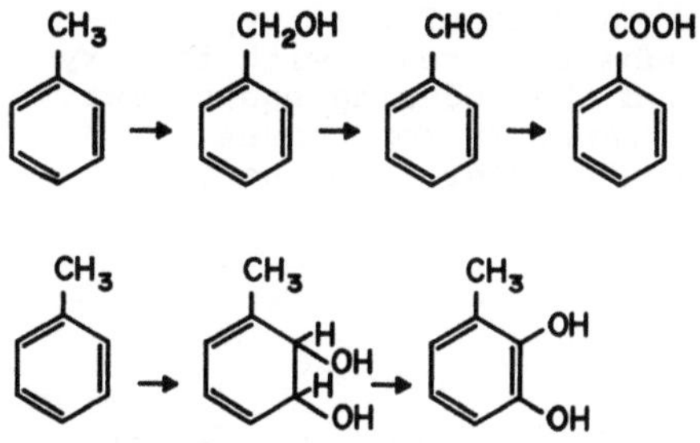

Fig. 8. The two degradation pathways of toluene in *Ps. aeruginosa* (*upper*) and in *Ps. putida* (*lower*)

Chloridazon

In the herbicide chloridazon as well as in the structurally related analgetica, Antipyrine and Pyramidon, the phenyl ring is attached to a heterocycle. Again in the initial degradation step, a dihydrodiol is formed, which is then dehydrogenated to the catechol (Haug et al. 1973). The *ortho* cleavage of this compound yields a product that is not further metabolized (Blobel et al. 1976). The cleavage in *meta*-proximal position yields a product that is not stable and decomposes into the heterocycle and 2-hydroxymuconic acid-δ-lactone, both dead end products. From extracts of chloridazon degrading bacteria, however, an enzyme could be isolated, which inhibits the spontaneous decomposition and instead hydrolyzes the *meta*-cleavage product to the heterocycle and free 2-hydroxymuconate (Müller et al. 1977) (Fig. 9). The hydroxymuconate can then be decarboxylated as described before. The heterocycle, however, cannot be degraded further and contributes to environmental pollution.

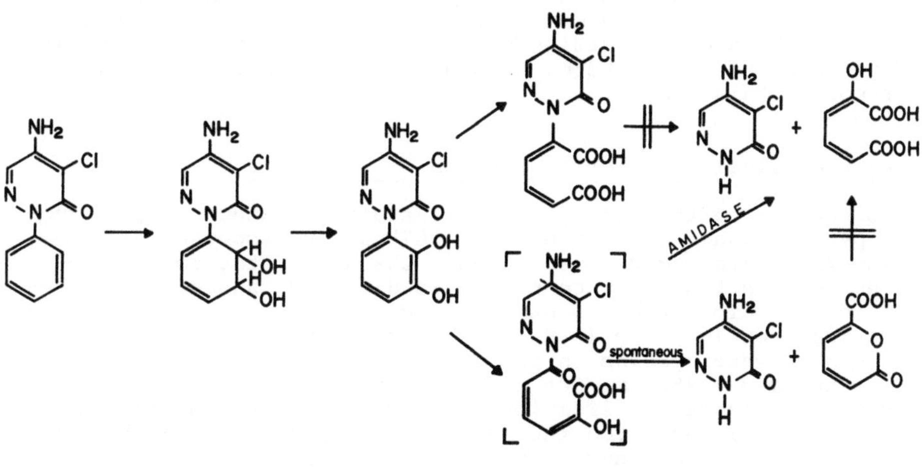

Fig. 9. The degradation of the herbicide chloridazon

Oxygenases Involved in the Degradation of Aromatic Compounds

Dioxygenases Forming Dihydrodiols

Several enzymes or enzyme systems are reported that catalyze the formation of *cis-ortho*-dihydrodiols from various aromatic compounds, including benzene, benzoate, toluene, anthranilate, naphthalene, and chloridazon. Tracer experiments showed that both oxygen atoms are derived from the same oxygen molecule. All enzymes of this type comprise several components. Oxygen and the substrate react with the oxygenase component to form the dihydrodiol and the reducing equivalent required is delivered from NADH via a reductase. In some cases, a third component, which serves as an electron carrier from the reductase to the oxygenase, is required (Fig. 10).

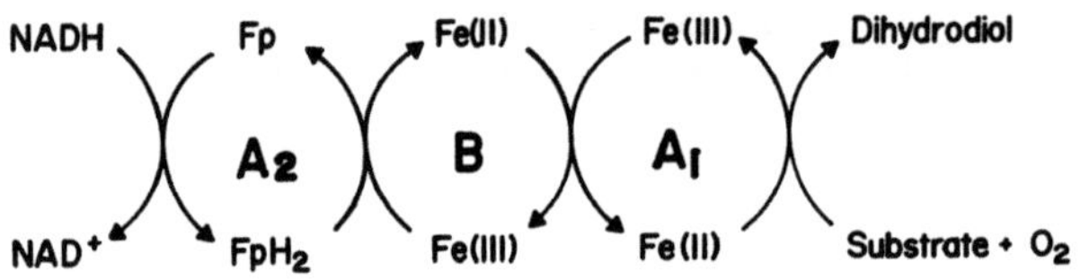

Fig. 10. Reaction scheme proposed for the three component system of chloridazon dioxygenase

The oxygenase and the electron carrier components are invariably iron sulphur proteins. The reductase contains FAD as prosthetic group. For chloridazon dioxygenase activity, three components are required when the bacterial cells were disrupted by extensive ultrasonic treatment (Sauber et al. 1977). Cell disruption under mild conditions, i.e., lysozyme treatment and subsequent solubilization of the bacterial membranes with the detergent triton X 100 led to new results on the composition of the dioxygenase complex. In extracts prepared by this method, only two components were found to be necessary for enzymatic activity: the terminal oxygenase and reductase. Whether under these conditions the electron carrier is included in the oxygenase is not clear at the moment. The oxygenase component isolated this way gave a positive staining for lipids, indicating that this component is originally membrane-bound. About the mechanism of this type of enzymes and the binding of oxygen or the substrate, nothing is known at present and waits for further investigation. Recently, another type of dioxygenase from *Corynebacterium renale* was reported by Dua and Meera (1981). This naphthalene oxidizing enzyme consists of a single component and contains no chromophore. Nothing is known about any functional groups in this enzyme.

Dioxygenases Catalyzing the *Ortho*-Fission of Catechols

Mainly two enzymes of this group are well-studied: catechol 1,2-dioxygenase or pyrocatechase (Kojima et al. 1967) and protocatechuate 3,4-dioxygenase (Fujiwara and Hayaishi 1968). Characteristic for *ortho*-cleaving enzymes is that they comprise two different subunits of similar size in stoichiometric amounts. The composition of the intact enzymes varies from $(\alpha\beta)_2$ to $(\alpha_2\beta_2)_8$. They all contain iron in the trivalent state, which yields an ESR signal at around $g = 4.3$ and a red color with $\lambda_{max} = 450$ nm. Que and Heistand (1979) suggested from Raman resonance spectra the ligation of a tyrosin-OH to the iron and Felton et al. (1982) suggested the reversible ligation of two histidines to the iron. In the enzyme substrate complex, the substrate seems to bind to the Fe^{3+}. The amino acid sequence of protocatechuate-3,4-dioxygenase has been reported (Kohlmiller and Howard 1979; Iwaki et al. 1981) and preliminary data on the X-ray structure are available (Satyshur et al. 1980), so that in near future the problem of how iron and the substrate are bound in this enzyme may be solved.

Dioxygenases Catalyzing the *Meta*-Fission of Catechols

Several enzymes of this group have been studied, including catechol 2,3-dioxygenase or metapyrocatechase (Kojima et al. 1961), protocatechuate 4,5-dioxygenase (Dagley et al. 1968) 3,4-dihydroxyphenylacetate 2,3-dioxygenase (Kita et al. 1965), and chloridazon-catechol dioxygenase (Müller et al. 1982). Most of these enzymes are extremely unstable in the presence of air. However, low concentrations of organic solvents like acetone or ethanol protect the enzymes.

They are composed of 3-6 identical subunits with a mol. w. of 35,000. Unlike intradiol dioxygenases, they are colorless and show neither significant absorption in the visible range nor any ESR signal. Iron is the sole co-factor and seems to be in the ferrous form. In the presence of substrate, however, in some cases an ESR signal at around $g = 4.3$ was observed. This signal disappears after the exhaustion of oxygen or substrate, indicating a reversible valence change of the iron during the catalytic cycle. Chloridazon-catechol dioxygenase is a *meta*-cleaving enzyme crystallized (Müller et al. 1982) and characterized (Müller et al. 1977) in our laboratory. This enzyme requires Fe^{2+}-ions as co-substrate. The particular mechanism of this enzyme is illustrated in Fig. 11. First, the inactive (probably Fe^{3+}) enzyme has to be activated by exogenous Fe^{2+} to the active (probably Fe^{2+}) form. This reacts with the substrate and oxygen and is thereby converted back to the inactive form. One electron is required for one reaction cycle. It seems as if in this enzyme, the transfer of an electron back to the enzyme at the end of the reaction cycle is uncoupled, which makes this enzyme interesting for further mechanistic studies. Where the electrons come from in intact cells of *Phenylobacteria* we don't yet know.

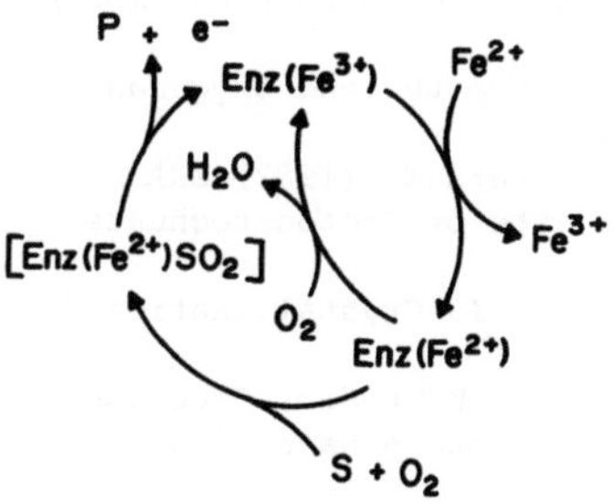

Fig. 11. Proposed reaction mechanism for Chloridazon-catechol dioxygenase

Que et al. (1981) recently reported the purification of a *meta*-cleaving enzyme from *Bacillus brevis*, which contains Mn^{2+} as cofactor instead of iron. It would be interesting to know more about the mechanism of this enzyme, since a valence change of the manganese does not seem very likely. Fujiwara et al. (1975) reported the *meta*-cleavage of 3-methyl-catechol by an intradiol dioxygenase from *Pseudomonas arvilla*. This Fe^{3+}-containing enzyme cleaved 3-methyl-catechol in *ortho*-position. The ratio between the two reactions was about 17:1. This indicates that the reaction mechanism of the *ortho* and *meta* fission of catechols may not be so different and maybe there is a common intermediate for both reactions. However, to confirm this, we have to know much more about the active site of these enzymes, especially about the role of the iron and the binding of substrate.

Discussion

Mason: Are these enzymes of preliminary metabolism induced or constitutive, and if induced, what is the mechanism of development of specifity?

Müller: Some of the enzymes described, like chloridazon-catechol dioxygenase, are constitutive, some are inducible. If they are induced, usually the first enzyme of a reaction sequence is induced by its substrate, but the enzymes of the pathways after ring-cleavage are induced coordinately by the catechol derivatives. The enzymes induced are specific for the inducer used. An exceptional case is the regulation of the *ortho*-pathway in *Pseudomonas putida*, where an endproduct, β-ketoadipate, is responsible for the induction of the *ortho*-pathway.

References

Blobel F, Eberspächer J, Haug S, Lingens F (1976) Enzymatische Bildung eines Muconsäurederivates mit Hilfe Pyrazon-abbauender Bakterien. Z Naturforsch 31c:756

Buck R, Eberspächer J, Lingens F (1979) Abbau und Biosynthese von L-Phenylalanin in Chloridazon-abbauenden Bakterien. Hoppe Seyler's Z Physiol Chem 360:957-969

Catelani D, Sorlini C, Treccani V (1971) The Metabolism of Biphenyl by Pseudomonas putida. Experientia 27:1173-1174

Catelani D, Colombi A, Sorlini C, Treccani V (1973) Metabolism of Biphenyl, 2-Hydroxy-6-oxo-6-phenylhexa-2,4-dienoate: The meta-cleavage product form 2,3-dihydroxybiphenyl by Pseudomonas putida. Biochem J 134:1063-1066

Dagley S (1971) Catabolism of Aromatic Compounds by Micro-Organisms. Adv Microbial Physiol 6:1-46

Dagley S, Geary PJ, Wood JM (1968) The Metabolism of Protocatechuate by Pseudomonas testosteroni. Biochem J 109:559-568

Dua RD, Meera S (1981) Purification and Characterization of Naphthalene Oxygenase from Corynebacterium renale. Eur J Biochem 120:461-465

Felton RH, Barrow WL, May SW, Sowell AL, Goel S, Bunker G, Stern EA (1982) EXAFS and Raman Evidence for Histidine Binding at the Active Site of Protocatechuate 3,4-Dioxygenase. J Am Chem Soc 104:6132-6134

Fujisawa H, Hayaishi O (1968) Protocatechuate 3,4-Dioxygenase I. Crystallization and Characterization. J Biol Chem 243:2673-2681

Fujiwara M, Golovleva LA, Saeki Y, Nozaki M, Hayaishi O (1975) Extradiol Cleavage of 3-Substituted Catechols by an Intradiol Dioxygenase, Pyrocatechase, from a Pseudomonad. J Biol Chem 250:4848-4855

Gibson DT, Cardini GE, Maseles FC, Kallio RE (1970a) Incorporation of Oxygen-18 into Benzene by Pseudomonas putida. Biochemistry 9:1631-1635

Gibson DT, Hensley M, Yoshioka H, Mabry TJ (1970b) Formation of (+)-*cis*-2,3-Dihydroxy-1-methylcyclohexa-4,6-diene from Toluene by Pseudomonas putida. Biochemistry 9:1626-1630

Haug S, Eberspächer J, Lingens F (1973) Enzymatic and Chemical Preparation of 5-Amino-4-chloro-2-(2,3-dihydroxyphen-1-yl)-3(2H)-pyridazinone. Biochem Biophys Res Commun 54:760-763

Iwaki M, Kagamiyama H, Nozaki M (1981) The Primary Structure of the β-Subunit of Protocatechuate 3,4-Dioxygenase from Pseudomonas aeruginosa. Arch Biochem Biophys 210:210-223

Jerina DM, Daly JW, Witkop B, Zalzman-Nirenberg P, Udenfriend S (1968) Role of the Arene Oxide-Oxepin system in the Metabolism of Aromatic Substrates. Arch Biochem Biophys 128:176-183

Kita H, Kamimoto M, Senoh S (1965) Crystallization and Some Properties of 3,4-Dihydroxyphenylacetate 2,3-Oxygenase. Biochem Biophys Res Commun 18:66-70

Kitagawa M (1956) Studies on the Oxidation Mechanism of Methyl Group. J Biochem 43:553-563

Klages U, Lingens F (1979) Degradation of 4-Chlorobenzoic acid by a Nocardia species. FEMS Microbiol Lett 6:201-203

Klages U, Lingens F (1980) Degradation of 4-Chlorobenzoic Acid by a Pseudomonas sp. Zbl Bakt Hyg I Abt Orig C1:215-223

Klamann E (1976) Abbau des (-)-Ephedrins durch Arthrobacter globiformis. Zbl Bakt Hyg I Abt Orig B 162:184-187

Klamann E, Lingens F (1980) Degradation of (-)-Ephedrine by Pseudomonas putida. Detection of (-)-Ephedrine: NAD^+-oxidoreductase from Arthrobacter globiformis. Z Naturforsch 35c:80-87

Köcher H, Lingens F, Koch W (1976) Untersuchungen zum Abbau des Herbizids Chlorphenpropmethyl im Boden und durch Mikroorganismen. Weed Res 16:93-100

Kohlmiller NA, Howard JB (1979) The Primary Structure of the α-Subunit of Protocatechuate 3,4-Dioxygenase. J Biol Chem 254:7309-7315

Kojima Y, Itada N, Hayaishi O (1961) Metapyrocatechase: A New Catechol-Cleaving Enzyme. J Biol Chem 236:2223-2228

Kojima Y, Fujisawa H, Nakazawa A, Nakazawa T, Kanetsuna F, Taniuchi H, Nozaki M, Hayaishi O (1967) Studies on Pyrocatechase, I. Purification and spectral properties. J Biol Chem 242:3270-3278

Müller R, Schmitt S, Lingens F (1982) A Novel Non-heme Iron-Containing Dioxygenase: Chloridazon-Catechol Dioxygenase from Phenylobacterium immobilis DSM 1986. Eur J Biochem 125:579-584

Müller R, Haug S, Eberspächer J, Lingens F (1977) Catechol-2,3-Dioxygenase aus Pyrazon-abbauenden Bakterien. Hoppe-Seyler's Z Physiol Chem 358:797-805

Que L Jr, Heistand RH II (1979) Resonance Raman Studies on Pyrocatechase. J Am Chem Soc 101:2219-2221

Que L Jr, Widom J, Crawford RL (1981) 3,4-Dihydroxyphenylacetate 2,3-Dioxygenase. A Manganese (II) Dioxygenase from Bacillus brevis. J Biol Chem 256:10941-10944

Reiner AM, Hegeman GD (1971) Metabolism of Benzoic Acid by Bacteria: Accumulatior of (-)-3,5-Cyclohexadiene-1,2-diol-1-carboxylic Acid by a Mutant Strain of Alcaligenes eutrophus. Biochemistry 10:2530-2636

Ruisinger S, Klages U, Lingens F (1976) Abbau der 4-Chlorbenzoesäure durch eine Arthrobacter-Species. Arch Microbiol 110:253-256

Satyshur KA, Rao ST, Lipscomb JD, Wood JM (1980) Preliminary Crystallographic Study of Protocatechuate 3,4-Dioxygenase from Pseudomonas aeruginosa. J Biol Chem 255:10015-10016

Sauber K, Fröhner C, Rosenberg G, Eberspächer J, Lingens F (1977) Purification and Properties of Pyrazon Dioxygenase from Pyrazon-Degrading Bacteria. Eur J Biochem 84:89-97

Stanier RY, Ornston LN (1973) The β-Ketoadipate Pathway. Adv Microbiol Physiol 9:89-151

Tittmann U, Wegst W, Blecher R, Lingens F (1980) Abbau von trans-Zimtsäure durch Chloridazon-abbauende Bakterien. Zbl Bakt Hyg I Abt Orig C1:124-132

Wegst W, Lingens F (1981) Abbau von L-Phenylalanin und von Phenylcarbonsäuren in Chloridazon-abbauenden Bakterien. Hoppe Seyler's Z Physiol Chem 362:1219-1227

Oxygen Radicals and Hydroperoxides in Mammalian Organs: Aspects of Redox Cycling and Hydrogen Peroxide Metabolism

E. Cadenas, R. Brigelius, Th. Akerboom, and H. Sies[1]

Introduction

The 1968 conference on the Biochemistry of Oxygen (Hess and Staudinger 1968) contained only formal reference to reduction states of oxygen such as one-electron, two-electron, or three-electron states. Time was surprisingly short between the initial recurrence of interest in the biological chemistry of oxygen metabolites about one or two decades ago and the recent efforts, largely successful, of applications of new knowledge on oxygen metabolites in clinical medicine (Bannister and Bannister 1980). In 1966, the peroxisome was described as a subcellular organelle occupied, in particular, with hydrogen peroxide metabolism (De Duve and Baudhuin 1966). In 1969, the field of oxygen free radicals received a major stimulus with the discovery of superoxide dismutase activity of erythrocuprein (McCord and Fridovich 1969). In 1970, a steady state production of hydrogen peroxide in mammalian organs was found to occur as a normal attribute of aerobic life (Sies and Chance 1970; Chance et al. 1979).

The advantage of aerobic life in terms of energy yield is thus linked to the risk of oxidative damage. Powerful defense systems against the so-called oxidative stress have evolved, both enzymatic and nonenzymatic. These include the activities of GSH peroxidase (Mills 1960), a seleno-enzyme (Se GSH-Px) (Rotruck et al. 1973; Flohé et al. 1973) and the non-selenium dependent GSH peroxidase activity (non-Se GSH-Px) (Lawrence and Burk 1976); catalase activity (Thénard 1818), the above mentioned superoxide dismutases, and other enzyme activities, notably various peroxidases. Nonenzymatic antioxidant defense includes vitamins E, C, and A and nutritional antioxidants, both naturally occurring and as chemical products. The biological basis of detoxication of oxygen free radicals (Sies and Cadenas 1983) and the compartmentation of detoxication reactions (Orrenius and Sies 1982) have recently been reviewed and will not be presented here in detail.

The balance between prooxidant and antioxidant challenges in biological systems is delicate and a number of imbalances are of medical interest (Autor 1982; Balentine 1982).

Recent studies on redox cycling in relation to oxygen metabolism have provided advances in knowledge on the toxicity of oxygen. This encompasses mainly reactions initially dependent on the generation of the superoxide anion radical. Further, the cellular metabolism of hydrogen peroxide has recently been studied in different types of cell; this includes the significance of peroxisomal fatty acid metabolism and the potential effects of hydrogen peroxide as a metabolic signal.

---

1 Institut für Physiologische Chemie I, Universität Düsseldorf, Moorenstraße 5, D-4000 Düsseldorf, FRG

34. Colloquium-Mosbach 1983
Biological Oxidations

These areas of research form but a segment of ingoing investigations on oxygen metabolites and they will be presented here in brief.

Estimated Steady-State Concentrations of Oxygen Metabolites

Most of the information on steady-state concentrations of oxygen metabolites has been obtained with the hepatocyte. However, except for conditions such as the respiratory burst in specialized cells, the data in Table 1 may serve as an orientation for other cell types. Clearly, oxygen concentrations may vary largely in different parts of the organism, with lung, skin, and heart on the upper end of the range and compartments such as intestine or bile more at the lower end. Further, the heterogeneity of the capillary bed and its dependence on the functional state as is known for skeletal muscle entails considerable fluctuations.

Table 1. Estimated steady-state concentrations of oxygen and oxygen metabolites. Data for rat liver, based on calculations for the "normal" physiological conditions

Oxygen	10^{-5} M	
Superoxide Radical ($O_2^{\cdot-}$, $HO_2^{\cdot}$)	10^{-11} M	Tyler 1975
Hydrogen Peroxide (H_2O_2)	10^{-9} M	Oshino et al. 1973
Singlet Oxygen ($^1O_2$)	10^{-15} M	Cadenas et al. 1983a

The heterogeneity of hepatocytes may be linked to the periportal to perivenous gradient in oxygen (Jungermann and Katz 1982). Periportal hepatocytes seem to be better equipped for oxidative metabolism. Reductive catabolism would take place in perivenous hepatocytes, which are largely endowed with glycolytic enzymes. It was suggested that activation of certain xenobiotics would predominate over other physiological mechanisms in hypoxic areas of the cell. Thus, in hypoxia carbon tetrachloride-induced lipid peroxidation (Kieczka and Kappus 1980) and the formation of the halothane adduct with cytochrome P-450 (Nasteinczyk et al. 1978) are enhanced. The particular importance of hypoxic conditions for deleterious oxygen effects is given by the concomitant flux through reductive as well as oxidative pathways (Bachur et al. 1979; Kappus and Sies 1981).

One-Electron Reduction of Oxygen: Superoxide Radical Production and Redox Cycling

The one-electron reduction of molecular dioxygen with formation of the superoxide anion radical takes place in cells due to a multiplicity of enzymatic and nonenzymatic reactions (Fig. 1). Cellular sources include the mitochondrial and microsomal respiratory chains and the bactericidal activity of leukocytes and macrophages. At a physiological level, the importance of these cellular source is difficult to

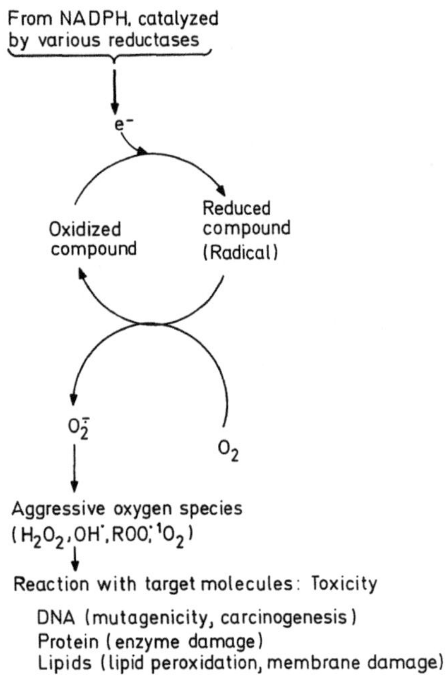

Fig. 1. Formation of reactive oxygen species upon redox cycling and some biological effects. (Modified from Sies and Cadenas 1983)

assess in a general fashion (although they might provide a large share in keeping the superoxide anion radical cellular steady-state concentration). Certainly, autoxidation reactions, including redox cycling, are growing in relevance for explaining the oxidative stress caused by several xenobiotics. The further metabolism of intermediates of oxygen appears to proceed by way of so-called dismutation reactions. As shown in Table 2, not only is the one-electron reduction state detoxified in a disproportionation reaction, but also the two-electron reduction products. Whereas dismutation of superoxide anion radical and hydrogen peroxide by the respective enzymes yields ground state oxygen, that of

Table 2. Dismutation reactions of oxygen metabolites

$2\ O_2^{\cdot-} + 2\ H^+ \rightarrow H_2O_2 + O_2$	(Superoxide Dismutase)	(McCord and Fridovich 1969)
$2\ H_2O_2 \rightarrow 2\ H_2O + O_2$	(Catalase)	(Thênard 1818)
$2\ PGG_2 \rightarrow 2\ PGH_2 + {}^1O_2$	(PG Hydroperoxidase)	(Cadenas et al. 1983d)
$2\ ROOH \rightarrow 2\ ROH + {}^1O_2$	(Cytochrome P-450)	(Cadenas et al. 1983c)
$2\ PhI{=}O \rightarrow 2\ PhI + {}^1O_2$	(Cytochrome P-450)	(Cadenas et al. 1983c)

Abbreviations used: PGG_2, prostaglandin G_2; PGH_2, prostaglandin H_2; PhI=O, iodosobenzene

hydroperoxides as products of lipid peroxidation or prostaglandin G2 (which could be considered an enzymatic-controlled lipid peroxidation) can produce singlet oxygen.

Mitochondrial Electron Transport

Superoxide anion radical and hydrogen peroxide production by mitochondria constitutes a physiological event under aerobic conditions, if one transposes data from isolated organelles to their function within the intact cell. It accounts for 1%-4% of the total oxygen consumption and depends on the metabolic state, being higher in the absence of phosphate acceptor (State 4) and lower when the respiratory carriers are largely oxidized (State 3). Two sites of superoxide radical generation exist in the mitochondrial inner membrane, ubiquinone-cytochrome c reductase and the NADH dehydrogenase flavin, accounting for 85% and 15% of the total superoxide radical production, respectively (Boveris and Cadenas 1982). Ubisemiquinone was found to be the major mitochondrial autoxidizable component yielding superoxide radical. Recently, evidence was presented for a heterogeneous pool of ubisemiquinone radicals in the neutral and anionic form and bound to a protein in the NADH-ubiquinone segment of the mitochondrial inner membrane (Suzuki and King 1983).

Hyperoxia enhances superoxide anion radical generation and hydrogen peroxide release by isolated lung mitochondria and microsomes, thus supporting the hypothesis that oxygen toxicity could be a consequence of increased intracellular rates of superoxide anion radical and hydrogen peroxide production in the lung (Turrens et al. 1982a,b).

Microsomal Electron Transport

The microsomal electron transport chain generates both the superoxide radical (Bartoli et al. 1977; Kuthan et al. 1978) and hydrogen peroxide (Gillette et al. 1957; Thurman et al. 1972; Hildebrandt and Roots 1975) upon autoxidation of cytochrome P-450 and the flavoprotein NADPH-cytochrome P-450 reductase. However, it remains still to be clarified whether in the intact endoplasmic reticulum cytochrome P-450 is capable of generating superoxide radical and/or hydrogen peroxide as suggested by Jones et al. (1978). Evidence against the occurrence of the release of hydrogen peroxide by cytochrome P-450 in the intact organ has been brought forward with perfused liver (Oshino et al. 1975; Sies et al. 1978; Sies and Graf 1982): both the GSSG release as an indicator for cytosolic hydrogen peroxide metabolism and the steady-state level of catalase compound I were similar in phenobarbital-pretreated and control rat livers.

NADPH-Cytochrome P-450 Reductase: Lipid Peroxidation, Activation of Xenobiotes

The flavoprotein NADPH-cytochrome P-450 reductase participates as initial electron donor in the activation of different xenobiotics generally through the formation of an initial radical adduct with the further generation of an activated oxygen species.

The superoxide anion radical is not a very reactive species as compared to its protonated form, the perhydroxyl radical (Sawyer and Valentine

1981) and secondary more reactive radicals might be formed from the superoxide anion radical in order to account for the initiation of lipid peroxidation. The superoxide-supported formation of the hydroxyl radical in the presence of iron salts finds circumstantial evidence for its occurrence in vivo (Halliwell 1982), although it seems that the requirement needed for an adequate catalyst is not met in all cases; moreover, evidence is available that the superoxide radical might be substituted by, for example, ascorbic acid (with a wide distribution in biological systems) in generating hydroxyl radicals (Winterbourn 1982). Hydrogen peroxide, formed during the cytochrome P-450 oxidase activity, seems not to be involved in the NADPH-dependent lipid peroxidation (Bast et al. 1983).

An alternative route for the formation of more aggressive species derived from the superoxide radical would be the nonenzymatic dismutation of superoxide radical yielding singlet molecular oxygen (identified as the monomol emission band of chemiluminescence; Khan 1981). This evidence disagrees with the lack of formation of a 5-α-hydroperoxy adduct of cholesterol, as a trap for singlet oxygen, in a superoxide radical-generating system (Foote et al. 1980). Singlet oxygen is formed in comparatively low amounts and it has become of interest because of its involvement at a specific stage of lipid peroxidation; singlet oxygen can add to the double bond of polyunsaturated fatty acids with formation of lipid hydroperoxides.

Alkoxy and peroxy radicals are formed in the decomposition of lipid hydroperoxides and are important in the maintenance of radical chain reactions in addition to the lipid radical R. These radicals seem to be of localized occurrence due to the control exerted by protection systems. Sources of these oxygen-centered radicals include the fragmentation of peroxides or their reactions with endogenous radicals or superoxide radicals. The decomposition of lipid hydroperoxides by metal ions renders alkoxy radicals, which are fairly reactive and through secondary free radical formation could lead to mutagenic or carcinogenic effects. Propagation reactions of an autoxidation chain of unsaturated fatty acids also yield peroxy radicals (Pryor 1976).

The mechanism of action of several antitumor drugs and that of therapeutic agents against infections with protozoa involve intracellular redox cycling with formation of oxygen free radicals upon autoxidation of the drug free radicals. NADPH-cytochrome P-450 reductase flavin plays a major role in the reductive activation of antitumor agents. In addition to this activity in endoplasmic reticulum, similar activities in mitochondrial membranes and nuclei of different tissues can carry out this initial reduction step (Low and Sim 1977). Enzymes other than the NADPH-cytochrome P-450 reductase, which catalyze the one-electron reduction of several xenobiotics to a free radical metabolite, include NADH-cytochrome b5 reductase, NADH dehydrogenase, xanthine dehydrogenase, aldehyde oxidase, ferredoxin-NADP reductase (Mason and Chignell 1982). Oxygen is a requirement for the cytotoxicity observed by certain antineoplastic drugs as well as other substances which undergo redox cycling; the radical form of such drugs - largely accumulated in anaerobiosis - is unlikely to be directly responsible for the cytotoxicity observed (Bozzi et al. 1981), but rather the oxygen-dependent radical formation. This free radical metabolism is a diversion of normal electron transport, where the xenobiotic competes with the endogenous cytochrome for electrons.

Compounds used to treat trypanosomiasis, such as naphthoquinone and nitrofuran derivatives undergo similar reductive activation and have been shown to lead to the formation of the superoxide radical and hydrogen peroxide by redox cycling, the latter oxygen metabolite being most effective against Trypanosoma cruzi which lacks catalase (Docampo and Moreno 1983). These compounds set an example on the role of oxygen free radicals in host-invader relations.

In addition, redox cycling is common to the intracellular activation of different compounds responsible for cytotoxic effects. Nitro-compounds used as radiosensitizers and aromatic amines can yield an aromatic nitro radical anion (Mason 1982) and nitroxide radical products (Floyd 1980), which would further react with molecular oxygen. Likewise, the herbicide paraquat (methyl viologen) is known to generate oxygen radicals upon intracellular redox cycling. Menadione as a model quinone for studies of oxidative stress has been employed with isolated hepatocytes and perfused liver (Thor et al. 1982; Wefers and Sies 1983) and the significance of DT-diaphorase in quinone reduction has been elucidated (Lind et al. 1982).

Microsomal Peroxidase Reactions

Cytochrome P-450. As mentioned above, in the initial step of NADPH-supported lipid peroxidation, an electron acceptor is reduced by NADPH subsequently yielding radical species after oxidation, which in turn sustain lipid peroxidation in the propagation phase. Cytochrome P-450 has a function in this propagation phase by catalyzing the homolytic or heterolytic scission of the formed lipid hydroperoxides to produce either free radicals or activated oxygen species, respectively.

In the homolytic scission (reaction 1), the formation of a $Fe\text{-}OH^{\cdot}$ complex and alkoxy radical are used as a tentative model (White and Coon 1980).

$$Fe^{3+} + ROOH \rightarrow Fe^{3+}\text{-}OH^{\cdot} + RO^{\cdot} \qquad (1)$$

Formation of singlet oxygen in such a reaction would not be a primary consequence of cytochrome P-450 activity, but might rather be attributed to further free radical interactions (Cadenas and Sies 1982). Previous observations in model systems have shown that oxidized hemoproteins can catalyze hydroperoxide breakdown with formation of oxygen in the activated state (Cadenas et al. 1980) or peroxyl free radicals (Kalyanaraman et al. 1983). In the latter case, oxygen would play a subsidiary role on the formation of peroxy radicals from hemoprotein/hydroperoxide mixtures.

A heterolytic scission of the O-O bond of hydroperoxides, under formation of a transient $(FeO)^{3+}$ species, relies on the oxene transferase activity of cytochrome P-450, which allows the transfer of an oxygen atom from oxene donors like hydroperoxides to substrates of the monooxygenase system (reaction 2) (Ullrich 1977).

$$Fe^{3+} + ROOH \rightarrow [FeO]^{3+} + ROH \qquad (2)$$

In this case, this "active oxygen complex" could either hydroxylate or epoxidize a substrate (RH) (reaction 3a) or may react with a second hydroperoxide molecule and yield some active form of oxygen, such as singlet oxygen (reaction 3b).

$$[FeO]^{3+} + RH \rightarrow Fe^{3+} + ROH \qquad (3a)$$

$$[FeO]^{3+} + ROOH \rightarrow Fe^{3+} + ROH + {}^{1}O_2 \qquad (3b)$$

Spectral analysis of low-level chemiluminescence has shown that this reaction is associated with singlet oxygen formation (Cadenas et al. 1983c). The occurrence of chemiluminescence also under anaerobic conditions suggests that the photoemissive species formed upon interaction of a microsomal component with hydroperoxides could originate from the oxygen atom present in the hydroperoxide molecule.

In the presence of hydroxylatable substrates, where reaction (3a) would be operative, the transfer of the activated oxygen to the substrate results obviously in the quenching of the chemiluminescence, which reflects singlet oxygen formation. Furthermore, optimal conditions to observe the formation of an activated oxygen species require the oxidized state of cytochrome P-450: in the presence of reducing equivalents to cytochrome P-450 (NADPH or NADH), a peroxidase type reaction activity is supported with the reduction of the organic hydroperoxide to the corresponding alcohol (Bidlack 1980) (reaction 4).

$$NAD(P)H + H^{+} + ROOH \rightarrow NAD(P)^{+} + H_2O + ROH \qquad (4)$$

This reaction is equivalent to the uncoupling process which involves two-electron reduction to water of the activated oxygen complex generated in the presence of nonhydroxylatable substrates (Staudt et al. 1979). In the case of reactions 3a and 4, the destruction of cytochrome P-450 is diminished significantly.

Prostaglandin Synthase. The oxidation of arachidonic acid by PG-endoperoxide synthase involves two distinct activities in a single protein, cyclooxygenase and prostaglandin hydroperoxidase. During the hydroperoxidase-catalyzed conversion of PGG2 to PGH2 a potent oxidant is formed (Marnett et al. 1974). This oxidant is not only responsible for the rapid destruction of the enzyme, but also participates in the cooxidations of a large number of xenobiotics, which might develop potent toxic and/or carcinogenic compounds. Some of these hydroperoxidase-catalyzed cooxidations render free radical intermediates, as for the case of aromatic amines and hydrazines (Mason and Chignell 1982). Then, the possibility exists that prostaglandin synthase, like the P-450-dependent monooxygenase, might be involved in the oxidative activation of xenobiotics.

Several reports attempted to identify this oxidizing equivalent as an oxygen-centered radical (Kuehl et al. 1980), more specifically the hydroxyl radical (Pangamala et al. 1976), a hemoprotein free radical (Kalyanaraman et al. 1982) or more recently, singlet oxygen (Cadenas et al. 1983d).

Since prostacyclin synthase has the characteristics of cytochrome P-450 and can catalyze a heterolytic cleavage of the 9,11-endoperoxide of 15-hydroxy-arachidonic acid (Ullrich et al. 1981) - and in accordance with the commonly accepted hydroperoxidase reaction - it can be proposed that formation of oxidizing equivalents in the PG hydroperoxidase reaction is also the result of two successive steps, similar to reactions (2) and (3b). The formation of a Fe=O complex (reaction 5) would be followed by its reaction with a second molecule of available PGG2 (reaction 6).

$$Fe^{3+} + PGG_2 \rightarrow [FeO]^{3+} + PGH_2 \qquad (5)$$

$$[FeO]^{3+} + PGG_2 \rightarrow Fe^{3+} + PGH_2 + {}^{1}O_2 \qquad (6)$$

Plasma Membrane Electron Transport

The plasma membrane of several mammalian cell types possesses electron transferring enzymes (Goldberg 1982), the activity of which is also associated with the generation of oxygen radicals.

In leukocytes, the respiratory burst observed during phagocytosis is accompanied by the formation of active oxygen species, superoxide radical, and hydrogen peroxide, for which NADH and NADPH can serve as electron donors (Badwey and Karnovsky 1980; Klebanoff 1982), the latter being more widespread (Nakamura et al. 1981). Also, a cytochrome b-type might play an essential role in the NAD(P)H oxidase system (Segal and Jones 1979; Michell 1983). The microbicidal or "killing" activity of leukocytes and myeloperoxidase relies on the production of superoxide radical and hydrogen peroxide; this is an example of a beneficial action of oxygen radicals. An impairment of microbicidal activity would thus occur upon a deficiency of NAD(P)H oxidase or lack of cofactors. However, activated oxygen species are released from the leukocyte to the extracellular milieu, accounting for the tissue damage that accompanies the inflammatory process (McCord 1974), as well as for alterations in microcirculation caused by changes in permeability. Although superoxide radical and hydrogen peroxide are well-established intermediates participating in the microbicidal activity of leukocytes, other oxygen species, such as hydroxyl radical and singlet oxygen were also detected and might have a potential role in the bactericidal mechanism (Badwey and Karnovsky 1980; Klebanoff 1982).

Adipocytes contain a plasma membrane NADPH oxidase, the activity of which is linked to hydrogen peroxide generation (May and de Haen 1979) and stimulated by preincubation with insulin (Mukherjee and Lynn 1977). It was postulated that hydrogen peroxide could act as a second messenger for the observed insulin effects, i.e., stimulation of glucose transport, glucose C-1 oxidation, and also pyruvate dehydrogenase activity, (Paetzke-Brunner and Wieland 1980; Wieland 1983). The generation of the superoxide radical and hydrogen peroxide at the cell border could thus be an effective means for the regulation of metabolic events occurring not only at the plasma membrane, but also in the remainder of the cell. Hydrogen peroxide exerts in adipocytes (May 1981) and phagocytosing human neutrophils (Voetman et al. 1980), a decrease in intracellular glutathione. Membrane-protein sulfhydryl redox transitions, which are under the influence of the reactive oxygen species evolved at the plasma membrane, can be transmitted to more distant enzyme systems via glutathione, the most abundant low molecular weight thiol present in the cell (disulfides as a "third messenger"; Gilbert 1982). Also, thioredoxin plus thioredoxin reductase may play an important role in the control of the protein sulfhydryl status and of hormone action (Holmgren 1979b).

Two-Electron Reduction of Oxygen: Direct H_2O_2 Formation

In addition to the hydrogen peroxide produced by the superoxide dismutase reaction, hydrogen peroxide can be formed directly in a number of oxidase reactions, shown in Fig. 2 (Sies 1974). Peroxisomal hydrogen peroxide production via acyl-CoA oxidase, the initial enzyme of the peroxisomal beta-oxidation system (Lazarow and de Duve 1976) is of particular interest (Kindl and Lazarow 1982). However, other enzymes, such as D-amino acid oxidase, glycolate oxidase, or urate oxidase,

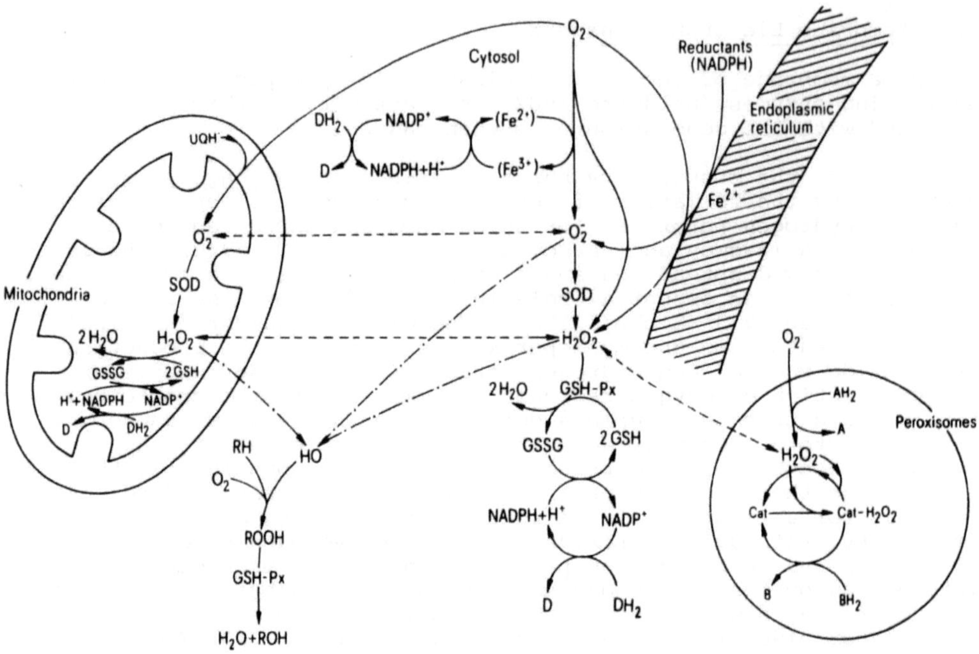

Fig. 2. Scheme to illustrate, in a simplified manner, the linkage of Fe^{2+} (or drug-mediated redox cycling; see Fig. 1) to cellular reductant systems, notably to the $NADPH/NADP^+$ system and to cellular oxidant systems as well as defense reactions. NADPH is involved in $O_2^{\overline{\cdot}}$ radical production and in hydroperoxide reduction by GSH peroxidase reactions. In addition to its reduction by GSH peroxidase, GSSG is partially utilized for the formation of mixed disulfides. Redox cycling is represented here in detail only in the cytosol, but is known to occur at membranes of endoplasmic reticulum and mitochondria possibly at higher rates. (Modified from Kappus and Sies 1981)

produce hydrogen peroxide as well. The hydrogen peroxide formation from fatty acids occurs in intact cells notably with octanoate or oleate (Oshino et al. 1973). Interestingly, the metabolism of palmitate is comparatively minor in peroxisomal hydrogen peroxide formation (Mannaerts et al. 1978; Foerster et al. 1981). Thus, the partitioning between the mitochondrial and peroxisomal pathways of beta-oxidation of fatty acids in intact cells is predominantly in favor of the mitochondrial compartment for long-chain saturated fatty acids, with medium-chain or long-chain monounsaturated having the highest contribution, but also this is low compared to the mitochondrial rates. Chain-shortening of very long-chain monounsaturated acids (C22-C26) seems to be a function of the peroxisomal system (Bremer 1977).

The functions of hydrogen peroxide in mammalian cells are not yet fully elucidated. Apart from toxic effects, which have been recently evaluated (Krinsky et al. 1981; Nathan et al. 1979; Simon et al. 1981), there are some putative roles of interest in metabolism. The most important and widespread antimicrobicidal enzyme system consists of myeloperoxidase/hydrogen peroxide/halide (Badwey and Karnovsky 1980; Klebanoff 1982). Furthermore, it was suggested that hydrogen peroxide is a second messenger in the insulin response in adipocytes (see above).

Vitamins E, C, and A and Selenium

The antioxidant effect of vitamin E was demonstrated in vivo using pentane expiration as an index of lipid peroxidation (Tappel 1980). Recently, it was concluded that the vitamin E is the only chain-breaking antioxidant in human blood (Burton et al. 1983). Vitamin E is a two-electron donor; the reactivity of the hydrogen atom from the OH group in the benzene ring has been shown to be higher than that of other possible electron donors in the molecule. Vitamin E free radicals, of low reactivity even within the frame of a radical-radical reaction, have been identified by ESR (Simic 1981a). The hydroxyl radical group of the benzene ring acts as a reductant upon reaction with free radicals and generates the vitamin E free radical (reaction 7).

$$ROO^{\cdot} + Vit\text{-}EOH \rightarrow ROOH + Vit\text{-}EO^{\cdot} \qquad (7)$$

This agrees with the hypothesis that antioxidants generally function by scavenging lipid-derived, oxygen-centered radicals. However, tocopherol, like some other phenols, can chemically react with and quench singlet oxygen; the latter seems to predominate, although the ratio varies with the solvent used (Foote 1976). This might constitute an additional mechanism of protection of vitamin E against oxidative damage.

The position of vitamin E on the membrane surface is important in its role as antioxidant, because it can both intercept free radical initiators and make possible their reduction by polar donors, such as ascorbic acid or the self-regenerating GSH system involving $GS^{\cdot}$ (reactions 8 and 9).

$$Vit\text{-}EO^{\cdot} + AH_2 \rightarrow AH^{\cdot} + Vit\text{-}EOH \qquad (8)$$

$$Vit\text{-}EO^{\cdot} + GSH \rightarrow GS^{\cdot} + Vit\text{-}EOH \qquad (9)$$

Ascorbate-dependent recovery of vitamin E can take place in aqueous solutions (Packer et al. 1979). In addition to its role in regeneration of vitamin E, ascorbic acid (in the ascorbate form) is reactive with peroxy radicals (reaction 10) ($K = 2.0 \times 10^8\ M^{-1}s^{-1}$) (Simic 1981b).

$$ROO^{\cdot} + AH^{\bar{\cdot}} \rightarrow ROOH + A^{\bar{\cdot}} \qquad (10)$$

Glutathione and other thiol groups can react nonenzymatically with radicals to form thiyl radicals ($RS^{\cdot}$) (reaction 11), which yield dimers, such as GSSG, GSSR, RSSR (reaction 12) (Kosower and Kosower 1978).

$$R^{\cdot} + GSH \rightarrow RH + GS^{\cdot} \qquad (11)$$

$$GS^{\cdot} + GS^{\cdot} \rightarrow GSSG \qquad (12)$$

Radical interactions other than that of reaction (12) are possible; in that case, a decrease in GSH by radical reactions will not necessarily be paralleled by an increase of GSSG through reaction (12). Under comparable concentrations of oxygen and glutathione, $R^{\cdot}$ would react preferentially with oxygen (reaction 13), because the formation of peroxy radicals is faster than that of thiol radicals.

$$R^{\cdot} + O_2 \rightarrow ROO^{\cdot} \qquad (13)$$

The inhibition of microsomal lipid peroxidation by GSH (Burk 1983) can be regarded as vitamin E dependent (Reddy et al. 1982); this synergism between vitamin E and GSH - probably through the regeneration of vitamin E by GSH - is mediated by a microsomally-associated factor.

It was recently observed that nonradical products of lipid peroxidation, such as 4-hydroxy-2,3-trans-nonenal (Esterbauer 1982), increase cellular low-level chemiluminescence and volatile hydrocarbon release, processes that are abolished by pretreatment of the animals with vitamin E (Cadenas et al. 1983b). Vitamin E was suggested to be metabolically beneficial by protecting light-exposed tissues, such as those in the eye, against oxidative damage caused by light-induced formation of oxygen free radicals. This finding could have certain relevance in the age-associated formation of cataracts (Varma et al. 1982).

The antioxidant activity exerted by carotenoids (vitamin A) has been related to their capacity to quench species such as singlet molecular oxygen; the mechanism seems to be a physical quenching reaction that does not affect chemically the structure of the pigment (Krinsky 1982). It has been suggested that dietary vitamin A could have a protective influence against cancer development (Peto et al. 1981).

The selenium status of the cell has been shown to be important regarding the antioxidant capacity of mammalian tissues (Levander 1982). The biochemical basis for this phenomenon is provided by glutathione peroxidase, a selenoprotein which is responsible for the reduction of hydroperoxides. Selenium deficiency was found to be responsible for Keshan disease, a congestive cardiomyopathy affecting persons living in rural selenium-deficient areas in Red China (Keshan Disease Group 1979). Administration of sodium selenite was shown to prevent the disease. A case of cardiomyopathy in a patient with a diet-induced selenium deficiency was described (Johnson et al. 1981). In addition to glutathione peroxidase, there are other selenoproteins that may become of interest (Wilhelmsen et al. 1981; Burk and Gregory 1982).

Two nonselenium glutathione peroxidase activities have been identified, one located in the soluble fraction of the cell (Lawrence and Burk 1976; Prohaska and Ganther 1977), the other in the microsomal compartment (Reddy et al. 1981). The latter enzyme might be important for the reduction of membrane lipid hydroperoxides, which may result from lipid peroxidation in vivo, whereas hydrogen peroxide is not reduced (Sies et al. 1982).

Metabolic Significance of Superoxide Radical and Hydrogen Peroxide

The provision of reducing equivalents from NADPH is of critical importance in maintaining the effectiveness of glutathione in the detoxication of hydroperoxides and also in supporting the reduction of certain compounds, which undergo intracellular redox cycling (see above).

In addition to the formation of oxygen radicals, this dual and continuous utilization of reducing equivalents during redox cycling might in itself have toxicological significance. Paraquat produces an oxidation in the cellular NADPH/NADP system in lung (Witschi et al. 1977; Forman et al. 1982), and therefore, may interfere with the production of surfactant. In liver cells, the NADPH/NADP ratio is also decreased by paraquat as well as by nitrofurantoin as it was shown for the perfused liver (Brigelius et al. 1983a) and by menadione (Eklöw et al. 1981; Wefers and Sies 1983).

The cellular GSSG level increases at the expense of GSH as observed during redox cycling of paraquat in rat liver (Brigelius et al. 1982) and nitrofurantoin in rat lung (Dunbar et al. 1981) and liver (Akerboom et al. 1982). Concomitantly, an increase in biliary GSSG release (Sies et al. 1983) is observed during organ perfusion (Fig. 3) with paraquat (Brigelius et al. 1981), nitrofurantoin (Akerboom et al. 1982), and Nifurtimox (Dubin et al. 1983).

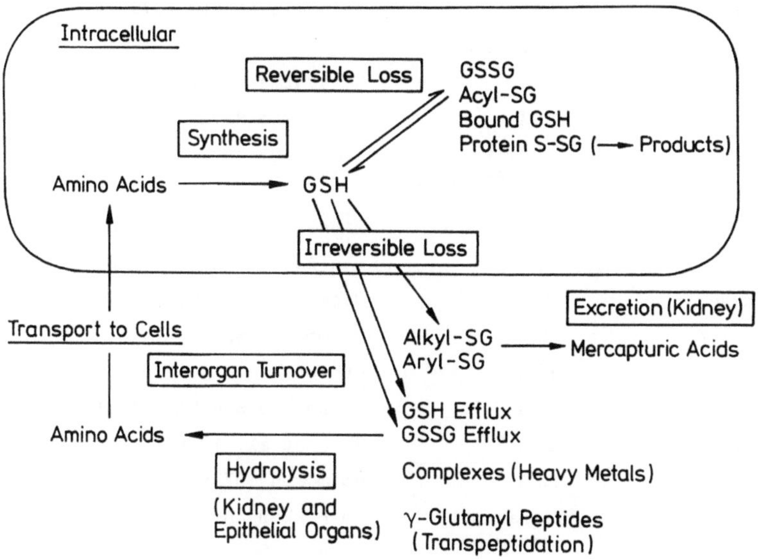

Fig. 3. Processes (chemical and translocation) affecting intrahepatic and extrahepatic glutathione status. (Modified from Sies et al. 1983)

A further consequence of the oxidation in the glutathione system is the formation of mixed disulfides between GSSG and protein thiol groups, catalyzed by the thioltransferase (Mannervik 1980). An increase in total mixed disulfides upon paraquat redox cycling occurs in rat liver (Brigelius et al. 1982) and lung (Keeling et al. 1982). An increase is also observed in hepatic mixed disulfides derived from either CoASH and glutathione upon organic hydroperoxide metabolism (Crane et al. 1982) or protein and glutathione upon paraquat, nitrofurantoin, or hydroperoxide metabolism (Brigelius et al. 1983b).

Alterations in the glutathione redox state relate to the membrane sulfhydryl redox potential and thiol groups in enzymes. In fact, the activity of several enzymes can be modified by formation of a mixed disulfide between an enzyme thiol group and a low molecular weight thiol. Glutathione may play a major role in such thiol/disulfide exchange reactions. The enzymes listed in Table 3 are modified by either mixed disulfide formation or the oxidation state of the glutathione system. Thus, thiol-disulfide exchange participates in regulating whole metabolism, flux of glucose-6-phosphate is channelled into the pentose phosphate pathway when glutathione is oxidized (Eggleston and Krebs 1974). This would make sense, since NADPH is needed for the reduction of GSSG as well as for sustaining the redox cycling process. Indeed, a stimulation of pentose phosphate pathway occurs in rat lung treated with paraquat (Rose et al. 1976) or when rat liver (Sies and Summer 1975) or rat heart (Zimmer et al. 1981) are perfused with t-butylhydroperoxide.

Table 3. Enzyme activities modified by thiol/disulfide exchange

	Activation by	*References*
Glucose-6-P-dehydrogenase	GSSG	Eggleston and Krebs 1974
Collagenase (Leukocytes)	GSSG	Tschesche and Macartney 1981
Acid Phosphatase (Spinach)	GSSG	Buchanan et al. 1979
FDPase	Cystamine	Pontremoli et al. 1967
	CoA	Nakashima et al. 1969
δ-Aminolaevulinate Synthetase	GSSG	Tuboi and Hayasaka 1972
	Inhibition by	
Pyruvate Kinase	GSSG	Mannervik and Axelsson 1980
		van Berkel et al. 1973
Phosphorylase Phosphatase	GSSG	Usami et al. 1980
Phosphofructokinase	GSSG	Gilbert 1982
Glycogen Synthase D	GSSG	Ernest and Kim 1973
HMG-CoA-Reductase	GSSG	Dotan and Shechter 1981
Adenylate Cyclase (Brain)	GSSG	Baba et al. 1978
Ribonucleotide Reductase (E.coli)	GSSG	Holmgren 1979a
Hexokinase	Tetraethyl-cystamine	Nesbakken and Eldjarn 1963
Tyrosine Aminotransferase	Cystine	Federici et al. 1978
PDH Kinase	DTNB	Pettit et al. 1982
Fatty Acid Synthetase	DTNB	Stoops and Wakil 1982
α-Glutamylcysteine Synthetase	Cystamine	Griffith et al. 1977
Papain, Trypsin	Dimethyl-disulfide	Steven et al. 1981

Abbreviations used: FDPase, Fructose diphosphatase; HMG-CoA-Reductase, 3-hydroxy-3-methylglutaryl-Coenzyme A reductase; DTNB, 5,5'-dithio-bis(2-nitrobenzoic acid); PDH kinase, pyruvate dehydrogenase kinase

In addition, thiol-disulfide conversions can play an important role in diverse membrane related processes, such as transport of metabolites like sugar (Czech 1976), amino acids (Sips and Van Dam 1981; Young 1980), or solutes like calcium (Sies et al. 1981) and magnesium (Siliprandi et al. 1978) and in hormone-receptor interactions (Wright and Drummond 1983).

Under conditions of oxidative stress, membrane thiols are oxidized either by forming intramolecular disulfide bridges or mixed disulfides (Haest et al. 1979; Kosower et al. 1982). Also, the extracellular glutathione redox state may be of importance for certain membrane processes. An extracellular glutathione pool of 3-25 μM exists (Tietze 1969; Häberle et al. 1979; Anderson et al. 1980), which may interact with membrane thiols. Recently, it was reported that the uptake of drugs like methrotrexate by isolated liver cells is dependent on the presence of extracellular gluthatione (Leszczynska and Pfaff 1982), affecting the redox state of the -S-S-/-SH groups of the cellular plasma membrane.

While the inner surface of the plasma membrane is accessible to both NADPH and GSH/GSSG, the outer surface could be reached from the inside indirectly by transfer of the reducing equivalents of NADPH via NADPH oxidase in a fashion similar to that known for handling of NADPH in glutathione reductase (Schirmer et al. 1983).

The membrane sulfhydryl status is also believed to play a role in the release of insulin by pancreatic islets. Diamide, a well-known oxidant of thiols and t-butyl hydroperoxide are able to suppress glucose-induced insulin release (Ammon et al. 1977), a process in which adenylate cyclase may play a crucial role (Ammon et al. 1982). The selective toxicity of alloxan towards the pancreatic B-cell involves the formation of active oxygen species via redox cycling (Malaisse 1982), apparently sustained by the thioredoxin/thioredoxin reductase system (Grankvist et al. 1982).

Acknowledgements. Fruitful cooperation with Prof. V Ullrich and Dr. W Nastainczyk, University of Saarland, Homburg and with A Müller and H Wefers of this Institute and helpful suggestions by Prof. L Flohé, Aachen, are gratefully acknowledged.

Work coming from the Authors' laboratory was supported by the Deutsche Forschungsgemeinschaft, Schwerpunktsprogramm "Mechanismen toxischer Wirkungen von Fremdstoffen" and by the Ministerium für Wissenschaft und Forschung, Nordrhein-Westfalen.

Discussion

Hamprecht: You mentioned that at high intracellular concentrations of glutathione disulfide, this disulfide is going to be released from the cell. Is there a transport system mediating extrusion?

Sies: Yes, several types of cells have been shown to mediate transport of GSSG. In the red cell, the transport system has been characterized in detail by the group of Beutler; it is ATP-dependent and there is a low-K_m and a high-K_m system as was shown with inside-out vesicles from erythrocyte ghosts. Recent work carried out by Dr. Akerboom together with Dr. Inoue with isolated biliary canalicular membrane vesicles from rat liver also has identified a transport system for glutathione disulfide.

Hamprecht: Are there transport systems in other cells that catalyze uptake of glutathione disulfide into those cells?

Sies: The kidney is the major organ utilizing extracellular glutathione disulfide, as a part of an interorgan turnover system. Whether the tubular cell takes the molecule up in intact form has so far not been demonstrated. In our work together with Dr. Häberle (1979) (FEBS Lett 108:335-340), it was shown that renal uptake exceeds the glomerular filtration fraction, so that significant uptake occurs at the basolateral membrane of the tubular cell.

Other epithelial organs are capable of taking up glutathione; the lung is a good candidate, but has not been studied in sufficient detail.

Buckel: Glutathione peroxidase and glutathione reductase, do they both contain selenium?

Sies: Only glutathione peroxidase is a selenoenzyme, GSSG reductase is not.

Buckel: Are other selenium-containing proteins involved in your system?

Sies: Selenoproteins or selenopeptides other than GSH peroxidase are known to exist. For example, skeletal muscle contains a low molecular weight selenoprotein and rat testis does likewise. The metabolic significance of these and other known seleno-compounds is yet to be elucidated.

Helmreich: You mentioned that hydrogen peroxide has been considered as a second messenger of insulin action. In this context, you implied that adipocytes may be exceptionally capable of accumulating hydroperoxides. Why is that so?

Sies: May and DeHaen (1979) (J Biol Chem 254:2214-2220) reported on insulin-stimulated intracellular hydrogen peroxide production in rat epididymal fat cells and they suggested that H_2O_2 may act as a second messenger for the observed effects of insulin. Whether adipocytes are exceptionally capable of accumulating hydroperoxides is unknown and I did not mean to imply that.

Recklinghausen: Is redox cycling and/or lipid peroxidation an important mechanism in anthracycline anticancer drug-induced cardiomyopathy?

Sies: The cardiotoxicity of drugs like adriamycin is presumably associated with redox cycling, which produces the superoxide anion radical at the expense of reducing equivalents from NADPH. Whether lipid peroxidation is critically involved in mediating toxicity is controversial in the literature.

Recklinghausen: Why are radical scavengers so far not effective in preventing anthracycline-induced cardiomyopathy?

Sies: Although I'm not sure whether this point has been proven (work with Vitamin E and Q_{10} has been controversial), one reason could be that systemic application of such scavengers will not reach the target sites near the mitochondrial inner membrane in sufficient concentration in order to locally intercept the radicals.

Wulff: At which wavelength are you measuring the singlet oxygen chemiluminescence?

Sies: If one talks about singlet oxygen dimol emission, the appropriate wavelenghts are 634 nm and 703 nm and we employ interference filters at these wavelengths plus one at 668 nm, which yields comparatively lower signals. The latter is used to exclude that a broad band of red chemiluminescence is contributing [Deneke and Krinsky (1977) Photochem Photobiol 25:299-304]. This seems up to now to be the most reasonable criterion to support singlet oxygen participation in biological systems.

However, unfortunately when light emission is very low, interference filters might not be usable because of their low transmittance. Instead, cut-off gelatin filters are employed, having about 90% transmission. In this way, the distinction between red and green chemiluminescence can be made and even the spectral resolution is possible in part [Boveris et al. (1981) Fed Proc 40:23-26].

Wulff: The 634 nm emission originates from the collision pair. Is there a high enough probability that these pairs are formed in solution and hence contribute significantly to the emission?

Sies: From the physicochemical standpoint, the probability in homogeneous aqueous solution indeed is very low. However, in the biological samples, we have a multiphase system and it is not clear whether in different phases or at the interfaces between them the probability may not be significantly increased. Nevertheless, we are detecting the two bands in suspensions of cells and preparations of subcellular organelles, so that the detection limits are surpassed even though the signals are sometimes as low as 100 counts/s.

References

Akerboom TPM, Bilzer M, Sies H (1982) The relationship of biliary glutathione disulfide efflux and intracellular glutathione content in perfused rat liver. J. Biol. Chem 257:4238-4252

Anderson ME, Bridges RJ, Meister A (1980) Direct evidence for inter-organ transport of glutathione and that the non-filtration renal mechanism for glutathione utilization involves gamma-glutamyl transpeptidase. Biochem Biophys Res Commun 96:848-853

Ammon HPT, Akhtar MS, Niklas H, Hegner D (1977) Inhibition of p-chloromercuribenzoate- and glucose-induced insulin release in vitro by methylene blue, diamide, and t-butyl hydroperoxide. Mol Pharmacol 13:598-605

Ammon HPT, Heinzl S, Abdel-Hamid M, Kallenberger HM, Hagenloh I (1982) Effect of diamide and reduced glutathione on the elevated levels of cyclic AMP in rat pancreatic islets exposed to glucose, p-chloromercuribenzoate and aminophylline. Naunyn-Schmiedeberg's Arch Pharmacol 319:243-248

Autor AP (ed) (1982) Pathology of Oxygen . Academic, New York

Baba A, Lee E, Matsuda T, Kihara T, Iwata H (1978) Reversible inhibition of adenylate cyclase activity of rat brain caudate nucleus by oxidized glutathione. Biochem Biophys Res Commun 85:1204-1210

Bachur NR, Gordon S, Gee M, Kon H (1979) NADPH-cytochrome P-450 reductase activation of quinone anticancer agents of free radicals. Proc Natl Acad Sci USA 76: 954-957

Badwey JA, Karnovsky ML (1980) Active Oxygen species and the functions of phagocytic leukocytes. Ann Rev Biochem 49:695-726

Balentine JD (1982) Pathology of Oxygen Toxicity. Academic, New York

Bannister WH, Bannister JV (eds) (1980) Biological and Clinical Aspects of Superoxide and Superoxide Dismutase. Vol 11B, Elsevier/North Holland, New York

Bartoli GM, Galeotti T, Palombini G, Parisi G, Azzi A (1977) Different contribution of rat liver microsomal pigments in the formation of superoxide anion and hydrogen peroxide during development. Arch Biochem Biophys 184:276-281

Bast A, Brenninkmeijer JW, Savenije-Chapel EM, Nordhoek J (1983) Cytochrome P-450 oxidase activity and its role in the NADPH-dependent lipid peroxidation. FEBS Lett 151:185-188

Bidlack WR (1980) Microsomal peroxidase activities - effect of cumene hydroperoxide on the pyridine nucleotide reduced cytochrome b5 steady state. Biochem Pharmacol 29:1605-1608

Boveris A, Cadenas E (1982) Production of superoxide radicals and hydrogen peroxide in mitochondria. In: Oberley LW (ed) Vol II, CRC Florida, pp 15-30

Bremer J (1977) Carnitine and its role in fatty acid metabolism. Trends Biochem Sci 2:207-209

Brigelius R, Hashem A, Lengfelder E (1981) Paraquat-induced alterations of phospholipids and GSSG release in the isolated perfused rat liver and the effect of superoxide dismutase-active copper complexes. Biochem Pharmacol 30:349-354

Brigelius R, Lenzen R, Sies H (1982) Increase in hepatic mixed disulfide and glutathione disulfide levels elicited by paraquat. Biochem Pharmacol 31:1637-1641

Brigelius R, Akerboom TPM, Sies H (1983a) Hepatic superoxide production by redox cycling. Paraquat and nitrofurantoin effects on NADPH, glutathione and mixed disulfides. In: Greenwald RA and Cohen G (eds) Superoxide and superoxide dismutase III, vol II, Elsevier, New York, pp 59-64

Brigelius R, Muckel C, Akerboom TPM, Sies H (1983b) Identification and quantitation of glutathione in hepatic protein mixed disulfides and its relationship to glutathione disulfide. Biochem Pharmacol 32:2529-2534

Bozzi A, Mavelli I, Mondovi B, Strom R, Rotilio G (1981) Differential cytotoxicity of daunomycin in tumor cells is related to glutathione-dependent hydrogen peroxide metabolism. Biochem J 194:369-372

Buchanan BB, Crawford NA, Wolosiuk RA (1979) Plant Sci Lett 14:245-251

Burk RF (1983) Glutathione-dependent protection by rat liver microsomal protein against lipid peroxidation. Biochim Biophys Acta 757:21-28

Burk RF, Gregory PE (1982) Some characteristics of 75Se-P, a selenoprotien found in rat liver and plasma, and comparison of it with selenoglutathione peroxidase. Arch Biochem Biophys 213:73-80

Burton GW, Joyce A, Ingold KU (1983) Is vitamin E the only lipid-soluble, chain-breaking antioxidant in human blood plasma and erythrocyte membranes? Arch Biochem Biophys 221:281-290

Cadenas E, Sies H (1982) Low-level chemiluminescence of liver microsomal fractions initiated by t-butyl hydroperoxide. Relation to microsomal hemoprotein, oxygen dependence and lipid peroxidation. Eur J Biochem 124:349-356

Cadenas E, Boveris A, Chance B (1980) Chemiluminescence of lipid vesicles supplemented with cytochrome c and hydroperoxide. Biochem J 187:131-140

Cadenas E, Boveris A, Chance B (1983a) Low-level chemiluminescence of biological systems. In: Pryor WA (ed) vol VI, Free Radicals in Biology, Academic, New York, in press

Cadenas E, Müller A, Brigelius R, Esterbauer H, Sies H (1983b) Effects of 4-hydroxynonenal on isolated hepatocytes. Studies on low-level chemiluminescence, alkane production, and glutathione status. Biochem J 214:479-487

Cadenas E, Sies H, Graf H, Ullrich V (1983c) Oxene donors yield low-level chemiluminescence with microsomes and isolated cytochrome p-450. Eur J Biochem 130: 117-121

Cadenas E, Sies H, Nastainczyk W, Ullrich V (1983d) Singlet oxygen formation detected by low-level chemiluminescence during enzymatic reduction of prostaglandin G2 to H2. Hoppe-Seyler's Z Physiol Chem 364:519-528

Chance B, Sies H, Boveris A (1979) Hydroperoxide metabolism in mammalian organs. Physiol Rev 59:527-605

Crane D, Häussinger D, Sies H (1982) Rise of coenzyme A-glutathione mixed disulfide during hydroperoxide metabolism in perfused rat liver. Eur J Biochem 127:575-578

Czech, MP (1976) Differential effects of sulfhydryl reagents on activation and deactivation of the fat cell hexose transport system. J Biol Chem 251:1164-1170

De Duve C, Baudhuin P (1966) Peroxisomes (microbodies and related particles). Physiol Rev 46:323-357

Docampo R, Moreno SNJ (1983) Free Radical intermediates in the atniparasitic action of drugs and phagocytic cells. In: Pryor WA (ed) Free Radicals in Biology vol VI, Academic, New York, in press

Dotan I, Shechter I (1981) Thiol-disulfide-dependent interconversion of active and latent forms of rat hepatic 3-hydroxy-3-methylglutaryl-coenzyme A reductase. Biochim Biophys Acta 713:427-434

Dunbar JA, DeLucia AJ, Bryant LR (1981) Effects of nitrofurantoin on the glutathione redox status and related enzymes in the isolated, perfused rabbit lung. Res. Commun Pathol Pharmacol 34:485-492

Dubin M, Moreno SNJ, Martino EE, Docampo R, Stoppani AOM (1983) Increased biliary secretion and loss of hepatic glutathione in rat liver after nifurtimox treatment. Biochem Pharmacol 32:483-487

Eggleston LV, Krebs HA (1974) Regulation of the pentose phosphate cycle. Biochem J 138:425-435

Eklöw L, Thor H, Orrenius S (1981) Formation and efflux of glutathione disulfide studied in isolated rat hepatocytes. FEBS Lett 127:125-128

Ernest MJ, Kim KH (1973) Regulation of rat liver glycogen synthetase. Reversible inactivation of glycogen synthetase D by sulfhydryl-disulfide exchange. J Biol Chem 248:1550-1555

Esterbauer H (1982) Aldehydic products of lipid peroxidation. In: McBrien DCH, Slater TF (eds) Free Radicals, Lipid Peroxidation, and Cancer. Academic, New York, pp 102-108

Federici G, Di Cola D, Saccheta P, Di Ilio C, Del Boccio G, Polidoro G (1978) Reversible inactivation of tyrosine aminotransferase from guinea pig liver by thiol and disulfide compounds. Biochem Biophys Res Commun 81:650-655

Flohé L, Günzler WA, Schock HH (1973) Glutathione peroxidase: a seleno-enzyme. FEBS Lett 22:132-134

Floyd RA (1980) Free radicals in arylamine carcinogenesis. In: Pryor WA (ed) Free Radicals in Biology vol IV, Academic, New York, pp 187-208

Foerster E-C, Fährenkemper T, Rabe U, Graf P, Sies H (1981) Peroxisomal fatty acid oxidation as detected by hydrogen peroxide production in intact perfused rat liver. Biochem J 196:705-712

Foote CS (1976) Photosensitized oxidation and singlet oxygen. In: Pryor WA (ed) Consequences in biological systems vol II, pp 85-133

Foote CS, Shook FC, Akaberli RB (1980) Chemistry of superoxide anion. 4. Singlet oxygen is not a major product of dismutation. J Am Chem Soc 102:2503-2504

Forman HJ, Aldrich TK, Posner MA, Fisher AB (1982) Differential paraquat uptake and redox kinetics of rat granular pneumocytes and alveolar macrophages. J Pharmacol Exp Ther 221:428-433

Gilbert HF (1982) Biological disulfides: the third messenger? Modulation of phosphofructokinase activity by thiol/disulfide exchange. J Biol Chem 257:12086-12091

Gillette JR, Brodie BB, Ladu BN (1957) The oxidation of drugs by liver microsomes: on the role of TPNH and oxygen. J Pharmacol Exp Ther 119:532-543

Goldberg H (1982) Plasma membrane redox activities. Biochim Biophys Acta 694:203-223

Grankvist K, Holmgren A, Luthman M, Taljedal I-B (1982) Thioredoxin and thioredoxin reductase in pancreatic islets may participate in diabetogenic free-radical production. Biochem Biophys Res Commun 107:1412-1418

Griffith OW, Larsson A, Meister A (1977) Inhibition of gamma-glutamyl cysteine synthetase by cystamine; an approach to therapy of 5-oxoprolynurea(pyroglutamic aciduria). Biochem Biophys Res Commun 79:919-925

Häberle D, Wählländer A, Sies H (1979) Assessment of the kidney function in maintenance of plasma glutathione concentration and redox state in anaesthetized rats. FEBS Lett 108:335-340

Haest CWM, Kamp D, Deuticke B (1979) Formation of disulfide bonds between glutathione and membrane SH groups in human erythrocytes. Biochim Biophys Acta 557:-363-381

Halliwell B (1982) Superoxide-dependent formation of hydroxyl radical in the presence of iron salts is a feasible source of hydroxyl radicals in vivo. Biochem J 205:461-462

Hess B, Staudinger H-J (eds) (1968) Biochemie des Sauerstoffs. 19. Kolloquium der Gesellschaft für Biologische Chemie. Springer, Berlin Heidelberg New York

Hildebrandt AG, Roots I (1975) Reduced nicotinamide dinucleotide phosphate-dependent formation and breakdown of hydrogen peroxide during mixed function oxidation reactions in liver microsomes. Arch Biochem Biophys 171:385-397

Holmgren A (1979a) Glutathione-dependent synthesis of deoxyribonucleotides. Characterization of the enzymatic mechanism of Escherichia coli glutaredoxin. J Biol Chem 254:3672-3678

Holmgren A (1979b) Reduction of disulfides by thioredoxin. Exceptional reactivity of insulin and suggested functions of thioredoxin in mechanism of hormone action. J Biol Chem 254:9113-9119

Johnson RA, Baker SS, Fallon JT, Maynard EP, Ruskin JN, Wen Z, Ge K, Cohen HJ (1981) An occidental case of cardiomyopathy and selenium deficiency. N Engl J Med 304:1210-1212

Jones DP, Thor H, Andersson B, Orrenius S (1978) Detoxification reactions in isolated hepatocytes. Role of glutathione peroxidase, catalase, and formaldehyde dehydrogenase in reactions relating to N-demethylation by the cytochrome p-450 system. J Biol Chem 253:6031-6037

Jungermann K, Katz N (1982) Metabolic heterogeneity of liver parenchyma. In: Sies H (ed) Metabolic Compartmentation. Academic, New York, pp 411-435

Kalyanaraman B, Mason RP, Tainer B, Eling TE (1982) The free radical formed during the hydroperoxide-mediated deactivation of ram vesicles is hemoprotein derived. J Biol Chem 257:4764-4768
Kalyanaraman B, Mottley C, Mason RP (1983) A direct ESR spin-trapping investigation of peroxyl free radical formation by hematine/hydroperoxide systems. J Biol Chem, 258:3855-3858
Kappus H, Sies H (1981) Toxic drug effects associated with oxygen metabolism: redox cycling and lipid peroxidation. Experientia 37:1233-1241
Keeling PL, Smith LL (1982) Relevance of NADPH depletion and mixed disulfide formation in rat lung to the mechanism of cell damage following paraquat administration. Biochem Pharmacol 31:3243-3245
Keshan Disease Research Group of the Chinese Academic of Medical Sciences (1979) Observations on effect of sodium selenite in prevention of Keshan disease. Chin Med J (Peking, Engl Ed) 92:477-482
Khan AU (1981) Direct spectral evidence for the generation of singlet molecular oxygen in the reaction of potassium superoxide with water. J Am Chem Soc 103: 6516-6517
Kieczka H, Kappus H (1980) Oxygen dependence of carbon tetrachloride-induced lipid peroxidation in vivo and in vitro. Toxicol Lett 5:191-196
Kindl H, Lazarow PB (eds) (1982) Peroxisomes and Glyoxysomes. Annals of the New York Academy of sciences, Vol 386, New York
Klebanoff SJ (1982) Oxygen-dependent cytotoxic mechanisms of phagocytosis. In: Gallin JI and Fanci AS (eds) Advances in Host Defense Mechanisms. vol I, Raven, New York, pp 111-162
Kosower NS, Kosower EM (1978) The glutathione status of the cells. Intern Rev Cytol 54:109-160
Kosower NS, Zipser Y, Faltin Z (1982) Membrane thiol-disulfide status in glucose-6-phosphate dehydrogenase-deficient red cells. Relationship to cellular glutathione. Biochim Biophys Acta 691:345-352
Krinsky NI (1982) Photobiology of carotenoid protection. In: Regan JD and Parrish JA (eds) The Science of Photomedicine. Plenum, New York, pp 397-407
Krinsky NI, Sladdin DG, Levine PH (1981) Effect of oxygen radicals on platelet functions. In: Rodgers MAJ and Posers EL (eds) Oxygen and Oxy-Radicals in Chemistry and Biology. Academic, New York, pp 153-160
Kuehl FA Jr, Humes JL, Ham EA, Egan RW, Dougherty HW (1980) Inflammation: The role of peroxidase-derived products. In: Samuelsson B, Ramwell PW, Paoletti R (eds) Prostaglandin and thromboxane research, vol VI, Raven, New York, pp 77-86
Kuthan H, Tsuji H, Graf H, Ullrich V (1978) Generation of superoxide anion as a source of hydrogen peroxide in a reconstituted monooxygenase system. FEBS Lett 91:343-345
Lawrence RA, Burk RF (1976) Glutathione peroxidase in selenium-deficient rat liver. Biochem Biophys Res Commun 71:952-958
Lazarow PB, DeDuve C (1976) A fatty acyl-CoA oxidizing system in rat liver peroxisomes: enhancement by clofibrate, a hypolipidemic drug. Proc Natl Acad Sci USA 73:2043-2046
Leszczynska A, Pfaff E (1982) Activation by reduced glutathione of methotrexate transport into isolated rat liver cells. Biochem Pharmacol 31:1911-1918
Levander OA (1982) Selenium: Biochemical actions, interactions, and some human health implications. In: Clinical, Biochemical, and Nutritional Aspects of Trace Elements. Alan R. Liss, New York, pp 345-368
Lind C, Hochstein P, Ernster L (1982) DT-diaphorase as a quinone reductase: a cellular control device against semiquinone and superoxide radical formation. Arch Biochem Biophys 216:175-185
Lown JW, Sim S (1977) The mechanism of bleomycin induced cleavage of DNA. Biochem Biophys Res Commun 77:1150-1157
Malaisse WJ (1982) Alloxan toxicity to the pancreatic B-cell. A new hypothesis. Biochem Pharmcol 31:3527-3534
Mannaerts GP, Debeer LJ, Thomas J, De Schepper PJ (1978) Mitochondrial and peroxisomal fatty acid oxidation in liver homogenates and isolated hepatocytes from control and clofibrate-treated cells. J Biol Chem 254:4585-4595
Mannervik B (1980) Thioltransferases. In: Jakoby WB (ed) Enzymatic Basis of Detoxication, vol II, Academic, New York, pp 229-244

Mannervik B, Axelsson K (1980) Role of cytoplasmic thiol transferase in cellular regulation by thiol-disulfide interchange. Biochem J 190:125-139

Marnett LJ, Wlodawer P, Samuelsson B (1974) Light emission during the action of prostaglandin synthetase. Biochem Biophys Res Commun 60:1286-1292

Mason RP (1982) Free radical intermediates in the metabolism of toxic chemicals. In: Pryor WA (ed) Free Radicals in Biology, vol V, Academic, New York, pp 161-221

Mason RP, Chignell CF (1982) Free Radicals in Pharmacology and Toxicology - Selected Topics Pharmcol Rev 33:189-211

May JM, de Haen C (1979) Insulin-stimulated intracellular hydrogen peroxide production in rat epididymal fat cells. J Biol Chem 254:2214-2220

May JM (1981) The role of glutathione in rat adipocyte pentose phosphate cycle activity. Arch Biochem Biophys 207:117-127

McCord JM (1974) Free radicals and inflammation: Protection of synovial fluid by superoxide dismutase. Science 185:529-535

McCord JM, Fridovich I (1969) Superoxide dismutase. An enzymic function for erythrocuprein (hemocuprein). J Biol Chem 244:6049-6055

Mitchell B (1983) The lethal oxidase of leukocytes. Trends Biochem Sci 8:117-118

Mills GC (1960) Glutathione peroxidase and the destruction of hydrogen peroxide in animal tissues. Arch Biochem Biophys 86:1-5

Mukherjee SP, Lynn WS (1977) Reduced nicotinamide adenine dinucleotide phosphate oxidase in adipocyte plasma membrane and its activation by insulin. Possible role in the hormone's effects on adenylate cyclase and the hexose monophosphate shunt. Arch Biochem Biophys 184:69-76

Mukherjee SP, Lynn WS (1979) Role of cellular redox state and glutathione in adenylate cyclase activity in rat adipocytes. Biochim Biophys Acta 568:224-233

Nakamura M, Baxter CR, Masters BSS (1981) Simultaneous demonstration of phagocytosis-connected oxygen consumption and corresponding NAD(P)H oxidase activity: direct evidence for NADPH as the predominant electron donor to oxygen in phagocytizing human neutrophils. Biochem Biophys Res Commun 98:743-751

Nakashima K, Pontremoli S, Horecker BL (1969) Activation of rabbit liver fructose diphosphatase by coenzyme A and acyl carrier protein. Proc Natl Acad Sci USA 64:947-951

Nastainczyk W, Ullrich V, Sies H (1978) Effect of oxygen concentration on the reaction of halothane with cytochrome P-450 in liver microsomes and isolated perfused liver. Biochem Pharmcol 27:387-392

Nathan CF, Silverstein SC, Brukner LH, Cohn ZA (1979) Extracellular cytolysis by activated macrophages and granulocytes. II. Hydrogen peroxide as a mediator of catotoxicity. J Exp Med 149:100-113

Nesbakken R, Elkjarn L (1963) The inhibition of hexokinase by disulfides. Biochem J 87:526-532

Orrenius S, Sies H (1982) Compartmentation of Detoxification Reactions. In: Sies H (ed) Metabolic Compartmentation, Academic, New York, pp 485-520

Oshino N, Chance B, Sies H, Bucher T (1973) The role of hydrogen peroxide generation in perfused rat liver and the reaction of catalase compound I and hydrogen donors. Arch Biochem Biophys 154:117-131

Oshino N, Jamieson D, Sugano T, Chance B (1975) Optical measurment of the catalase-hydrogen peroxide intermediate (compound I) in the liver of anaesthetized rats and its implication to hydrogen peroxide production in situ. Biochem J 146:67-77

Packer JE, Slater TF, Willson RL (1979) Direct observation of a free radical interaction between vitamin E and C. Nature (London) 278:737-738

Paetzke-Brunner I, Wieland OH (1980) Activation of the pyruvate dehydrogenase complex in isolated fat cell mitochondria by hydrogen peroxide and t-butyl hydroperoxide. FEBS Lett 122:29-32

Pangamala RV, Sharma HM, Heikkila RE, Geer JC, Cornwell DC (1976) Role of hydroxyl radical scavengers dimethyl sulfoxide, alcohols, and methional in the inhibition of prostaglandin biosynthesis. Prostaglandins 11, pp 599-607

Peto R, Doll R, Buckley JD, Sporn MB (1981) Can dietary beta-carotene materially reduce human cancer rates? Nature (London) 290:208-210

Pettit FH, Humphrey J, Reed LJ (1982) Regulation of pyruvate dehydrogenase kinase activity by protein thiol-disulfide exchange. Proc Natl Acad Sci USA 79:3945-3948

Pontremoli S, Traniello S, Enser M, Shapiro S, Horecker BL (1967) Regulation of fructose diphosphatase activity by disulfide exchange. Proc Natl Acad Sci USA 58:286-293

Prohaska JR, Ganther HE (1977) Glutathione peroxidase activity of glutathione-S-transferases purified from rat liver. Biochem Biophys Res Commun 76:437-445

Pryor WA (1976) The role of free Radical reactions in biological systems. In: Pryor WA (ed) Free Radicals in Biology, vol I, Academic, New York, pp 1-49

Reddy CC, Tu CPD, Burgess JR, Ho C-Y, Scholz RW, Massaro EJ (1981) Evidence for the occurrence of selenium-independent glutathione peroxidase activity in rat liver microsomes. Biochem Biophys Res Commun 101:970-978

Reddy CC, Scholz RW, Thomas CE, Massaro EJ (1982) Vitamin E-dependent reduced glutathione inhibition of liver microsomal lipid peroxidation. Life Sci 31:571-576

Rotruck JT, Hoekstra WG, Pope AL, Ganther HE, Swanson AB, Hafeman DG, Hoekstra WG (1973) Relationship of selenium to GSH peroxidase. Fed Proc 31:691

Rose MS, Smith LL, Wyatt J (1976) The relevance of pentose phosphate pathway stimulation in rat lung to the mechanism of paraquat toxicity. Biochem Pharmacol 25:1763-1767

Sawyer DT, Valentine JS (1981) How super is superoxide? Acc Chem Res 14:393-400

Schirmer RH, Schulz GE, Untucht-Grau R (1983) On the geometry of leukocyte NADPH-oxidase, a membrane flavoenzyme. Inferences from the structure of glutathione reductase. FEBS Lett 154:1-4

Segal AW, Jones OTG (1979) The subcellular distribution and some properties of the cytochrome b component of the microbicidal oxidase system of human neutrophils. Biochem J 182:181-188

Sies H (1974) Biochemistry of peroxisome in the liver cell. Angew Chem Int Ed Engl 13:706-718. (Biochemie des Peroxysoms in der Leberzelle. 1974 Angew Chem 86: 789-901

Sies H, Cadenas E (1983) Biological basis of detoxication of oxygen free radicals. In: Caldwell J, Jakoby WB (eds) Biological Basis of Detoxication. Academic, New York, pp 181-211

Sies H, Chance B (1970) The steady state level of catalase compound I in isolated hemoglobin-free perfused rat liver. FEBS Lett 11:172-176

Sies H, Graf P (1982) Redox relations associated with drug metabolism in perfused liver. In: Sato R, Kato R (eds) Microsomes, Drug Oxidations and Drug Toxicities. Jap Sci Soc, Tokyo, Wiley-Intersci, New York, pp 613-620

Sies H, Summer KH (1975) Hydroperoxide metabolizing systems in rat liver. Eur J Biochem 57:503-512

Sies H, Gerstenecker C, Menzel H, Flohé L (1972) Oxodation in the $NADP^+$ system and the release of GSSG from hemoglobin-free perfused rat liver during peroxidatic oxidation of glutathione by hydroperoxides. FEBS Lett 27:171-175

Sies H, Bartoli GM, Burk RF, Waydhas C (1978) Glutathione efflux from perfues rat liver after phenobarbital treatment during drug oxidations, and in selenium deficiency. Eur J Biochem 89:113-118

Sies H, Graf P, Estrela JM (1981) Hepatic calcium efflux during cytochrome P-450-dependent drug oxidations at the endoplasmic reticulum in intact liver. Proc Natl Acad Sci USA 78:3358-3362

Sies H, Wendel A, Burk RF (1982) Se and Non-Se glutathione peroxidases: enzymology and cell physiology. In: King TE et al. (eds) Oxidases and related redox systems. Pergamon, New York, pp 169-189

Sies H, Brigelius R, Akerboom TPM (1983) Intrahepatic glutathione status. In: Larsson A, Orrenius S, Holmgren A, Mannervik B (eds) Functions of Glutathione: Biochemical, Physiological, Toxicological, and Clinical Aspects. Raven, New York, pp 51-64

Siliprandi N, Siliprandi D, Bidoli A, Toninello A (1978) Effect of oxidation of glutathione and membrane thiol groups on mitochondrial functions. In: Sies H, Wendel A (eds) Functions of Glutathione in Liver and Kidney. Springer, Berlin Heidelberg New York, pp 139-147

Simic MG (1981a) Vitamin E radicals. In: Rodgers MAJ, Powers EL (eds) Oxygen and Oxy-Radicals in Chemistry and Biology. Academic, New York, pp 109-118

Simic MG (1981b) Free radical mechanisms in autoxidation processes. J Chem Educ 58:125-131

Simon RH, Scoggin CH, Patterson D (1981) Hydrogen peroxide causes the fatal injury to human fibroblasts exposed to oxygen radicals. J Biol Chem 256:7181-7186
Sips HJ, Van Dam K (1981) Amino acid dependent sodium transport in plasma membrane vesicles from rat liver. J Membr Biol 62:231-237
Smith LL, Kulig MJ (1976) Singlet molecular oxygen from hydrogen peroxide disproportionation. J Am Chem Soc 98:1027-1029
Staudt H, Lichtenberger F, Ullrich V (1979) The role of NADH in uncoupled microsomal monoxygenations. Eur J Biochem 46:99-106
Steven FS, Griffin MM, Smith RH (1981) Disulphide exchange reactions in the control of enzymic activity. Evidence for the participation of demethyl disulphide in exchanges. Eur J Biochem 119:75-78
Stoops JK, Wakil SJ (1982) The reaction of chicken liver fatty acid synthetase with 5,5'-dithiobis(2-nitrobenzoic acid). Biochem Biophys Res Commun 104:1018-1024
Suzuki H, King TE (1983) Evidence of an ubisemiquinone radical(s) from NADH-ubiquinone reductase of the mitochondrial respiratory chain. J Biol Chem 258:351-358
Tappel AL (1980) Measurement of and protection from in vivo lipid peroxidation. In: Pryor WA (ed) Free Radicals in Biology, Vol IV, Academic, New York, pp 1-47
Thênard LJ (1818) l'Academie des Sciences, Paris; cf Thenard LJ: Traite de Chimie, 6th Edn., vol I, Ed Crochard, Paris 1834, p 529
Thor H, Smith MT, Hartzell P, Bellomo G, Jewell SA, Orrenius S (1982) The metabolism of menadione (2-methyl-1,4-naphthoquinone) by isolated hepatocytes. A study of the implications of oxidative stress in intact cells. J Biol Chem 257:12419-12425
Thurman RG, Ley HG, Scholz R (1972) Hepatic microsomal ethanol oxidation, hydrogen peroxide formation and the role of catalase. Eur J Biochem 25:420-430
Tietze F (1969) Enzymic method for quantitative determination of nanogram amounts of total and oxidized glutathione: applications to mammalian blood and other tissues. Analyt Biochem 27:502-522
Tschesche H, Macartney HW (1981) A new principle of regulation of enzymic activity. Activation and regulation of human polymorphonuclear leukocytes collagenase by disulfide-thiol exchange is catalyzed by the glutathione cycle via peroxidase-coupled reaction to glucose metabolism. Eur J Biochem 120:183-190
Tuboi S, Hayasaka S (1972) Control of delta-aminolevulinate synthetase activity in Rhodopseudomonas spheroides. II. Requirement of a disulfide compound for the conversion of the inactive form of fraction I to the active form. Arch Biochem Biophys 150:690-697
Turrens JF, Freeman BA, Levitt JG, Crapo JD (1982a) The effect of hyperoxia on superoxide production by lung submitochondrial particles. Arch Biochem Biophys 217:401-410
Turrens JF, Freeman BA, Crapo JD (1982b) Hyperoxia increases hydrogen peroxide release by lung mitochondria and microsomes. Arch Biochem Biophys 217:411-421
Tyler DD (1975) Polarographic assay and intracellular distribution of superoxide dismutase in rat liver. Biochem J 147:493-504
Ullrich V (1977) The mechanism of cytochrome P-450 action. In: Ullrich V, Roots, I, Hildebrandt A, Estabrook RO, Conney AH (eds) Microsomes and Drug Oxidations. Plenum, New York, pp 192-210
Ullrich V, Castle L, Weber P (1981) Spectral evidence for the cytochrome P-450 nature of prostacyclin synthase. Biochem Pharmacol 30:2033-2036
Usami M, Matsushita H, Shimazu T (1980) Regulation of liver phosphorylase phosphatase by glutathione disulfide. J Biol Chem 255:1928-1931
van Berkel JC, Koster JF, Hülsmann WC (1973) Two interconvertible forms of L-type pyruvate kinase from rat liver. Biochim Biophys Acta 293:118-124
Varma SF, Beachy NA, Richards RD (1982) Photoperoxidation of lens lipids: prevention by vitamin E. Photochem Photobiol 36:623-626
Voetman AA, Loos JA, Roos D (1980) Changes in the levels of glutathione in phagocytosing human neutrophils. Blood 55:741-747
Wefers H, Sies H (1983) Hepatic low-level chemiluminescence during redox cycling of menadione and the menadione glutathione conjugate. Arch Biochem Biophys 224:568-578
White RE, Coon MJ (1980) Oxygen activation by cytochrome P-450. Annu Rev Biochem 49:315-356

Wieland O (1983) The mammalian pyruvate dehydrogenase complex: structure and regu lation. Rev Physiol Biochem Pharmacol 96:124-170

Wilhelmsen EC, Hawkes WC, Motsenbocker MA, Tappel AL (1981) Selenium-containing proteins other than glutathione peroxidase from rat tissue. In: Spallholz JE, Martin JL, Ganther HE (eds) Selenium in Biology and Medicine. AVI Publishing Company, Connecticut, pp 535-539

Winterbourn CC (1982) Superoxide-dependent production of hydroxyl radicals in the presence of iron salts. Biochem J 205:463

Witschi H, Kacero S, Hirai KJ, Cote MG (1977) In vivo oxidation of reduced nicotinamide-adenine dinucleotide phosphate by paraquat and diquat in rat lung. Chem Biol Interact 15:143-160

Wright M, Drummond GI (1983) Inactivation of the beta-adrenergic receptor in skeletal muscle by dithiols. Biochem Pharmacol 32:509-515

Young JD (1980) Effects of thiol-reactive agents on amino acid transport by sheep erythrocytes. Biochim Biophys Acta 602:661-672

Zimmer H-G, Bünger R, Koschine H, Steinkopff G (1981) Rapid stimulation of the hexose monophosphate shunt in the isolated perfused rat heart: possible involvement of oxidized glutathione. J Mol Cell Cardiol 13:531-535

Subject Index

S. Fliszár

Charge Distributions and Chemical Effects

A New Approach to the Electronic Structure and Energy of Molecules

1983. 15 figures. XIII, 205 pages
ISBN 3-540-90854-4

Contents:

This book offers a coherent theory of the relationships between electronic charge distributions and chemical binding energies. Theoretical charges (and correlations with NMR Shifts) allow simple energy calculations with experimental accuracy. While the bulk of the monograph is dedicated to the "backbone of organic chemistry", the hydrocarbons, some inroads into heteroatomic systems are described. The implications of this new theory on our present view of the nature of the chemical bond are delineated.

Springer-Verlag
Berlin
Heidelberg
New York
Tokyo

MIX
Papier aus verantwortungsvollen Quellen
Paper from responsible sources
FSC® C105338

If you have any concerns about our products,
you can contact us on
ProductSafety@springernature.com

In case Publisher is established outside the EU,
the EU authorized representative is:
Springer Nature Customer Service Center GmbH
Europaplatz 3, 69115 Heidelberg, Germany

Printed by Libri Plureos GmbH
in Hamburg, Germany